Intelligent Systems, Control and Automation: Science and Engineering

Volume 92

More information about this series at http://www.springer.com/series/6259

Erika Ottaviano · Assunta Pelliccio
Vincenzo Gattulli
Editors

Mechatronics for Cultural Heritage and Civil Engineering

Springer

Editors
Erika Ottaviano
Department of Civil and Mechanical Engineering
University of Cassino and Southern Lazio
Cassino, Frosinone
Italy

Vincenzo Gattulli
Department of Structural and Geotechnical Engineering
Sapienza University of Rome
Rome
Italy

Assunta Pelliccio
Department of Civil and Mechanical Engineering
University of Cassino and Southern Lazio
Cassino, Frosinone
Italy

ISSN 2213-8986 ISSN 2213-8994 (electronic)
Intelligent Systems, Control and Automation: Science and Engineering
ISBN 978-3-319-68645-5 ISBN 978-3-319-68646-2 (eBook)
https://doi.org/10.1007/978-3-319-68646-2

Library of Congress Control Number: 2017957183

Printed on acid-free paper

This Springer imprint is published by Springer Nature
The registered company is Springer International Publishing AG
The registered company address is: Gewerbestrasse 11, 6330 Cham, Switzerland

Preface

This book presents recent advances in mechatronic and integrated monitoring and management systems with application to architecture, archaeological survey, construction management and civil engineering. It comprises 16 chapters, authored by recognized experts in a variety of fields, including dynamics, signal processing, inverse modeling, robotics and automation, applied here to the design and construction of civil structures and architectural surveys, and the monitoring and maintenance of cultural heritage assets, structures and infrastructure.

Topics include image processing for automated visual inspection, fiber-optical sensor technology, embedded systems for cables and wireless sensor monitoring, bridge inspection and monitoring of tunnel infrastructures, design tools for construction engineering, and smart cities. Direct and inverse modeling of multibody systems and robots contributes to the development of applications for civil engineering and smart cities, for infrastructure inspection and condition assessment, unmanned aerial vehicle (UAV) systems and direct data processing. Digital technology and mechatronic systems change the way of looking at restoration of historical and archeological sites, analysis, inspection, visualization, management systems and sensor networks for human–machine interfaces (HMI). The combined use of geographic information systems (GIS), laser scanning, remote sensing, digital thermography and drones as integrated systems is also discussed.

The book serves as a valuable reference volume for scientists, architects, engineers, researchers and practitioners in engineering and architecture, as the integrated development of new technologies for the design and management of existing and new infrastructure may give rise to a new market for services and products designed to provide safe and economically optimized infrastructure management.

Through the dissemination of advanced research concepts in mechatronics and integrated management systems, our objective is to promote the exchange of ideas and collaboration among researchers of different disciplines, and to contribute to further advancements in the rapidly growing field at the intersection of robotics, automation and information technology, for the integrated management of structures and infrastructure.

The organization of the book into three main parts—"Robotics and Automation", "Digital Technologies for Cultural Heritage" and "Civil Structural Health Monitoring"—evidences the different backgrounds of the editors, unified by a common interest. This perspective, which is expressed in the first chapter, reveals the authors' views regarding the integration of these three key aspects. A summary of the main contributions of each part follows.

Robotics and automation are undergoing wide dissemination and effective use throughout the industrial environment, with advances in related disciplines such as mechatronics, informatics and mechanics, which are rapidly expanding in scope in application for assisting, entertaining and interacting with humans.

Very recent applications can be seen in the use of robots and automated systems in construction. In the MECH-Section, two chapters report new trends in the use of cable-driven parallel robots in the field of civil engineering ICT and construction. These robots have a moving platform suspended and operated by cables connected to the base and relying on the force of gravity. Used in a broad range of applications of manipulative tasks, they enable operation in an uncluttered working environment. In addition, the large workspace and ability to work in a 3D environment make them suitable for a wide range of applications. In this context, a joint collaboration in the fields of mechanical and civil engineering has produced a contribution discussing upcoming transformations in shell production by comparing the conventional construction process with that of the future, influenced by cable-driven parallel robots, including a discussion of conceptual design and economic feasibility analysis.

The second contribution deals with the analysis and design of reconfigurable cable robots for use in construction. The reconfiguration of the robots allows better positioning capability and workspace for the proposed case study, which deals with sandblasting and painting of a 3D tubular structure.

The algebraic and numerical techniques used extensively in robotics and computational kinematics can be successfully applied to other disciplines. In this regard, a contribution links distance geometry, traditionally used for position and motion analysis of mechanisms, to novel applications dealing with tensegrity and deployable structures. The design of novel structures with variable geometry relies on a complete characterization of their valid configurations, defined by a system of equations encoding the assembly, task or contact constraints. Different formulations encode the equations presented.

Recent trends, also demonstrated by the outgoing H2020 calls, show growing interest in smart and sustainable cities to better integrate environmental, transport, energy and digital networks in the EU's urban environments. In this context, a contribution proposes a systematic approach for simulating a complex urban traffic scenario. This can be extended and further implemented, being computationally efficient, for the optimization of an existing traffic-road system or the design of a new one.

Home automation is the residential extension of building automation, and involves the interoperability of smart sensors and security. It represents a promising market in the context of both technology and engineering techniques that exploit the integration of mechatronics and information and communications technology (ICT) for the development of smart buildings. The contribution presents a methodology for obtaining and analyzing requirements associated with residential automation systems design.

In recent decades, the use of ICT for the interpretation, conservation and preservation of world heritage has grown exponentially. In light of the many tools and methodological approaches used by researchers for the analysis of natural and cultural heritage and for communicating the results, UNESCO in 2003 was compelled to define the significance of "digital heritage" and to establish its value and the need for its protection.

For the analysis of the territory, geographic information systems (GIS) are particularly powerful and effective tools, as they allow for modeling ranging from large- to small-scale representations, with which a large amount of data is associated and can be processed and correlated with the digital model.

Indeed, in the Arch-Section, the first contribution discusses the use of GIS to explore the effects of subsidence in the city of Bologna, as the valuable historic building heritage and churches and monuments of the city are exposed to that risk. Based on geostatistical methods, the system is able to analyze the spatial and temporal evolution of subsidence phenomena and to reveal the roles of different factors and their eventual mutual correlation.

In the architectural heritage field, numerous digital technologies and tools, each different from the other, are available for surveying, visualizing and modeling. In fact, the second contribution relates the use of thermal diagnostics as a non-destructive testing method for the analysis of historical buildings. Using analysis based on infrared thermography (of external walls), it is possible to correlate the thermal properties with defects in the walls of historic buildings, and to then apply a numerical calculation based on the finite element method on the results obtained. This method is particularly well suited for the analysis of a special type of historical building, known as "workers' housing estates", in Upper Silesia (Poland), due to the typology, the material used and the location of those buildings.

In the specific branch of metric survey, laser scanning technology is the most commonly used. The third contribution reports a detailed analysis of the current technologies; specifically, the use of laser scanners is especially suitable for metric survey of the structures and analysis of the materials of each individual element, using the property of reflectance. Recent survey tools include 3D image modeling used to obtain orthophotos for metric analysis, and the use of a multicopter with an onboard Nikon camera to represent the architectural object as a cloud of millions of points. The integration of all of these technologies is helpful for understanding very complex architectural phenomena both in Italy and worldwide. Always more often, monitoring to evaluate the structural plastic deformation of elements of historic buildings is increasingly important in restoration. For this purpose, the integration of various technologies is a key element. The fourth contribution shows how an integrated system of 3D modeling, photo-modeling and laser scanning technology is able to produce 3D computational modeling to simulate the static performance of vaulted roofs and constructive elements of historic architectures. In the area of parametric modeling as applied to survey, the building information modeling (BIM) concept is used in various approaches for cultural heritage. In the fifth contribution, "H-BOM" is defined as BIM applied to a single object which makes up the cultural heritage. The system is capable of detailed and accurate measurement for semantic analysis of the building components based on 3D model with different levels of architectural detail.

Research themes and interest in the integration of automation and information technologies for inspection, maintenance, rehabilitation and management of existing structures and infrastructures has garnered increasing attention in recent years. In addition, in the construction of new facilities, the embedding of technologically advanced systems and methodologies can improve service life and reduce life-cycle cost. In the CIV-Section, the first contribution presents a general methodology for achieving so-called smart heritage, which can be obtained through properly designed SHM systems, and when connected to an automated diagnostic system can even self-evaluate retrofitting needs. After the general concept is illustrated, three subsequent contributions with specific themes related to this concept. The first provides insights into the potentiality of statistical learning algorithms, both supervised and unsupervised, in SHM for real-time condition assessment of civil infrastructure systems. This approach allows for the construction of a surrogate model from the acquired data as a substitute for a high-fidelity model in the context of data-driven models. The methodology is promising in view of large deployment of distributed sensing for example in seismic urban areas. A second, similar contribution deals with automated vibration-based structural health monitoring as a useful alternative or complement to visual inspection or local non-destructive testing performed manually.

The key challenge addressed here is in providing a robust damage diagnosis from the recorded vibration measurements, for which statistical signal processing methods are required. In this chapter, a damage detection method is examined through comparison of vibration measurements from the current system to a reference state in a hypothesis test, where data-related uncertainties are taken into account. The chapter presents an updated version of the current methodologies in the specific field.

The interconnection between strengthening and sensing in the retrofitting of ancient and modern construction provides an interesting contribution.

The chapter identifies a full-coverage reinforcement in the form of a polymeric multi-axial textile methodology based on mesoscopic and multiscale representations within the context of characterization, identification and performance assessment. Finally, the monitoring and maintenance of non-conventional structures is investigated, including water tanks and the steel frames inside them, by considering the underground structures in service at the Gran Sasso National Laboratory (LNGS).

Cassino, Italy Erika Ottaviano
Cassino, Italy Assunta Pelliccio
Rome, Italy Vincenzo Gattulli

Contents

Mechatronics in the Process of Cultural Heritage and Civil Infrastructure Management

Vincenzo Gattulli, Erika Ottaviano and Assunta Pelliccio

Abstract Automatized survey, construction, inspection, maintenance, restoration and reconstruction have become challenging activities conducted during the process of cultural heritage and civil infrastructure management, due to the revolutionary impact of mechatronics and information technology in routine operations. The complete process is summarized, considering different aspects related to the interconnection between classical engineering and architectural problems with the emerging technologies related to automation, robotics and information communication technologies (ICT). The impact of new technologies on data acquisition for survey, inspection and monitoring is, firstly, considered, drawing upon evidence of how the use of robotized systems and sensor networks determines new sets of available data to be processed by digital tools to build advanced models. The integration among different information and numerical models permits one to test the novelties related to the use of ICT technologies for creating an exhaustive description of the examined facility. Data and models can be then used to identify and to describe defects and degradation, especially in view of determining possible performance reduction in existing structures. All the acquired knowledge opportunely managed constitutes the input for automated or partially automated decision-making process useful in facilities and infrastructure management.

V. Gattulli (✉)
DISG—Department of Structural and Geotechnical Engineering, Sapienza University of Rome, Via Eudossiana 18, 00184 Rome, Italy
e-mail: vincenzo.gattulli@uniroma1.it

E. Ottaviano · A. Pelliccio
DICeM—Department of Civil and Mechanical Engineering, University of Cassino and Southern Lazio, via Di Biasio 43, 03043 Cassino (FR), Italy
e-mail: ottaviano@unicas.it

A. Pelliccio
e-mail: pelliccio@unicas.it

E. Ottaviano et al. (eds.), *Mechatronics for Cultural Heritage and Civil Engineering*, Intelligent Systems, Control and Automation: Science and Engineering 92,
https://doi.org/10.1007/978-3-319-68646-2_1

Introduction

In the interest of economic and environmental sustainability, cultural heritage and civil infrastructures increasingly require a defined process management system capable of containing multiple innovative procedures, and made possible by the introduction of mechatronics. This process results in the application of knowledge, skills, tools, techniques and systems to define, visualize, measure, control, report and improve new construction and restoration processes with the goal of meeting customer requirements profitably. In the process, several tasks are interconnected and likely to be enhanced by the introduction of new technologies. For example, the survey, inspection, assessment, maintenance and safe operation of existing civil infrastructure and historical sites mainly rely on large amounts of human activity, although this may be dangerous or unsafe. These activities are indispensable for the creation of information tools (geometric and numerical models), which permit accurate evaluation and creation of successive activities in the decision process and construction interventions. New construction projects follow similar processes, in which attention is devoted to optimization and control, in order to assure matching between design idea and as-built realization.

However, during new construction or improvement of existing facilities, conventional inspection, monitoring and survey are based primarily on visual investigation methods, which are technically complex and performed by well-trained staff. They are devoted to structures and structural components, including large-scale constructions such as tunnels, bridges, roads and pipelines, or buildings, caves, catacombs and brownfields, which may be hard to reach or inaccessible to operators. Trained staff or operators with specialized equipment, such as industrial climbers, large under-bridge units, elevating platforms, generally carry out these tasks [9]. In particular, existing structures need regular inspections in order to detect possible deterioration and to program their restoration and maintenance. Recursive inspection may be classified in initial inspection, routine inspection, damage inspection, in-depth inspection, fracture-critical inspection, underwater inspection and special inspection [40]. It can be planned or due to an emergency, and it may require an interruption of service of the infrastructure. The final goal is to extend the construction life or planning restoration intervals, taking into account that different sizes, geometries, complexity of conservation and environmental conditions require different approaches.

Urban search and rescue operations in disaster relief, conflicts and disasters in nuclear plants, for example, are extremely dangerous, particularly in buildings where no prior information may be available. For these specific applications, robotic inspections can be effectively used, as described in [51, 59]. Application of a mechatronic survey is proposed in [56].

With regard to *bridge inspections*, a large amount of funds is spent each year on non-destructive testing/evaluation (NDT/NDE) technologies. This is a critical issue for bridges that were built not only in Europe but also worldwide in the mid-20th century or earlier, because these bridges are reaching their life expectancy, leaving questions about their structural integrity and deterioration levels. Thus, routine inspection and maintenance are necessary to ensure sustainability. Commonly adopted procedures on structural components still rely on prior visual inspection to identify and evaluate deterioration, damage and faults. Automatic survey and data acquisition are carried out by NDT permanent monitoring systems. Results can be organized in a bridge management system (BMS), as reported in [24], in order to evaluate bridge conditions and manage maintenance. Moreover, effective tools and delivery systems for NDT inspection rely heavily on an inspector's subjective or empirical knowledge, which can lead to false evaluations. Inspection can be also performed with specialized equipment like large under bridge units, large trucks, special elevating platforms or scaffolding on structures, which are, in most cases, expensive and require expensive logistical efforts as well as high personnel costs for the specially trained machine operators. These units can cause disturbances in the operational conditions of the structures. Specially trained staff, like industrial climbers, can get access to certain parts of the structure, but they can rarely evaluate the influences detected damage can have on these structures. Therefore, they can only take photos or videos of the relevant part of the structure, which must then be analyzed by civil engineers after data recording.

Underlying these activities is the *survey*, which is defined as the procedure of acquiring all information, metrics, geometrics, about materials and technology necessary to evaluate the sites for new constructions or the existing objects of study (architectural or infrastructural). The survey is based on four phases: the first one involves the knowledge of the object or the site with the acquisition of the documentary, iconographic and photographic material, which is fundamental to understand its evolutionary phases. The second step is based on the choice of the method for taking the measures, direct by using a single measuring instruments, or instrumental with technologically advanced tools. The third phase includes performing the sampling of the measures, based on discretization according to the specified accuracy. The last phase involves the design of the basic technical drawings (e.g. layout, section) and 2D or 3D graph models, using standards, symbol graphic and metrological analysis. The survey practice follows the technological development of measuring instruments. In recent decades, the laser scanning, digital photogrammetry, total station or thermal imaging have simplified the procedure of data acquisition, but improving complexity of the post-processing. These survey technologies are widely used in the field of cultural heritage and archeology, because they rapidly acquire metric/geometric information and geo-referenced data. Furthermore, they discretize the objects in millions of points (point cloud) in an ontologically undifferentiated way, with the only variation according to the numerical coordinates and color information. In addition, they acquire color information about the materials using laser scanner and thermography technologies. A further advantage of using these technologies is the capability of interfacing with management, modeling and data visualization environments such as geographic information systems (GIS) and building information models (BIMs). With regard to historical sites, preservation and management of cultural heritage very often require a complex strategy based on data acquisition, processing and programmed intervention. In recent years the idea of *programmed conservation* has been investigated. *Programmed conservation* is based on the continuous monitoring and planning of intervention, *risk analysis*, and *decision-making* [71]. From this point of view, the first step is developing efficient strategies for monitoring, which can be planned as a function of several factors. In most cases, those sites are difficult to access or dangerous for operators and require additional systems, such as unmanned aerial vehicle (UAV). Therefore, the implementation of efficient technologies for automation facilitates surveying and monitoring, which is preliminary to further actions, as reported in Fig. 1. Indeed, the large amount of data potentially acquirable should be precisely selected to develop a specific objective of the process. Consequently, automation and robotics may help one to acquire data, possibly in an automatic manner, in order to generate information-based models, such as by finite element method (FEM), BIM, GIS.

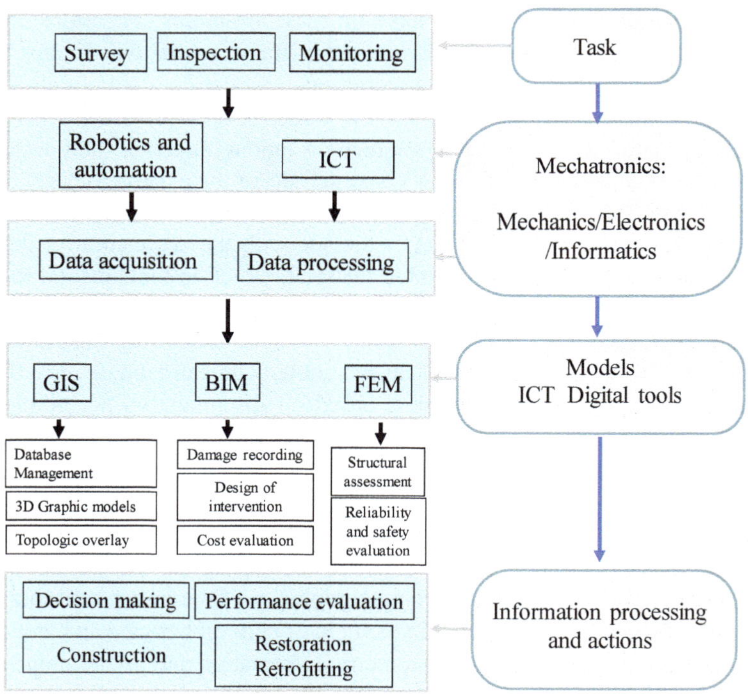

Fig. 1 Block diagram of the process for new and existing facility management enhanced by mechatronics and automation

BIMs are digital representations of facilities that encode all the relevant information about their life cycle from construction to demolition, including 3D design drawings, schedules, material characteristics, costs and safety specifications.

This approach is emerging due to its capacity to interconnect with other information environments, such as finite element method (FEM) or integrating directly the data coming from structural health monitoring (SHM). While creating as-designed BIMs (i.e. BIMs generated in the design stage of a facility) is a straightforward process, and is becoming increasingly common, generating as-built BIMs (i.e. BIMs that reflect a facility in its as-built condition) is a challenging, but necessary process for facilities not equipped with an as-designed BIM [52]. These potentialities open up a set of innovative research activities related to the integration of data coming from different processes within a BIM environment. Among these is information related to vision inspection, such as images that need to be accurately treated to identify or detect damage and stored and referred into a BIM. Defects and anomalies present in construction have to be hierarchized, in order to evaluate their effects on the reliability and safety of the structure with respect to all possible hazards or their potential impact on aesthetics and conservation of the property. Consequently, the information process for the performance evaluation of alternatives and decision-making procedures permits the design of interventions based on the economic evaluation of costs. Finally, actions are taken for restoration, retrofitting, programmed conservation and maintenance, taking into consideration the possible impact of innovative automation in all of these activities.

Following the proposed hierarchy of activities in the process management of new and existing facilities, the Chapter describes the emerging impact of mechatronics in each component of a unified process.

Automatic Data Acquisition in Survey, Inspection and Monitoring

Automatic data acquisition in survey, inspection and monitoring can be integrated using novel technologies such as robotics and automation. In the last decades, novel applications of automatic data acquisition for surveying, combining digital technologies based on laser scanning, photogrammetry, thermography, and remote sensing (e.g. aided by robotic and automatic systems) have been developed. In particular, 2D and 3D graph modeling techniques reconstruct complex objects. Thermography is used to assess the physicochemical behavior of conservation treatments, such as stone cleaning and consolidation, and mortar repair as well as to disclose any substrate features, such as tesserae on plastered mosaic surfaces. In this regard, mechanical or mechatronic devices can be used as guided or autonomous systems to carry equipment for an automatic relief of images and frames, especially when the site of interest is difficult or dangerous to access.

In the context of civil engineering, SHM deals with the observation of a system over time using sensors and the extraction of damage-sensitive features from measurements, and the statistical analysis of these features to determine the current state of system health. Recently, categories of SHM studies recall one (or a combination) of the following: (1) anomaly detection, (2) sensor deployment studies, (3) model validation, (4) threshold check, and (5) damage detection. This framework should aid engineers specifying monitoring systems to determine what should be measured and why, hence allowing them to better evaluate what value may be delivered to the relevant stakeholders for the monitoring investments. In this context, new technologies have being used in SHM activities among which two key emerging ones are, namely, imaging and computer vision and microelectromechanical systems (MEMS) sensors. Moreover, the use of guided or autonomous systems carrying suitable sensors for inspection and monitoring has become a key issue for operation management, especially when the site is dangerous or difficult to access.

Therefore, robots and automatic systems can be used in the above-mentioned fields to perform inspection and survey tasks, carrying sensors to acquire information and monitoring structural elements arranged horizontally (using ground robots) or vertically (climbing robots) or from above (aerial robots).

Automatic Data Acquisition for Survey of Cultural Heritage Sites

The relief of a part of territory or architecture consists in a method, based on a series of preliminary operations and measurements, able to return a 2D/3D graph model of the metrically correct observed object.

The development of *laser scanning* techniques is relatively new, though it is growing in the field of cultural heritage restoration, as well as in many others, like architecture, engineering, geology. These tools allow one to determine, quickly and easily and in an automated manner the position in space of a large number of points of the surveyed object. The detection consists in measuring distances by means of electromagnetic waves, also known as light detection and ranging (LIDAR). The device transforms electrical, optical, thermal or chemical energy into a laser beam that hits the object, generating a point cloud that describes the outer surface of the object. These devices are mainly divided into two large families: time-of-flight (TOF) and triangulation systems. In a TOF system, the laser at each point measures three parameters, namely the distance between the center of the grip of the device and the object hit by the laser beam (d), the angle of inclination (φ) of the beam in relation to the vertical axis of the instrument, and the azimuth angle (θ) of the beam emitted in relation to a horizontal axis. Most laser scanners reach accuracies between 6 mm and 100 mm of the measured distance. In the laser scanners based on triangulation, the laser beam is deflexed according to an incremental step, and it strikes the object according to a regular scan. The reflected signal (*point* or *laser line*) is captured by a solid-state image sensor (CCD), placed at a predetermined and calibrated distance (base) in relation to the point from which the laser is sent to the object. According to triangulation, a point is detected as the intersection of two straight lines of a known direction in space. Recently, several applications in the field of cultural heritage have been developed using models based on integrated technologies of surveying, such as the laser scanning and photogrammetry, with the aim of obtaining a virtual model. These integrated systems create a virtual model thanks to the geometric accuracy of the survey, photorealistic representations suited for interactive navigation and manipulation in digital framework [29].

Digital photogrammetry is another new approach in cultural heritage surveying. It allows one determine the shape and position of objects metrically, starting from at least two separate images that represent the same object (stereoscopic couple). This technique is used in cartography, topography and architecture. The development of computers that can handle large amounts of data and computer graphics has enabled increased speed, lower costs and ease of use. New software programs are available able to manage all photogrammetric procedures (orientation and plotting). These programs run on standard PCs and offer user-friendly interfaces, allowing them to be used even by unskilled operators [42]. For this reason, digital photogrammetry currently represents one of the most reliable, economic and precise acquisition technique of data. The first outcome of the photogrammetric survey is a point cloud, which can be integrated with 3D clouds, generated by other technologies, such as laser scanners or total stations. The procedure is structured into four phases: (1) acquisition of the image; (2) numerical transformation of each single point from its position in the projected space on the picture to a 2D position in orthogonal projection, defined as the image straightening; (3) graphics processing of over-mapping on the rectified image, which provides the measurements and formal characteristics of the object; and (4) redeployment of each straightened frame in the 2D space. The procedure makes use of software that can process the acquired data and "dress model". In particular, structure from motion (SFM) is a numerical technique that allows one to reconstruct the shape of objects through automatic collimation of points from a set of photos. Based on computer vision algorithms, SFM extracts significant points from photos and, acquiring photographic parameters, finds matching points across multiple photos, finding their coordinates in space.

Image alignment allows for the automatic orientation of the images in space and the extraction of the key points that are useful in the process of elaboration of the points and the mesh cloud. The use of different software then allows for the exporting of the point cloud in the most common extension files enabling one to post-process the data in other frameworks. The point cloud, processed with SFM, can be transformed into a photorealistic 3D model. In fact, thanks to the use of UV texture maps, one can associate the image texture with the pixels of the image.

Based on the acquisition of infrared images, *thermography*, one of the major non-destructive techniques, uses a thermo-camera to detect radiation emitted in the infrared band from bodies subjected to thermal stress. Known as radiant energy, this is a function of the surface temperature of the materials, in turn conditioned by thermal conductivity and the specific heat, which is expressed in quantitative terms as the ability of the materials to transmit or restrain heat. With this feature, this non-destructive technique is used for the assessment of several traditional historical materials and structures after they had been conserved, restored or repaired using, depending on the case, different treatments [6]. More generally, the primary applications in the field of building are the insulation test, verification of waterproofing, analysis of deterioration due to moisture, searching for causes of water seepage and search for hidden structural elements.

Emerging Technologies for Structural Health Monitoring

Permanent observation of the structural behavior of constructions actuated by the installation of a sensor is commonly referred to as SHM. This approach can be viewed as a possible source of data for further analyses. In this respect, the technical development of sensor technology has been rapid. Several new techniques are available on the market and provide higher performance and more reliable measurements for SHM. Fiber-optic sensors, advanced NDT methods such as acoustic emission and radar-based techniques, in addition to MEMS technology together with advanced data acquisition techniques and database automation supply have benefited SHM.

In the last few years, vibration based monitoring has enjoyed a rapid and extensive development due to the performance provided by the accelerometers. Among the most utilized accelerometers are the piezoelectric, servo and force-balance accelerometers and MEMS. The piezoelectric accelerometers take advantage of the piezoelectric material properties to generate an electric charge when they are subjected to a variable force. They are robust and reliable sensors, having stable characteristics over time, but have difficulties in the measuring low frequencies (below 1 Hz). In general, the servo-accelerometers are extremely precise, but are expensive, bulky and heavy. MEMS are sensors of a different nature (mechanical, electrical and electronic) whose principle is based on the variations of the electrical capacity that is produced by the acceleration imposed to the sensor.

Recent developments in MEMS, wireless communications, and digital electronics are producing low-cost, power-saving sensor nodes that are small, communicate untethered across short distances and possess multifunctional features [1]. In comparison, wired monitoring systems are typically implemented in buildings and infrastructures with a relatively high cost per channel. Such a unitary cost can quickly add up along with the logistical and operational difficulties inherent in the network deployment. Smart sensors, which can include onboard communication, processing and memory features, can combine the low-cost deployment and installation procedures with the richness of the communicated information [43, 61]. In the last decade, wireless communications, low-power computing and sensing technology have overcome several technical problems, permitting the monitoring of different typologies of civil infrastructures using wireless sensor networks (WSNs) [60].

Regarding static SHM, in contrast to the traditional measurement technique based on mechanical or electrical devices, innovative sensing systems based on optical fiber sensing have been developed [33, 41]. Optical fiber sensing has many important advantages compared to typical strain gauges: optical sensors are small and very lightweight, they are immune to electromagnetic interference and corrosion, and importantly, they possess embedding capability [27]. The basic principles regarding the monitoring of civil engineering structures using optical fiber sensors is illustrated in [2]. A general system needed for a network of fiber optical sensors is composed of a light transmitter, a receiver, an optical fiber, a modular element and a signal-processing unit. One of the most widely used optical sensors is the fiber Bragg grating (FBG), especially for the civil SHM, [11, 64]. A rough manner to describe the mechanism of an FBG sensor is the following: when a broadband light coming from a light transmitter passes through the grating of the sensor, a specific wavelength, called a Bragg wavelength, is reflected; this wave corresponds a grating period. If the grating is subjected to some deformation, the period of the wavelength reflected will change and so it is possible to connect the deformation with the variation of the Bragg wavelength. It is worth mentioning that the total strain is given by the sum of the mechanical and thermal strains. For this reason, optical measurements have to be purged by thermal effect. The complete set of the SHM sensors can communicate and be connected to the central acquisition and the signal-processing unit in two ways: wired or wireless. The rapid growth of ICT-based solutions has led in the last few years to the important development of WSNs [22, 26]. A typical sensor node contains a microcontroller unit, a radio unit, some kind of long-term stable storage (e.g. Flash memory, SD card) and I/O capabilities to support sensors. A recent review of SHM actions performed in Italy has pointed out differences between permanent installations and short-term experimental campaigns, with a particular focus on applications devoted to cultural heritage sites, which have an important historic, strategic and economic value for the country [25].

In most common cases in which one-day testing is operated instead of SHM permanent monitoring, movable monitoring tools for specific testing on site are traditionally positioned in planned interventions by specialized operators. In such cases, automation process can play a relevant part in the process.

Automatic and Robotic Inspection

Robotics and automation are undergoing a widespread application in many sectors. They were originally developed to be used in industrial environments. Then with the advances of related disciplines, such as mechatronics, informatics and mechanics, they have rapidly expanded in scope and application. Most recent applications involve urban search and rescue, remote exploration [48], service and simulation [49], military intelligence and defense [69], manufacturing, cleaning, healthcare and motion aiding [15]. The original scope was developed to substitute humans in repetitive and dangerous tasks perfectly applies to the specific application we are dealing with, that is monitoring and inspection of site of interest in a broad sense. More specifically, robotic or automatic systems can serve an intelligent agent to carry necessary instrumentation for the survey, monitoring or inspection, due to planned or emergency response operations. According to the task, the site or component may be reached and inspected by *ground*, *serpentine*, *climbing* or *aerial* systems. In the following section, the main types of robotic and automatic systems are classified.

Unmanned aerial vehicles (*UAVs*) are aircrafts without human pilots and can fly remotely controlled by an operator, controlled autonomously by a computer or semi-autonomously as a combination of both capabilities. This technology was originally developed for military purposes, but in recent years, it has been applied in different scenarios, such as in agriculture, where it is used for spraying pesticides, to urban search and rescue or inspection [31]. UAVs may be classified according to the flight type, as reported in Table 1. The winged drone with a multi rotor is the most appropriate for inspection tasks.

More specifically, this flying method, obtaining lift from one or two rotors rotated horizontally, similar to a helicopter, is classified as quadcopter, hexacopter and octocopter, according to the number of rotors. For the application of UAVs in the field of civil engineering and cultural heritage, these flight systems can be equipped with different kinds of high definition cameras and other equipment with many applications in the field of photography, photogrammetry, geology, geography, meteorology, agriculture and forestry. Recent advances report studies on the integration of UAVs, image processing and data acquisition procedures for crack detection and assessment of surface degradation in structures and infrastructure [39]. The UAV could follow a predetermined path or could move by visual servoing and detects by the means of image treatment the size and location of defects and cracks. The use of a UAV for inspection has multiple advantages over

Table 1 Classification of UAV according to flight type [39]

Flight mode	Description
Fixed wing	Relatively long-endurance due to small fuel consumption Runway or open space is needed
Rotary wing	Needs only a small space due to vertical takeoff and landing As led fuel-efficient, long-endurance is limited
Tilt-rotor	Variable take-off and landing Difficulty of stability/reliability
Co-axial	Flight duration and the system are stable, easy Needs more drag Time is needed for high-speed flight entry
Multi-rotor type	Three or more multiple rotors are mounted Vulnerable to wind

traditional inspection methods based on expert personnel: (1) reducing work accident risk, (2) lowering costs due to less logistics and fewer working hours, (3) eliminating operating interruption required for the structure or infrastructure, and (4) opening the possibility of using nondestructive techniques for inspection.

Ground mobile robots generally use *wheeled/tracked*, *legged*, a combination of both types named as *hybrid mobile*, or hyper-redundant articulated systems called *snake* or *serpentine robots.*

In general, legged locomotion requires higher degrees of freedom (DOF) and therefore greater mechanical complexity than wheeled/tracked locomotion. Wheels, in addition to being simple, are extremely well suited to flat ground, but they suffer in overpassing obstacles that are greater than the radius (dimension) of the wheel (track).

In addition, as the surface becomes soft or irregular, wheeled locomotion accumulates inefficiencies, due to rolling friction whereas legged locomotion suffers much less, because it consists only of point contacts with the ground. Theoretically, legged robots can go where wheeled robots cannot, so they have a potentiality that justifies the extra effort required controlling their locomotion, but they suffer from some drawbacks, such as energy consumption, stability and locomotion speed, if either we are dealing with two, four, six or more legs. The key advantages of legged locomotion include adaptability and maneuverability in rough terrain. Because only a set of point contacts is required, the quality of the ground among those points does not matter as long as the robot can maintain adequate ground clearance [59].

A number of different leg configurations can be found in nature: mammals and reptiles have four legs, and insects have six or more legs. In some mammals, like humans, bipedal walking and even jumping on one leg are performed. Walking can be performed either with static or dynamic balance. In order to achieve static walking, a robot may have at least six legs [59]. In such a configuration, it is possible to design a gait in which three legs are statically stable in contact with the ground. The leg, which may have several DOF, must be capable of sustaining part of the robot's total weight, and in many robots must be capable of lifting and lowering the robot. A classification of each individual leg can be also given, as reported in Table 2.

A single DOF leg design is considered in [57]. In the general case of legged mobile robots, a minimum of two DOF is generally required to move a leg forward by lifting and swinging it forward. An additional DOF can be added for maneuvers that are more complex.

Biped robots can have a fourth DOF at the ankle joint enabling more consistent ground contact. In general, adding DOF to a leg increases its maneuverability, but the drawbacks are related to additional joints and actuators, energy, control, and mass.

Hybrid mobile systems were developed to exploit the terrain adaptability of legs in rough terrain and simpler control and high-speed operation with wheels. They can drive quickly and smoothly on flat terrain and can stably negotiate natural or artificial uneven terrain.

Table 2 Classification of mobile robots

	Type	Characteristics
Wheeled	3 or more wheels (active and/or passive)	Efficient motion on almost flat surface Low energy consumption Fast motion
Legged	1, 2, 4, 6, 8 legs Each leg: 1, 2, 3 DOF	Efficient motion on uneven terrain High energy consumption High number of DOF—Complex to control
Hybrid	Combination of legs and wheels/tracks either active and/or passive	Combination of locomotion types Efficient use in complex unstructured environment (exploration, nuclear, search and rescue)
Serpentine	Hyper redundant modular robots with wheels or legs	Efficient motion on uneven terrain Small size solution well suited for inspection
Climbing	Magnetic or pneumatic	Efficient for vertical climbing inspection

Hybrid robots can be classified into two main groups: first, legs (or sub-tracks) and wheels (tracks) work independently and articulated legs (sub-tracks) provide traction for obstacle-climbing when wheels (tracks) cannot; second, wheels are purely passive and used to stabilize the system [50]. The solution dealing with an independent sub-track is adopted in [48, 51, 69], in which robots are used for search and rescue, inspection of nuclear plants or hazardous environment. The second type of hybrid locomotion involves wheels (or tracks) installed on each leg. In particular, the articulated leg efficiently overcomes obstacles that cannot be surpassed with tracks or wheels only.

Serpentine or snake robots belong to hyper-redundant articulated mobile robots [28] and can be used in industrial environment, but also for applications such as pipe inspection, search and rescue, and military. They have the ability to negotiate rugged or uneven terrain, such as concrete floors cluttered with debris, or unfinished floors, such as those in constructions sites. They have the ability to climb up and over tall vertical steps and stairs, and to travel inside and outside of horizontal, vertical or diagonal pipes, such as electric conduits or water pipes. Serpentine robots can be divided into two groups based on two characteristic features: with active joints with legged, wheeled, or tracked propulsion, or active joints (wheels) and passive joints (between two adjacent body segments).

Inspecting structures and infrastructure means often inspecting the surfaces composing the different parts, and thus being in close and safe contact with the surface is crucial in any direction regarding gravity.

Adhesion can be achieved by magnetisms, being most of the industrial elements ferromagnetic, with either magnetic wheels or magnets in the body structure [13]. Magnetic wheels have the advantages to be in direct and close contact to the ferromagnetic underground and thus having a reduced air gap essential for strong adhesion.

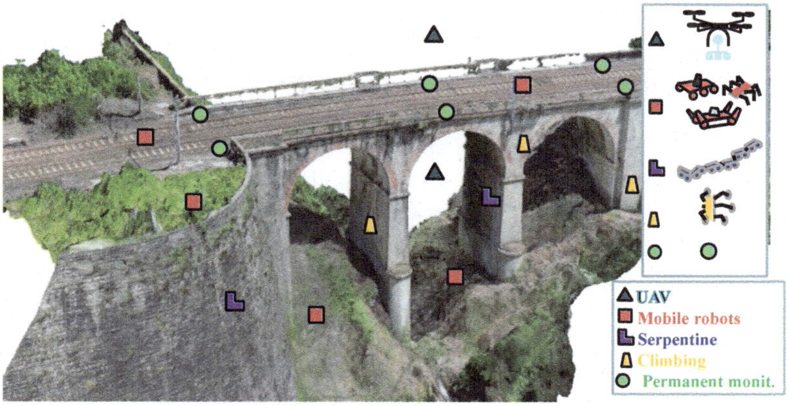

Fig. 2 Example of automatic/robotic access to an infrastructure

It is worth noting that, in the context of survey and monitoring in cultural heritage sites and civil engineering, several NDT/NDE techniques are ripe for automating. This can be pursued either by placing/installing/embedding sensors at some critical points of structures (*permanent monitoring*). Alternatively, sensors can be brought to the site of interest by an automatic/robotic system (survey or non-permanent monitoring).

Note that visual inspection, like radiographic testing and active infrared thermography, is based on the information that an image provides. In addition, if sensors are mounted on a robot to control their position and orientation during the acquisition, the use of suitable algorithms based on image processing is enabled/facilitated [70].

Figure 2 shows a scheme of possible use and access of robotic and automatic solutions for the monitoring and survey, which may be used in combination of the well-established permanent monitoring.

3D Graphic/Numerical Modeling and Information Digital Tools

The digital data acquired through the different technologies, summarized in the previous section, furnish generally and directly or by means of a pre-processing, meaningful input to different modeling and data storage environment. This section describes the most used digital tools, which take data from automatized acquisition system; the analysis of the different tools permits one to evidence the possibility to get information from independent model environments, which can possibly be merged and used for further decision and performance evaluation.

Automatic Processing of Data in GIS

GIS can be included within information technology communication (ITC), since it is based on the use of computer applications for the storage, retrieval, manipulation, analysis, display and share the geographically referenced data [63]. It was created in 1960, but only in recent decades research on both the methodology and application of GIS has greatly developed all over the world, as a very important and versatile tool for many different geographical fields, including natural, social and medical sciences as well as historical, archaeological, architectural and engineering [30]. The increasing use of these systems has prompted the international community to draw up a very well-structured set of standards to define, describe and manage geographic information with the purpose of facilitating the interoperability of GIS, including interoperability in distributed computing environments [34]. In detail, the ISO/TC211 specifies the methods, tools and services for data management, by acquiring, processing, analyzing, accessing, presenting and transferring such data between different users, systems and locations. These standards are organized into themes for the geographic information technology framework: architecture & framework, metadata, data content & definitions, core data model, data exchange formats, data interchange and services, data quality, spatial referencing and imagery.

Based on ISO TC211, the GIS design requires only two descriptors, what is present and where it is, to represent geographical phenomena of the real world [12]. Thus, GIS features depend on the specific application, and its use is defined via a link between the database and digital cartography (geo-localized). In particular, the alphanumeric and iconographic data are linked to vectorized cartographic units (point, line, polygon) or to raster elements (center of pixel) [54]. Briefly, these systems create maps of the territory with different themes to which it is possible to associate datasets (with attributes and locational information). The result is the possibility to establish a "cross-correlation" among the different data and return a full view of the results without sacrificing readability. In fact, through the process known as "overlaying," GIS enables the representation of a variety of information pertaining to the real world on the same map sheet. Therefore, thanks to the great flexibility of GIS, in combining the information from cartography sources (i.e. maps), earthbound surveys, remote sensing (i.e. aerial and satellite imagery) and transforming and manipulating interactively in one spatial structure [36], it is widely used for the management and monitoring of environmental and urban factors and the use of the land. With regard to territorial management, GIS is particularly used in the development of technological networks and the mapping of geological risks, such as those induced by earthquakes, hydro-geological events. In the scientific community there are in fact several applications in the field of *management*, such as the coupling of eco-hydrological modeling with remote sensing, landscape ecological analyses and GIS to develop a series of indicators of water catchment health as part of a geographical audit of environmental health [5] using multi-criteria analysis based on GIS as a model to support and to integrate the decision of the multiple actors of land management [35]. In several other instances, GIS analysis has proved to be a valuable tool in monitoring, such as the system applied to the monitoring of the effects of excavation in sensitive urban environments [21].

The integration of remote sensing techniques and GIS has an increasingly wide range of applications in many different fields of land survey and most earth science disciplines, including military, intelligence, commercial and economic planning, and humanitarian applications, due to the possibility to import the numerical data of aircraft or satellites into GIS. The application of Geo-information in current usage intended as remote sensing techniques, together with GIS, is also used to analyze and assess many urban phenomena: for instance, using multi-temporal Landsat Thematic Mapper data, the land use/cover change detection is carried out and by using a GIS-based modeling approach, urban growth patterns are analyzed. The integration of remote sensing and GIS is further applied to examine the impact of urban growth on surface temperatures [19].

With regard to enhancement, territory and cultural heritage management, two different tools have taken shape over time, depending on the scale of representation: the system currently known as GIS, to manage urban and territorial scale data and the system known as architectural information (AIS), which is concerned with architectural scale data. With the aim of integrating the two systems, the new hybrid system, based on 3D representation, has recently been created, capable of solving the problem of scale range [47]. The use of scale range is very important, for instance, in the analysis of historic small towns, which requires a system able to perform analyses in urban planning, but also in archaeological contexts or architectonic fields, respectively, by tuning the scale of representation from small, medium or large. Moreover, the benefits of GIS in the context of cultural heritage data management mainly relate to visualization and communication using the interfaces of the Web Information System [46]. Valorization, in the current international conception, is the knowledge and dissemination of the gathered information.

For this reason, informatics and telematics are becoming more and more fundamental in playing a critical role to understand, protect and manage cultural heritage. In any case, the design of a GIS should always take into account that "the complexity of the territory consists not only of metric parameters, such as dimensional and formal aspects or the management of digital models taken directly from photographs or derived with some manipulation, but also a large part of the complexity lies in the field of non-visible phenomena, which transform the territory and determine the quality of living and the environment" [44].

It is a mistake to think about information systems as only the equivalent of old, but technologically advanced maps, because they function as a cartography and as interactive information tools at the same time: by means of computer networks the system is able to evolve, to load new applications and finally to integrate data continuously. The features of GIS allow the investigation of the results of the analysis of new technologies applied in the field of survey. In recent decades the development of technologies, such as laser scanning, thermal imaging camera, aerial photogrammetry aided by drones, for the survey of architectures, structural aggregates, urban areas and so on, requires an increasing number of systems capable of making a logical link between the database management system (DBMS) and the relative raster or vector cartography, in specific spatial references. It is well known that 3D laser scanner technology can be used as a scanning operation in any complex environment or space. Moreover, the system can directly connect 3D data of scanned objects or scenes (in any large, medium and small scale, complex, irregular, standard or nonstandard, etc.) to computers, and then it quickly gives back 3D model and cartography data, such as line, surface and space [16]. The laser scanning operation acquires the object as a data point cloud, which can be post processed, such as mapping, measuring, analyzing, emulation, simulation, displaying, inspecting, VR, and so on. In laser scanning technology an object is hit by an artificial light ray, laser light (LIDAR), at a very high speed which, by measuring the reflection time of an electromagnetic radiation (monochromatic light wave), enables to determine the position (distance) of hundreds of thousands of points in a given metric and topological reliability. Furthermore, the recording wave intensity of return laser light allows detecting the reflectivity values of the materials, radiometric coefficient (intensity at last pick point) known as reflectance. In the laser scanning process, the reflectance values are associated with the individual points through the red, green, blue (RGB) color space. The reading and analysis of the RGB values of the detected points enables one to make attractive valuations, especially in the identification of the elements of the architectural object and any degradation.

GIS assists in post-processing the reflectivity data of laser scanning due to its topological features (geo-localized points). Taking advantage of the geo-referencing of the cloud, GIS can import the 3D model, or digital surface model (DSM)—that is the cloud points acquired with scanning—within information structures using ASCII files to process the coordinates of all points of the cloud and the extension of the drawing exchange format (DXF) file for readability. Once the cloud is imported in GIS, the system is able to recognize for each point P (x, y, z) the color value of the RGB color space. Moreover, with the aid of a few queries, it was possible to segment the entire data in into different layers depending on the range of the RGB value.

By comparing these ranges with the values of the last pick point (Info Point Cloud), it is possible to associate each range with a different object—a material such as an organic coating, for example (Fig. 3).

The structured information system thus allows the single or simultaneous reading of the identified layer, thus allowing the performance of metrics and quantitative evaluations as well as on the 3D digital model [45].

Recently, thermography technology, applied in the energy assessment of building or urban areas, has taken on a particularly significant role in the diagnosis of a historical building due to its non-destructive approach. This technology is increasingly used to detect, for example, any closures of openings in walls under plaster, the type of masonry, the presence of lintels or stone columns, geometry of

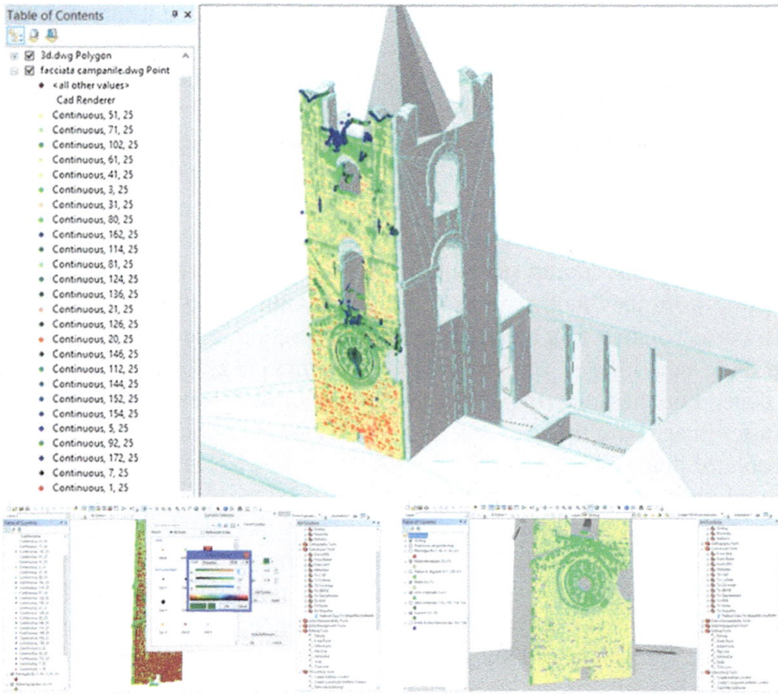

Fig. 3 Example of post-processing of point cloud, captured by laser scanner, by GIS

vaulted surfaces (false times), the degradation associated with the detachment of material, organic coatings or infiltration and thermal dispersion.

Thermography is based on the measurement of the infrared radiation (heat intensity) emitted by a body hit by natural light. The result of the thermographic survey is a raster thermal image of the object, or a map, in "false colors" as the "representation of the thermal state": the system associates the relative color with each different surface temperature. The variation of the color (from dark purple to yellow) depends on the temperature range chosen by the operator, according to the internal/external temperature variation of the investigating object. In brief, the thermographic camera detects and measures images in the frequency of the infrared band, allowing one to see different levels of thermal energy emitted from an object.

The energy levels are much greater, in terms of emitted infrared radiation, the higher the result in the temperature of the object itself. The variation of the energy levels depends not only on the chemical composition of the materials of the object investigated, but also on the presence of thermal bridges caused, for example, by water seepage. The thermographic investigation is qualitative and the results are enclosed in a raster image, in which the chromatic scale of the temperature range is selected by the operator. For this analysis as well, GIS is able to process not only vector images, but also the raster results of thermographic analysis. In fact, by reading the data from the center of the pixel, one can obtain colorimetric or Boolean information. Using the application Map Algebra, which operates with algebraic-own math functions on pixels of the images, one can process raster input data, obtaining new raster maps (Fig. 4). Starting from the raster image of the object, with the thermographic camera, GIS can rework the image and produce new ones, each of which contains separately the information about the single parts of the wall, such as bricks, mortar or mildew. The system achieves the new images by subtracting from the "mother image" the pixels with the RGB value, or range of values, corresponding to a specific material, generating a new raster image.

On these new images, one can perform quantitative assessments, which, however, depend on the property of the camera (size of the pixels and dpi image). The new raster images can then be "smeared" on 3D terrain or building models, as GIS is capable of doing [45].

Digital photogrammetry is thus able to create 3D surface models as a cloud of millions of points. Using specific image acquisition procedures, which can even be automated through the use of aerial (drones) and terrestrial (mobile robots) systems, one can obtain a 3D surface modeling of the object, including the geo-localization and a specific project of gripping points. In the photogrammetric processing, the digital images are enhanced by software able to generate 3D spatial data of a surface model upon which it is possible to apply the texture, generating photorealistic models. The density of the point cloud can be higher of lower depending on the number of points chosen by the operator. The possibility of analyzing the 3D model surface in GIS creates a very important tool because of the interconnection between the data set and the geometry of object. This tool is very important in the field of cultural heritage preservation, above all in archeology. Figure 5 shows the computer vision approach that can create cloud and mesh model of a typical vaulted surface in southern Italy (Lecce), and how one can import into the model in GIS, both as mesh and textured models.

Fig. 4 Example of post-processing by GIS of raster images, captured by thermography: creation of new maps

Fig. 5 Mesh and texture model surface imported in GIS: typical vault surface of Southern Italy (Lecce)

Automatic Data Acquisition for BIM

Building information modeling (BIM) is a graphics and alphanumeric data model regarding the life cycle of a building or a building complex, from the phase of designing to realization and maintenance. BIM technology thus constructs an accurate virtual model for planning, design, construction and operation of the facility. The main aims of the system are to help architects, engineers and constructors (AEC) to visualizing what is to be built in a virtual way, identifying any potential design, construction or operational issues and encourage further integration of the roles of all stakeholders on a project [7]. The greater efficiency of BIM is mostly in architectural design due to the ability of the system to communicate and integrate all tools relating to the metric list of price, the development of photorealistic images, land registries and energy certification. In structural fields, BIM decreases the time of communication between modeling packages and calculation programs, because it avoids re-modeling of the structure, reducing possible human errors, transferring information on the sections used (e.g. materials and the characteristics of the construction elements). In plant engineering the system facilitates dialogue and the integration of data (output of different software) relating to the sizing and placement of the implants, and the modeling and metric list price calculation. Infrastructure design generally requires a high degree of control of the project in each phase and the integration of related design disciplines. BIM helps avoid mistakes and information deficiencies, turning information into insight to optimize designs and help accelerate approvals, resulting in more effective and resilient infrastructure. In the constructive process, the system allows a more accurate control on the phases and on the sequence of the process. It controls the cuts and scribing of the pieces with assisted machines, defines project management that involves all stakeholders in the work and facilitates the management of the shipyard allowing the manager to follow at a distance for example the storage and positioning of materials. During the life of the building, a BIM system with all the data acquired allows the implementation of the facility manager, which is the new frontier for many of the maintenance processes.

In the context of this chapter, the interaction of a series of available automatized technology (UAV, robots, smart glasses, drones) able to acquire data that can be utilized to construct BIMs of existing constructions has garnered much interest. These models are named as-built BIM (AB BIM) emphasizing the fact that the models are reporting the features of an already constructed object. Creating an AB BIM requires two major steps: data collection, to capture the as-built conditions, and data modeling, to generate compact, but rich representations readily understandable by other processes.

Auspiciously, the 3D reconstruction field, related to computer vision and/or 3D laser scanning techniques, filled the technology gap, offering off-the-shelf tools for generating 3D models of scenes. Consisting basically of a set of 3D points endowed with 3D Cartesian coordinates and possibly color information, these models that are denoted as point clouds are useful for visualization or augmented reality purposes (see the bridge represented in Fig. 2). However, the data modeling side continues to be deficient, and the problem of converting the raw point cloud into a semantically rich BIM model is far from being settled [52]. The commercial and academic tools available at this point to perform this conversion require extensive human intervention, making them expensive and prone to error. Given the expected economic impact, automatically generating AB BIMs from point clouds is a key objective for the industrial, academic, and governmental parties involved in the architectural/construction and facility management industry (AEC/FM). In this respect, a set of images acquired by UAV and human-based vision permit one to reconstruct a cloud of points that can be converted in a BIM, such as in the case of bridges [55].

Integration of Sensors and Acquired Data in BIM

The benefits associated with the installation of SHM as recently outlined may have different applications [67]. Certainly, the raw data measured by the sensors must be processed and put into a geometric context within the infrastructure asset, which facilitates the interpretation and analysis of the data. This supports informed decision making, which in turn leads to effective actions. Recently, a new approach has been proposed that enables the modeling of a structural monitoring system in a BIM environment, thus permitting sensor data storage be visualized directly on BIM models [18]. A case study related to a pre-stressed concrete bridge demonstrates that by including and visualizing monitoring data directly on BIM models the acquired data gain geometric context within the built asset, which facilitates better interpretation, analysis and all the data-sharing benefits associated with the BIM approach. In the case, the data measuring dynamic strain changes at discrete locations obtained by 140-fiber Bragg grating sensors installed on the main structural elements of the bridge, which are visualized in the BIM of the bridge. Recently, the bridge and engineering firm Mabey has unveiled the industry's first "Live BIM" tool, which will bring real-time monitoring of stresses on buildings. Using Live BIM, 3D models of buildings, rail, roads and bridges are connected with real-time sensors (FBG, accelerometer) as permanent SHM. It combines 3D modeling with other data and creates digital graphics evidencing problems [37]. A real-life and novel integration between BIM and radio-frequency identification (RFID)-based wireless strain sensor technologies has been proposed to enable automated structural condition assessment [8]. In this case, BIM has been used as an ideal 4D graphical computing environment for integrating SHM with management practice. This system can rapidly identify and locate the spot where pre-set strain or crack has happened and the corresponding structure element of the BIM model is automatically highlighted to provide an alert signal. A custom developed middleware was developed to enable automated sensor data entry into a BIM environment and thus avoid possible human-related errors.

Structural Finite Element Model Updating by BIM Data

The process of automation in the construction of new properties or in the management of existing ones cannot avoid in the future the interconnection between BIM and finite element analysis (FEA). In new construction, full integration of BIM and FEA allows for the complete control of structural design during the entire development through initial design, manufacturing processes and construction through the structure's life cycle [23]. Moreover, the potential for reducing the environmental impact of structural systems through a more efficient use of materials has been developed through a holistic and integrated methodology that utilizes BIM capabilities combined with structural analysis and life-cycle assessment (LCA) [20].

The process of automatization in the construction of a finite element mesh used to describe the analyzed structure needs the knowledge of the geometry of the object, which can be defined both for BIM and for FEM, and it can be acquired either by laser scanner or by image processing techniques, such as those explained in the previous section. Sometimes the processes of automatic data acquisition are used for both BIM and FEM in an interconnected manner.

In other cases this possibility is explored just for one of the two modeling ambients, BIM or FEM. On the one hand, the case of a medieval tower, a new semi-automatic procedure to transform 3D point clouds of complex objects to 3D FEM has been recently presented and validated [14].

On the other hand, the landmark building of the World Expo in Shanghai, which has a free surface with single-layer reticulated shell structure, has been analyzed using BIM-FEM technology to model the Sun Valley, including architecture modelling and structure analysis.

The new approach is viable and effective in combination with different software such as Rhino, Revit, and Midas in solution of the complex-shaped surfaces' structure for modeling and calculation [68].

The area of integration between BIM modeling and FEA is still in its infancy, but holds immense potential, especially if defined within a common framework of the entire process, as discussed here.

Information Processing and Decision-Making

The integration among robotics and automation as well as ITC allows one to generate very reliable models and define digital tools for the decision-making phase. In this phase the information that should be processed comes from the transformation of the output of the acquired data that have been used to create models (GIS, BIM, FEM).

The output of the modeling can be further enhanced together with other information from a direct treatment of other data, as in the case of the damage detection, which is expressly considered in this section. Finally, the outcome of the information processing helps operators and local governments in decision making.

Techniques for Damage and Defects Detection

In the last two decades, *digital image processing* has been introduced for identification and measure of defects in civil infrastructures. In this regard, in recent years, many research endeavors have attempted to empower visual inspection and optical imaging by quantitative image analysis or to achieve automatic or semi-automatic damage detection [17]. It has to be mentioned that a large percentage of damages (e.g. cracking, spalling, deformation, or collapse induced debris) in buildings and bridges may be captured using a commercial digital camera. Therefore, visual inspection for damage detection can be linked to image analysis and processing. Digital image processing allows for the extraction of quantitative and qualitative information of a reproduced scene starting from one image and producing a modified version of that image through calculation by numerical methods or discrete units.

The two main tasks are (1) object recognition, and (2) damage recognition and quantification.

One of the most used techniques for object recognition is based on Haar wavelets. They are natural set basis functions, which encode differences in average intensities among different regions. This approach requires the training of an object detector by means of a process of learning [65], with a database of positive and negative images.

Another approach is related to finding correspondences between two images of the same scene. The search procedure for finding discrete image correspondences can be divided into three main steps: first, “points of interest” are selected at distinctive locations in the image, such as corners, blobs, and T-junctions. Second, it requires that a feature vector represents the neighborhood of every point of interest. This descriptor has to be distinctive and, at the same time, robust to noise, detection errors, and geometric and photometric deformations. Third, the descriptor vectors are matched among different images [10].

Image processing-based methods can be classified as based on color information, textural information or a combination of the two, to segment and extract regions of interest in images. Texture and color-based segmentation approaches are the primary modes of segmentation employed for image analysis. While both approaches have important applications in image processing methods, the color-based methods have been researched to a much greater extent. Texture may be considered as an innate property of surfaces, and this technique finds particular relevance in cases where the regions of interest are more separable from the background based on their texture than color.

Color detection techniques allow for fast processing and are highly robust to geometric variations of the object pattern and viewing direction [38]. The most used color space is RGB, where each color is represented by the components (red, green and blue), which are added together. In the last few decades, this method has been used for many applications, since this technique can avoid problems related to the illumination issues by applying a normalized color space. Damage and defect detection can be then based on the contour and area detection methods that have to be implemented in the framework for damage scene analysis, as it is described in [55].

Figures 6 and 7 show examples of DIP for damage detection on a building and a bridge. It is worth mentioning that in addition to monitoring and inspection, the process of damage identification also can be semi-automatized [55].

Performance Evaluation

The results reported in the previous sections have evidenced that it is possible to have an indirect impact of the use of automation and mechatronics in civil engineering on the performance evaluation of a given structure. This potential enhancement is directly related to design automation methodology in which the technical uncertainty of inadequately defined design problems may be formulated through the application of intelligent design systems [62]. Many leading companies are exploring the potential of cognitive computing and artificial intelligence (AI) where software learns and applies its new knowledge. In the future, AI could run a whole range of complex systems, learning under human supervision then

Fig. 6 Example of defect analysis by DIP: from the original image (1), the crop area (2) is selected identifying the defect and the boundaries (3)

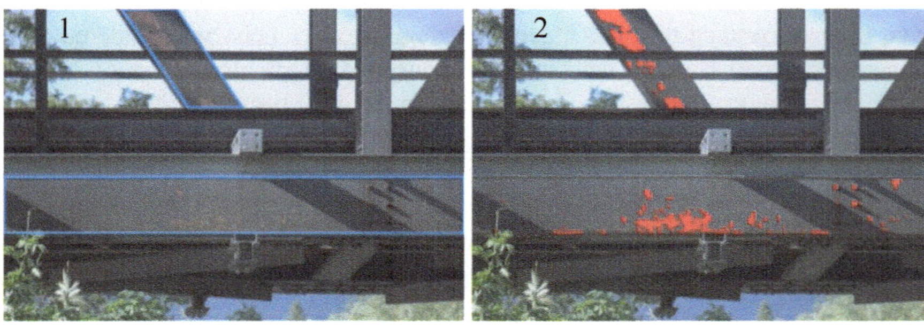

Fig. 7 Example of defect analysis by DIP: from the original image containing the selected components (1) the boundaries and defect areas are identified (2)

operating independently once it has achieved expert status using AI and learning algorithms.

Integrating technologies from the disciplines of robotics, sensing and process control are a decisive step on the road to automation. The oil industry is particularly interested in bringing new tools together to create a network of interacting, intelligent systems that will change how the oil and gas industry operates, delivering a step change in asset performance [32]. In bridge management, the impact of automation in inspection may affect directly the evaluation of the bridge condition and its correlation with bridge reliability [24]. Indeed, the use of image processing may reduce the uncertainties related to visual defect evaluation, which has been used to determine the condition of existing bridges. Moreover, the same processing of images may contribute to the determination of exact geometry of as-built construction, which may be favorably used to construct a 3D-point cloud model from which it is possible to generate a BIM model and consequently an FEM model, eventually updated by acceleration measurements [55]. This process may have a direct impact on the structural reliability evaluation, especially in reducing the scatter in the involved quantities. The potentiality of automated integrated procedure from data acquisition to structural analysis are far from being fully exploited, as is proved by the large on-going research in this area.

Decision-Making Procedures

The possibility of having decision-making (DM) procedure opens a vast area of applications and it plays an important role in the life of constructed objects, both historical and new ones. Decisions are required in the time span starting from conceptualization of new objects or planning for the renovation of old ones and covering design, construction, occupation, and decommissioning. Classical disciplines of engineering are also useful tool for facilitating decisions in civil engineering. However, all of these methods can be seen as means of providing input information for a formal DM.

The formal DM methods were intensively developed and applied to various engineering systems in recent decades. They are known as methods of multiple-criteria decision-making (MCDM). A recent special issue devoted to several applications of these methodologies has been published in 2015 [3]. Among these in our context, it is relevant that the framework developed for LCA and life-cycle cost assessment (LCCA) has been applied to a real building case to evaluate three possible alternatives for an external skin system. In the studied case, it took approximately 12 h to perform the BIM for the three alternatives for the external skin system, and the model provided information on the material quantities required and energy consumed, which are key items of information for conducting LCA and LCCA. The BIM made it possible to obtain the results of the quantity calculation immediately [58]. However, automated data collection can be pursued with different methods, and then introduces the automatic as-built BIM model creation process through a case study, which also successfully demonstrated the interoperability between the created as-built model and a typical energy simulation tool. At last, a discussion is made about the limitations and challenges of the current state of practice to enlighten the future direction [66]. It appears, then that BIM procedures are prone to be interconnected with DM methods to give a full support to the stakeholders. Indeed, BIM goes through a series of levels of development (LOD) which represent increased detailing that can be taken into account along with multiple-criteria decision-making (MCDM) [4].

Security and safety are activities, which are performed in order to keep people and organization safe from attack. Safety is achieved by protecting objects of civil engineering against fire, lightning, earthquakes, extreme winds, and contamination events, which can negatively affect indoor air quality. Effectiveness of security and safety can be characterized by at least two MCDM criteria (one for security and one for safety). Further criteria related to them may be costs of security and safety systems. We can add criteria, which measure impact of architectural and structural solutions on security and safety.

The presence of a number of criteria makes a case for the application of MCDM to building design with respect to security and safety. In an MCDM oriented towards security and safety, alternatives can also represent different architectural and structural solutions influencing security and safety. The architectural and structural solutions may seriously influence security and safety.

Making decisions in the design, construction and management of cultural heritage and civil infrastructure is a fully supported process through the integration of new and emerging tools customized to the problem and coming from competence related to mechatronics, automation and ICT in a continuing exchanging of multidisciplinary competence. In this context, the ability of GIS to acquire, for instance, update data, to manipulate them and to analyze the results statistically, showing the most suitable solution is helpful generally in the management. Moreover, the statistical analysis of events involved in how the monitoring system works can be realized, of course, by many methods whose applicability at each event is determined by the combination of parameters that characterize the process studied [67].

In cultural heritage management the "informative representation" (in which the physical, geographical and topological data are associated with the narrative and cultural characters of landscape) of GIS combines "e-government" with the new way of thinking about landscape protection, through the direct involvement of citizens (individual or collective) and through the promotion of transparency, dialogue and participation. In cultural heritage monitoring to protect for instance the historical or natural landscape, GIS can help the local government and operators to assess, manage and mitigate risk related to human actions (e.g. overexploitation of the soil) or physical vulnerability of the territory (e.g. seismic activity). The historical towns combine the homogeneous urban composition with the precious architectures and the natural environment: for these characteristics, they are defined as "environmental monuments". However, they have often arisen on particularly seismic territories, thus requiring a clear prevention policy.

A 3D graphical model of an urban area, in which is stored all GIS data for the city, can create a tool capable of defining the different levels of vulnerability (indexes). Through direct or instrumental survey (e.g. laser scanners) one can acquire all information regarding the urban settlement (city center), the structural aggregates and units, to identify indexes defining the attention of the environmental quality and the attitude to degradation. These indexes are subsequently linked to the structural unit and represented by different colors on the 3D model, thereby generating rankings of risk: the different levels of risk are characterized by a color scale (from a darker to a lighter color), according to the greater/lesser degree of criticality (Fig. 8). This procedure defines the scheduled maintenance by identifying buildings with a higher level of risk, thus suggesting priority actions to be taken on buildings before the occurrence of a seismic event. The ratio of cromia/indexes is automatically updated when the information system data are changed following ordinary and extraordinary maintenance interventions on buildings: the updating is also displayed through the chromatic variation on graph model [53].

The vulnerability of the natural landscape is often the result of human activity, as in the case of mining exploitation. In many areas in the world, throughout different

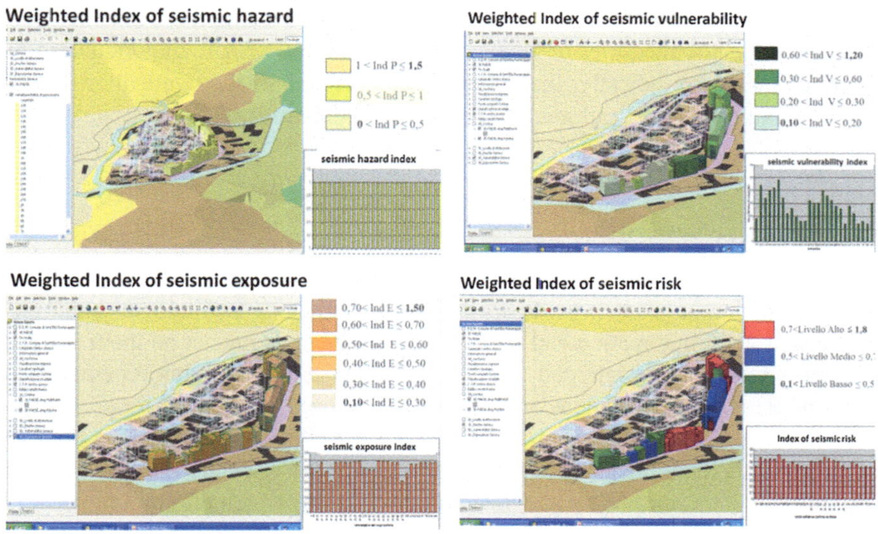

Fig. 8 Example of assessment of seismic risk by GIS through hazard, exposure and vulnerability indexes: historical town of S. Elia Fiumerapido (Southern Lazio)

periods, intense mining activity has "disfigured" the landscape: the processing methods have often altered the face of the mountain, removing the surface layer of vegetation and plant soil. This activity creates substantial visual discontinuity and accentuates the chromatic contrast between the light color of the bare rock and the surrounding environment characterized by different shades of green. The preservation of the landscape must always start with an exhaustive collection of knowledge regarding the deterioration: to this aim, the design of GIS for the governance of the territory may be of help.

At the base of the system, a data set relates the types of quarries (slope, summit, etc.), size in square meters and quarried volumes, types of processing, distance from the historical centers of the towns, and infrastructures and limitations by law of national and European rules (Fig. 9). This system is designed as a control tool for the excavation process and as a guide to defining the best interventions for recovery of the exploited area.

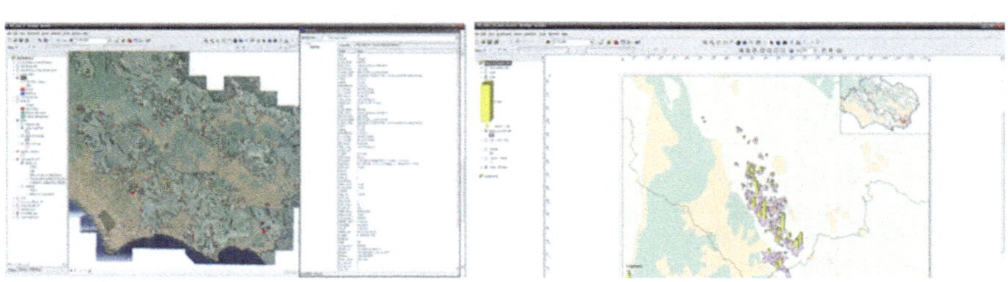

Fig. 9 Example of analysis of mining activities by GIS: Coreno Ausonio district (Southern Lazio)

Indeed, it provides local government and technicians a simplified reading of the data and criticality measures of land use. In addition, by periodic importation of the cloud of points periodically, the system is able to control the overexploitation of each monitored quarry, and identify the rules irregularities [18].

Conclusions

This chapter has aimed to provide insight into the emerging technologies in the area of mechatronics, automation and ICT, which will have an impact on the traditional processes related to design, construction and management of cultural heritage and civil infrastructure. The intent of the authors has been to frame a series of innovative procedures affecting typical activities of the civil construction and property and cultural heritage management market (e.g. survey, inspection, restoration, retrofitting) within a completely integrated process. This research effort is aligned with the concept of finding an interconnection among different disciplines in order to create innovative solutions for challenging problems emerging in new sustainable and environmentally compatible construction, and particularly from the perspective of cost optimization in the management of existing properties, which may constitute indispensable cultural heritage assets.

References

1. Akyldiz IF, Su W, Sankarasubramaniam Y, Cayirci E (2002) Wireless sensor networks: a suvey. Comput Netw 38:393–422
2. Ansari F (2007) Practical implementation of optical fiber sensors in civil structural health monitoring. J Intel Math Syst Struct 18(8):879–889
3. Antucheviciene J, Kala Z, Marzouk M, Rytas Vaidogas E (2105) Decision making, methods and applications in civil engineering. Math Probl Eng 2015. Article ID 160569
4. Antucheviciene J, Kala Z, Marzouk M, Vaidogas ER (2015) Solving civil engineering problems by means of fuzzy and stochastic MCDM methods: current state and future research. Math Probl Eng 2015. Article ID 362579
5. Aspinall R, Pearson D (2000) Integrated geographical assessment of environmental condition in water catchments: linking landscape ecology, environmental modelling and GIS. J Environ Manag Elsevier 59(4):299–319
6. Avdelidis NP, Moropoulou A (2004) Applications of infrared thermography for the investigation of historic structures. J Cult Heritage Elsevier 5(1):119–127
7. Azhar S (2011) Building information modeling (BIM): trends, benefits, risks, and challenges for the AEC Industry. Leadersh Manag Eng 11(3) American Society of Civil Engineers (ASCE)
8. Bai L (2013) RFID sensor-driven structural condition monitoring in integrated building information modeling environment. PhD Dissertation, University of Maryland, College Park
9. Balaguer C, Montero R, Victores JG, Martínez S, Jardón A (2014) Towards fully automated tunnel inspection: a survey and future trends. In: The 31st international symposium on automation and robotics in construction and mining (ISARC 2014), Sydney, pp 19–33

10. Bay H, Ess A, Tuytelaars T, Van Gool T (2008) Speeded-up robust features (SURF). Comput Vis Image Underst 110(3):346–359
11. Betz DC, Staudigel L, Trutzel MN, Kehlenbach M (2003) Structural monitoring using fiber-optic bragg grating sensors. Struct Health Mon 2(2):145–152
12. Burrough PA, McDonnell RA, Lloyd CD (1998, 2015) Principles of geographical information systems. Oxford University Press. ISSN 978-0-19-874284-5
13. Caprari C et al (2010) Highly compact robots for inspection of power plants. In: 2010 1st international conference on applied robotics for the power industry, Montreal, pp 1–6. https://doi.org/10.1109/CARPI.2010.5624412
14. Castellazzi G, D'Altri AM, Bitelli G, Selvaggi I, Lambertini A (2015) From laser scanning to finite element analysis of complex buildings by using a semi-automatic procedure. Sensors 15:18360–18380. https://doi.org/10.3390/s150818360
15. Castelli G, Ottaviano E, Rea P (2014) A cartesian cable-suspended robot for improving end-users' mobility in an urban environment. Robot Comput Integr Manuf 30(3):335–343
16. Chen XN, Xiab Q, Zhang SH, Zhou Y (2005) 3D laser scanner system for surveying and engineering. ISPRS.ecn.purdue.edu
17. Cruz-Ramırez SR, Mae Y, Arai T, Takubo T, Ohara K (2011) Vision-based hierarchical recognition for dismantling robot applied to interior renewal of buildings. Comput Aided Civil Infrastruct Eng 26(5):36–355
18. Davila Delgado J, Butler LJ, Gibbons N, Brilakis I, Elshafie MZEB, Middleton C (2016) Management of structural monitoring data of bridges using BIM. In: Proceedings of the institute of civil engineers, Bridge Engineering
19. Davis FW, Quattrochi DA, Ridd MK, Lam NSN, Walsh SJ, Michaelsen JC, Johnston CA (1991) Environmental analysis using integrated GIS and remotely sensed data. Some research needs and priorities. Photogram Eng Remote Sens 57(6):689–697
20. Eleftheriadis S, Mumovic D, Greening P, Chronis A (2015) BIM enabled optimisation framework for environmentally responsible and structurally efficient design systems. In: 32nd international symposium on automation and robotics in construction and mining (ISARC 2015) 15–18 June 2015, Oulu, Finland
21. Eramo N, Modoni G, Arroyo M (2012) Design control and monitoring of a jet grouted excavation bottom plug. In: Proceedings of the 7th international symposium on geotechnical aspects of underground construction in soft ground, TC28 IS Rome, Viggiani ed., Taylor & Francis Group London, 16–18 May 2011, pp 611–618. ISBN 978-0-415-66367-8
22. Federici F, Graziosi F, Faccio M, Colarieti A, Gattulli V, Lepidi M, Potenza F (2012) An integrated approach to the design of wireless sensor networks for structural health monitoring. Int J Distrib Sens Netw. Article ID 594842
23. Fedorik F, Makkonen T, Heikkilä R (2106) Integration of BIM and FEA in automation of building and bridge engineering design. In: 33rd international symposium on automation and robotics in construction (ISARC 2016), Auburn, USA, 18–21 July 2016
24. Gattulli V, Chiaramonte L (2005) Condition assessment by visual inspection for a bridge management system. Comput Aided Civil Infrastruct Eng 20:95–107
25. Gattulli V, Lepidi M, Potenza F (2016) Dynamic testing and health monitoring of historic and modern civil structures in Italy. Struct Monit and Maint 3(1):71–90
26. Gattulli V, Potenza F, Graziosi F, Federici F, Colarieti A, Faccio M (2014) Design of wireless sensor nodes for structural health monitoring applications. Proc Eng 87:1298–1301
27. Gattulli V, Potenza F, Toti J, Valvona F, Marcari G (2016) Ecosmart reinforcement for a masonry polycentric pavilion vault. Open Constr Build Technol J 10(Suppl 2: M7):259–273. https://doi.org/10.2174/1874836801610010259
28. Granosik G, Borenstein J, Hansen MG (2007) Serpentine robots for industrial inspection and surveillance. In: Low KH (ed) Industrial robotics: programming, simulation and applications. Published by pro-Literatur Verlag, Germany, pp 633–662
29. Guarnieri A, Pirotti F, Vettore A (2013) Cultural heritage interactive 3D models on the web: an approach using open source and free software. J Cult Heritage, Elsevier 11(3):350–353

30. Guisan A, Zimmermann NE (2000) Predictive habitat distribution models in ecology. J Ecol Model Elsevier 147–186
31. Hallermann N, Morgenthal G (2013) Unmanned aerial vehicles (UAV) for the assessment of existing structures. In: IABSE symposium, Kolkata
32. http://petrobotproject.eu/
33. Jacobs S, Matthys G, De Roeck G, Taerve L, de Waele W, Degrieck J (2007) Testing of a prestressed concrete girder to study the enhanced performance of monitoring by integrating optical fiber sensors. J Struct Eng 133(4):541–549
34. Jakobsson A, Giversen J (2013) Guidelines for implementing the ISO 19100 geographic information quality standards in national mapping and cadastral agencies. Eurographics
35. Joerin F, Musy A (2000) Land management with GIS and multicriteria analysis. Int Trans Oper Res Elsevier 7(1):67–78
36. Kamińska IA, Ołdak A, Turski WA (2004) Geographical information system (GIS) as a tool for monitoring and analysing pesticide pollution and its impact on public health. Ann Agric Environ Med 11(2):181–184. ISSN: 1232-1966 e-ISSN: 1898-2263
37. Kenny J (2016) Live BIM offers building stress monitoring in real-time, Dec 2016. http://www.bimplus.co.uk/technology/live-bim-monit6ors-bu5ilding-str9ess-real-time/
38. Khan FS (2012) Color attributes for object detection. IEEE Conf Comput Vision Pattern Recogn 3306–3313
39. Kim J-W, Kim S-B, Park J-C, Nam J-W (2015) Development of crack detection system with unmanned aerial vehicles and digital image processing, In: International congress on advances in structural engineering and mechanics, Incheon
40. Lee S, Kalos N (2015) Bridge inspection practices using non-destructive testing methods. J Civil Eng Manag 21(5):654–665
41. Li QB, Ansari F (2001) Circumferential strain measurement of high strength concrete in triaxial compression by fiber optic sensor. Int J Sol Struct 38(42–43):7607–7625
42. Lingua A, Piumatti P, Rinaudo F (2012) Digital photogrammetry: a standard approach to cultural heritage survey. In: The international archives of the photogrammetry, remote sensing and spatial information sciences, vol. XXXIV, Part 5/W12
43. Lynch JP, Loh K (2006) A summary review of wireless sensors and sensor networks for structural health monitoring. Shock Vib 38(2):91–128
44. Maurelli P (2006). I Sistemi Informativi Territoriali (SIT) come contesti di rappresentazione e interazione. In: Martone M (ed) Atti del Seminario La rappresentazione per la conoscenza dell'ambiente urbano e del territorio Rome Kappa Edizioni
45. Meschini A, Pelliccio A (2013) Il colore nel rilievo strumentale: laser scanner, termografia e postprocessing dei dati in un sistema GIS. In: Rossi M, Siniscalco A (eds), Colore e Colorimetria. Contributi multidisciplinari, Maggioli Editore S.P.A.: 70–81. ISBN 9788838762413
46. Meyer E, Grussenmeyer P, Perrin JP, Durand A, Drap P (2007) A web information system for the management and the dissemination of cultural heritage data. J Cult Heritage Elsevier 8 (4):396–411
47. Moscati A (2012) Integrated information systems for the enhancement of the urban/architectural heritage, including 3D GIS, AIS (architectural information systems) and web. PhD Dissertation, Rome URN: urn:nbn:se:hj:diva-28937OAI: oai:DiVA.org:hj-28937DiVA: diva2:892496
48. Nagatani K, Kiribayashi S, Okada Y, Otake K, Yoshida K, Tadokoro S, Nishimura T, Yoshida T, Koyanagi E, Fukushima M, Kawatsuma S (2013) Emergency response to the nuclear accident at the Fukushima Daiichi nuclear power plants using mobile rescue robots. J Field Robot 30:44–63
49. Ottaviano E, Ceccarelli M (2006) Application of a 3-DOF parallel manipulator for earthquake simulations. IEEE/ASME Trans Mechatron 11(2):240–246
50. Ottaviano E, Rea P (2013) Design and operation of a 2-DOF leg-wheel hybrid robot. Robotica 31(8):1319–1325

51. Ottaviano E, Rea P, Castelli G (2014) THROO: a tracked hybrid rover to overpass obstacles. Adv Robot 28(10):683–694. https://doi.org/10.1080/01691864.2014.891949
52. Patraucean V, Armeni I, Nahangi M, Yeung J, Brilakis I, Haas C (2015) State of research in automatic as-built modelling. Adv Eng Infor 29:162–171
53. Pelliccio A (2013) Informative representation for the vulnerability analysis of anthropic landscape. Two different areas in comparison: the historical center of St. Elia Fiumerapido (Fr) and the mining site of Coreno Ausonio. Disegnarecon, April 2013. ISSN 1828-5961
54. Pelliccio A, Cigola M (2015) Geographic information systems (G.I.S.) for the analysis of historical small towns. In: Khosrow-Pour M (ed) Encyclopedia of information science and technology, 3rd edn. IGI Global, pp 3128–3135. ISBN 978-1-4666-5888-2 (hardcover); ISBN 978-1-4666-5889-9 (ebook); ISBN 978-1-4666-5891-2 (print &perpetual access)
55. Potenza F, Castelli G, Gattulli V, Ottaviano E (2017) Integrated process of images and acceleration measurements for damage detection. In: X international conference on structural dynamics, EURODYN 2017, Procedia Engineering, in press
56. Rea P, Pelliccio A, Ottaviano E, Saccucci M (2017) The heritage management and preservation using the mechatronic survey. Int J Architect Heritage. https://doi.org/10.1080/15583058.2017.1338790
57. Saranli U, Buehler M, Koditschek DE (2001) RHex: a simple and highly mobile hexapod robot. Int J Robot Res 20(7):616–631. https://doi.org/10.1177/02783640122067570 [14]
58. Shin Y, Cho K (2015) BIM application to select appropriate design alternative with consideration of LCA and LCCA. Math Probl Eng 2015. Article ID 281640
59. Siegwart R, INourbakhsh R (2004) Introduction to autonomous mobile robots. MIT Press
60. Spencer BF, Chung-Bang Y (eds) (2010) Wireless sensor advances and applications for civil infrastructure monitoring, Newmark Structural Engineering Lab. Report Series, No.24. University of Illinois at Urbana-Champaign, Illinois. http://hdl.handle.net/2142/16434
61. Spencer BF, Ruiz-Sandoval Manuel E, Kurata N (2004) Smart sensing technology: opportunities and challenges. Struct Control Health Monit 11:349–368
62. Stapelberg RF (2009) Handbook of reliability, availability, maintainability and safety in engineering design. Springer
63. Tian Y, Wen C, Hong S (2008) Global scientific production on GIS research by bibliometric analysis from 1997 to 2006. J Inf Elsevier 2(1):65–74
64. Valvona F, Toti J, Gattulli V, Potenza F (2017) Effective seismic strengthening and monitoring of a masonry vault by using glass fiber reinforced cementitious matrix with embedded fiber bragg grating sensors. Comp Part B Eng 113:355–370
65. Viola P, Jones M (2001) Rapid object detection using a boosted cascade of simple features. In: Proceedings of the 2001 IEEE computer society conference on computer vision society conference on computer vision and pattern recognition, vol 1
66. Wang C, Cho YK (2105) Application of as-built data in building retrofit decision making process, Procedia Eng 118:902–908
67. Webb GT, Vardanega PJ, Middleton CR (2015) Categories of SHM deployments: technologies and capabilities. J Bridge Eng (ASCE) 20(11):04014118
68. Xiongjue W (2016) Analysis on complex structure stability under different bar angle with BIM technology. Perspect Sci 7:317–322
69. Yamauchi B (2004) PackBot: a versatile platform for military robotics. In: Proceedings of SPIE, vol 5422. Unmanned Ground Vehicle Technology VI, Orlando, FL
70. Yeum CM, Dyke SJ (2015) Vision-based automated crack detection for bridge inspection. Comput Aided Civil Infrastruct Eng. https://doi.org/10.1111/mice.12141
71. Zonta D, Pozzi M, Zanon P (2008) Managing the historical heritage using distributed technologies. Int J Archit Heritage 2(3):200–225

Part I
Robotics and Automation

Timed Cellular Automata-Based Tool for the Analysis of Urban Road Traffic Models

Camelia Avram, Adina Astilean and Eduardo Valente

Abstract The optimization process of urban transportation in smart cities is strongly connected to the elaboration of specific, efficient models. In this context, this chapter describes a versatile modelling formalism based on timed automata and implemented in the UPPAAL environment for different complex and easily changeable road traffic simulations. Microscopic models based on cellular automatons are analysed in order to simulate the behaviour of different vehicles in a specific group of urban streets. The proposed models integrate the main traffic elements present in urban traffic: streets with multiple traffic lanes; different types of vehicles, including automobiles, buses and trams; intersections controlled by traffic lights; bus and tram stops inside and outside of the traffic lane, pedestrian crosswalks; and parallel street parking. The basic concepts are detailed starting from scenarios which first merit to highlight possible modelling techniques and structures and to facilitate a comparative analysis of the limitations of the presented models. The models based on traffic cellular automata (TCA) have appropriate results inside of the urban traffic theory.

Keywords Urban traffic · Modelling · Simulation · Cellular automata
Formal verification

Urban Traffic Theory

Historically, traffic congestion has been regarded as a problem confined to major metropolitan areas. Over the years, the traffic problems that existed in densely developed urban areas have begun creeping into the suburbs. Automobile technology advances have allowed more people to drive and the feverish modern

C. Avram (✉) · A. Astilean
Technical University of Cluj-Napoca, Cluj-Napoca, Romania
e-mail: camelia.avram@aut.utcluj.ro

E. Valente
University of Minho, Guimarães, Portugal

E. Ottaviano et al. (eds.), *Mechatronics for Cultural Heritage and Civil Engineering*,
Intelligent Systems, Control and Automation: Science and Engineering 92,
https://doi.org/10.1007/978-3-319-68646-2_2

lifestyle has caused traffic congestion problems to appear even in small towns. This hectic scenario has given rise to the need for new solutions that improve traffic circulation, and traffic simulations have played a key role [1].

A good understanding of road traffic dynamics is fundamental to assist the choice between the strategies that are more efficient and more appropriate. In this context, simulations that reproduce the eventual effect of traffic change parameters may be extremely important for an improvement of road traffic circulation [2].

In this chapter, the fundamental concepts of the urban traffic theory are presented. Firstly, the main physical variables involved in the traffic road problems (flow, average speed, and density) are shown, as well as different approaches to express those variables. Afterward, relational diagrams between these variables are depicted and their theoretical behaviours are exposed. A presentation of some microscopic models for traffic road simulation closes this chapter [2].

Fundamental Concepts of the Urban Traffic Flow (Traffic Parameters)

The traffic behaviour can be evaluated by the following variables: flow (J), average speed (v) and density (ρ). These parameters are called traffic parameters, i.e., those variables that help to determine the road condition at a particular time. Traffic flow is defined as the quantity of vehicles that pass through a road section in a given period of time and its units are vehicles per time unit. The average speed is given by space units travelled by those vehicles per time unit. The density is determined by the number of vehicles per space unit [3, 4].

In this section different methods to calculate the urban traffic parameters are presented. The different expressions provide the necessary background for the utilization of urban traffic parameters in the course of this work [2].

The variables' behaviour is defined in the space-time diagrams, which represent each vehicle's trajectory in a period.

The traffic characteristics range in time and space. In order to simplify these variations, commonly average values for the traffic parameters are adopted. These average values may be temporal or spatial values. Thus, there are different expressions for the traffic variables: When one road is considered in one time interval, it is called temporal average; or a lane stretch in a time instant is called a spatial average [2, 5].

Spatial Average

The density for a lane stretch (*L*) in a time instant (*dt*) is the number of trajectories (vehicles) on this stretch at the time instant t_1 divided by the considered length (*L*) [2].

Equation 1 shows the mathematical expression of traffic density.

$$\rho = \frac{n}{L}, \tag{1}$$

where *n,* is de number of vehicles present in the traffic stretch *L.*

The vehicles' average speed on this traffic lane stretch can be expressed by the following equation:

$$\bar{v} = \frac{\sum_{i=1}^{n} v_i}{n}, \tag{2}$$

where v_i is the instant speed of the vehicle *i* in the considered stretch. For a permanent regime of speed, it can be calculated by:

$$v = \frac{J}{\rho}, \tag{3}$$

In other words:

$$J = \rho \times v, \tag{4}$$

As in this case is considered the vehicles speed on a section (v):

$$J = \rho \times \bar{v}, \tag{5}$$

The traffic flow equation can be written:

$$J = \frac{\sum_{i=1}^{n} v_i}{L}, \tag{6}$$

Temporal Average

In the previous case, the information is a function of a single time instant from the considered traffic lane stretch. Normally, when is necessary to obtain real data, motion detectors are installed in any traffic lane [2]. Therefore, it is necessary to define expressions that consider multiple measurements on the same traffic lane section x_k. The traffic flow is given by the number of vehicles that cross a determined lane section (m), in a certain time interval (T), i.e.

$$J = \frac{m}{T}, \tag{7}$$

The vehicle's average speed that crosses the lane section is expressed by:

$$\bar{v} = \frac{\sum_{i=1}^{n} v_i}{m}, \tag{8}$$

v_j, is the vehicle's speed that crosses this section. Substituting Eqs. 7 and 8 into Eq. 5, the average density of vehicles on that lane section is given by [2, 6]:

$$\rho = \frac{m^2}{T \times \sum_{i=1}^{n} v_i}, \tag{9}$$

Traffic Flow and Density

The traffic flow has three different phases [7, 8]:

- Low-density region, called free traffic flow. This phase, generically, allows the driver to achieve a desired speed, approaching the maximum speed permitted.
- Median density region, $c_1 < \rho < c_2$, where the traffic flow is not exclusively defined by density. The traffic configuration influences directly its flow, and may cause a free or congested flow. This region (the middle density) is also known as meta-stable region;
- High-density region, $\rho > c_2$, where the traffic flow drops as the density increases. The large concentration of vehicles causes them to cluster in traffic jams and a vehicle that leaves a place from the traffic jam will find congestion just ahead. This type of traffic is characterized by the behaviour of starting and stopping.

Models for Simulation of Vehicular Traffic

Mathematical models for traffic simulation can be divided into three different approaches: macroscopic, mesoscopic, and microscopic models [9, 5]. Figure 1 represents schematically these three different methodologies (Image adapted from [4]).

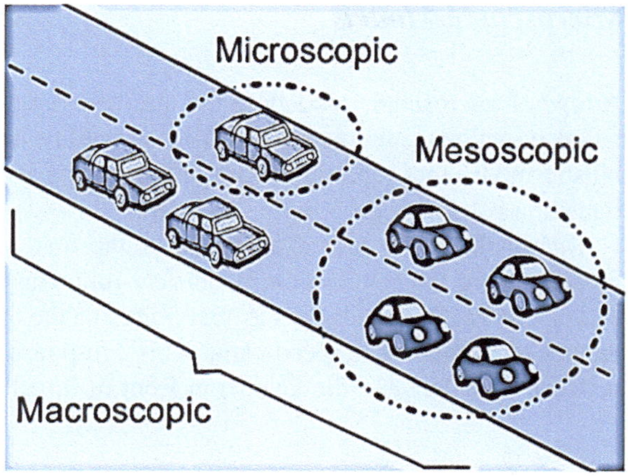

Fig. 1 Main strategies for traffic simulation

Macroscopic Models

The macroscopic analysis describes the global behaviour of traffic streams. Therefore, they relate density, flow and average speed parameters of the vehicles. To study its behaviour, the macroscopic approaches apply hydrodynamic rules. This is the reason why this approach is also known as traffic hydrodynamic analogy. For its features and considerations, macroscopic models are successfully applied in the study of high-density traffic, but they cannot provide accurate results easily for rarefied traffic situations [4, 10–13].

Mesoscopic Models

The mesoscopic models represent the behaviour of a group of vehicles, i.e., bases its traffic analysis in a group of vehicles that behaves according to some logical grouping criteria: an expedition, congestion behaviour, etc. [4, 10–14].

Microscopic Models

Microscopic models are the ones that focus the individual behaviour of each vehicle to obtain the global behaviour of a traffic road. They consider the interrelated parameters that determine the vehicle's dynamics. For example, knowing the acceleration of each vehicle at each time instant its position and speed can be known, after a time interval [4, 10–14].

Microscopic Models

A type of microscopic modelling that has been largely used in traffic simulation is based on cellular automata, due to its versatility and simplicity. The first probabilistic model that reproduced the basic traffic conditions with the use of simple transition rules was proposed in [15–18].

One of the most studied models in traffic road simulation are the microscopic models, using the persecution model (car-following), which was developed at the end of 1950s. This model has as first objective the translation of speed variation of the tracker vehicle. The speed variation is a response function of the speed stimulus between a vehicle and the vehicle in front of it, called leader vehicle [2].

Urban Road Traffic Modelling Issues

The formalism adopted for modelling the systems considered in this work (road traffic systems) is timed automata (TA) [12, 13, 19–22, 17, 18]. This formalism is widely used on several domains for modelling behaviour of physical systems due to its non-determinism characteristic [16, 23, 24, 25, 9].

In addition, because the UPPAAL simulation and verification environment is used for performing the simulation and formal verification analysis techniques, the edition of those models was implemented on the editor of this software tool, as performed in [26–30].

Each example presented in this chapter defines a concrete problem and the model's evolution until the creation of the final model that contains all the features needed to implement in the case study.

The main aim of this work is to create a systematic approach for traffic road models. This systematic approach can only be obtained using modular structures. This model also must be easily extendable, reused and contain a large level of detail, in order to be possible implementing the complex scenario proposed. The solution implemented to create such versatility features introduced matrices that provide a compact environment for discretization and modelling the physical environment.

The models presented in this chapter are divided into two levels of complexity. The models of low complexity consist of the following:

- A model with one traffic lane and one automobile.
- Two models are presented to describe a traffic lane with several automobiles travelling simultaneously.
- The complexity increases model by model. The models with a high complexity represent the following situations:
- Four traffic lanes connected by an intersection and the automobiles chooses the next street to continue its journey.

At the end of each subchapter, we present the features, advantages and/or limitations of each proposed model.

Low Complexity Modes

The first models created contain the simplistic interactions between a street and vehicles. These two models were the first approach and the level of detail and degree of realism are not the desired for this work. They were useful for understanding which modelling techniques and structures that have potential to be further developed in order to reach the versatility required.

Model with One Traffic Lane and One Automobile

The first model created used one automobile moving in a street freely, without interactions with other automobiles. The absence of interactions with other automobiles is because only an automobile can travelling inside of the street.

At the beginning of the street and also at the end of a street, motion sensors detect the presence of a vehicle approaching from the beginning of the street (Si1) or leaving the end of the street (So1). The initial conditions of this problem also consider that an automobile enters in the street with velocity different from zero.

In Fig. 2 the physical environment of the first model is schematically represented.

To implement this scenario in UPPAAL, it was necessary develop two automata: the automaton street and the automaton automobile.

The automaton street has five places:

- OUT_of_the_STREET_START: When an automobile is travelling outside of the beginning of the street.
- STREET_START: When the sensor Si1 has detected an automobile at the beginning of the street.
- OCCUPIED: When an automobile is on the street.

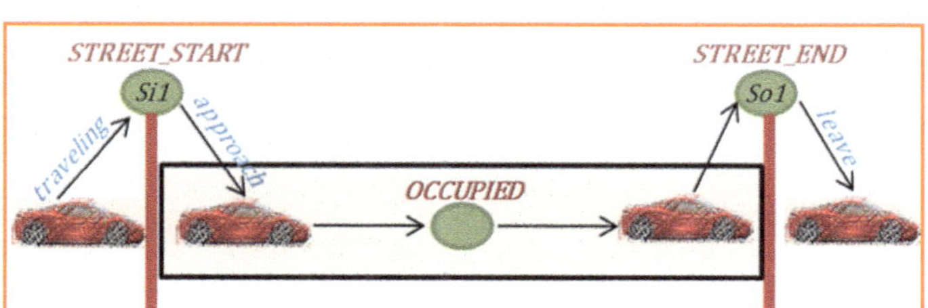

Fig. 2 A single traffic lane with one travelling automobile

- STREET_END: When an automobile was detected by the sensor So2 at the end of the street
- OUT_of_the_STREET_END: When an automobile is travelling outside of the end of the street.

These are the five states that a street can have when interacting with an automobile. To coordinate the interactions between the street and the automobile, it was necessary to create three synchronization channels. The signal that the car is travelling is sent by the automobile's automaton to the street's automaton, and it means that an automobile starts to move in the direction of the street. When sensor Si1 is activated, the automaton street sends and activates the channel *approach* to the automaton automobile, and it means that one automobile is at the street's start changing the automaton street to OCCUPIED. When the sensor So1 is activated, the automaton street sends and activates the channel *leave*, and it means that the automobile travelled the entire street and now is out of the street end. The street is again without any automobile and ready to receive another automobile.

Figure 3 shows the automaton street with the synchronization channels and respective places.

An automobile only has two possible states: *STOPPED* or *MOVING*. Sensors at the beginning and at the end of the street detect whether the automobile is in or out of the traffic lane. At the beginning of the simulation, the automobile is stopped, and the channel *travelling* will put the automobile moving in the direction of the start of the street, which has the sensor Si1. After being detected at the beginning of the street by the sensor Si1, the automobile will continue its trip until the end of the traffic lane. In order to create a permissive model, the automobile, after it finishes its trip in a street, can be forced to stop and go again in the direction of the street start or can continue moving in the direction of the street start.

Figure 4 represented the automaton automobile with the synchronization channels and the two places previously explained.

In this model, time features are not included and is only the first model created to represent the simplest interaction between an automobile and a street. Once, interactions between vehicles are not considered, the concept of neighbourhood is not applicable. The street is not discretized in cells, and this is the main reason for the absence of interactions between vehicles. If several automobiles were travelling in this model, its spatial position inside of the street would not be defined. The transition rule is that only one automobile can circulate in a street.

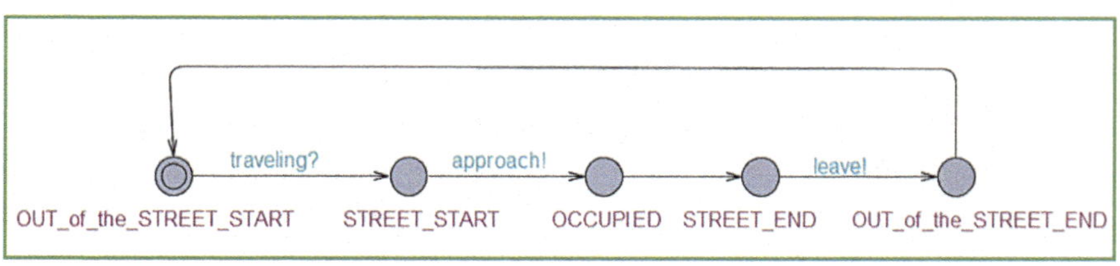

Fig. 3 *Street* automaton of the model with one automobile and one traffic lane

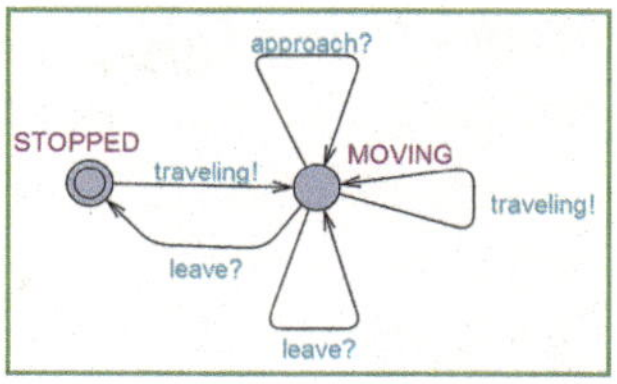

Fig. 4 Automaton *automobile* for the model with one automobile and one traffic lane

This model has the following limitations:

- It is a very simplified model, one only considers a type of vehicles (automobiles) and only one automobile can be inside of the street travelling.
- To implement in several streets with several automobiles, one would need to create each automaton automobile and street. Extending a scenario with a large number of automobiles and streets is complex.

Model with One Traffic Lane and Several Automobiles

In this new definition of the modelling problem, the complexity increases: one street with several automobiles moving inside it and interacting is simulated. To solve this problem two models were created. The first model presents the interaction between the vehicles moving cell by cell inside of the street if the cell in front of it is free using only variables to coordinate the movement. In the second model the interaction between the automobiles creates a queue at the end of the street if is not possible for them to circulate freely the entire length of the street using functions and variables to coordinate these movements.

Figure 5 schematically presents the physical environment of the model for one traffic lane and several automobiles. The automobiles are allocated inside of the street in the same order that initiated the movement.

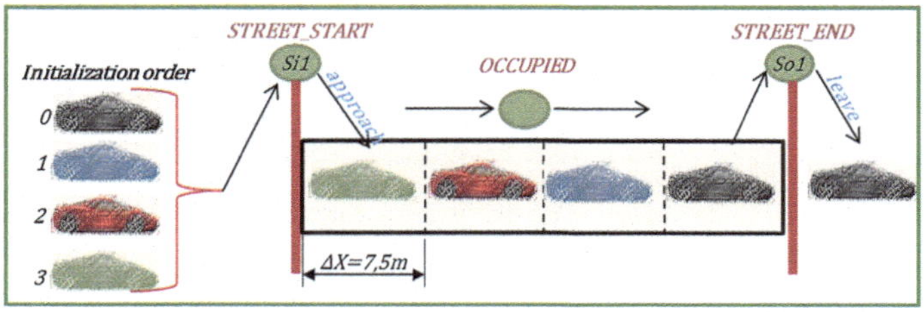

Fig. 5 Single traffic lane with several travelling automobiles

Model Using Only Variables to Coordinate the Interactions Between the Automobiles Travelling in a Traffic Lane

First, four control variables for each automobile were globally declared. The first digit of each variable refers to the number of the automobile, and the second digit corresponds to the respective street cell. The variable V_CELL11 is responsible when automobile Nr. 1 is in the first cell of the street; the variable V_CELL12 is the variable dedicated to the automobile Nr. 1 when it is in the second cell of the street. Similarly, for the others automobiles, control variables dedicated for each street cell were created.

In this example, the street has four cells and four automobiles will circulate. To implement this scenario four automata automobile and one automaton street were added.

Five channels were generated. The channel *approach_cell1* is activated when an automobile enters into the first cell of the street; the channel *approach_cell2* is activated when an automobile approaches the second street cell; the channel *approach_cell3* is activated when an automobile enters into the third cell of the street; the channel *approach_cell4* is activated when an automobile enters into the fourth street cell. The channel *leave* is activated when an automobile leaves the fourth and the last cell of the street. The channel *travelling* present in the previous model was erased, and it is assumed at the beginning of the simulation that all of the automobiles start moving inside the street with some velocity.

Figure 6 presents the global declaration for the model with one street with several automobiles travelling controlled with variables.

The automaton street is a loop of channels, i.e., it only contains the channels to coordinate the movement of the automobiles. The conditions for an automobile to move are in the structure of the automata automobiles through a combination of control variables. The sequence of the automobiles that will initialize its movement

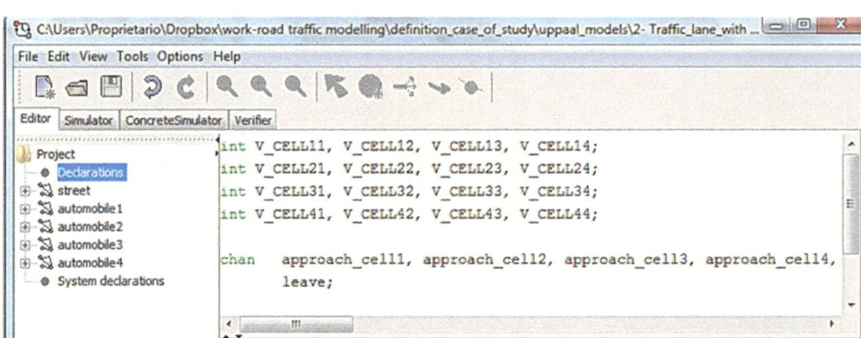

Fig. 6 Global declaration for the model with one traffic lane with several automobiles, using only variables to control the movement

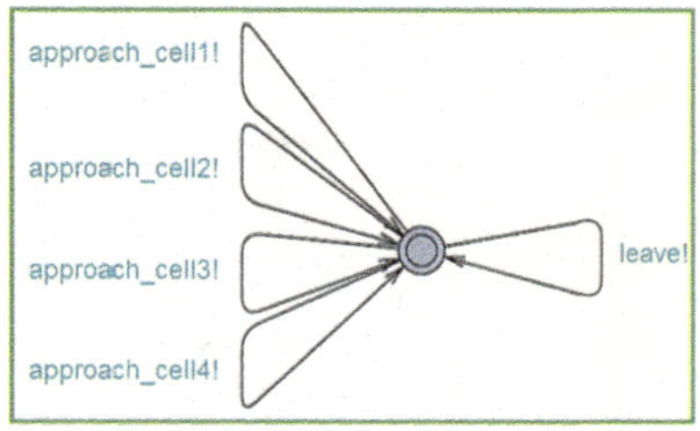

Fig. 7 Automaton *street* for the model with one traffic lane and several automobiles, using only variables to control the movement

can be implemented into the automaton street. The sequence of the automobiles was not considered.

Figure 7 represents the automaton street for the model with one street and several automobiles controlled only by variables.

The automaton automobile was reconfigured and has eight places:

- OUT_of_the_STREET_START_MOVING: When an automobile is moving in the direction of the beginning of the street
- CELL1: When an automobile has been detected by sensor Si1, which means it is inside of the first street cell travelling.
- CELL2: When an automobile is travelling in the second street cell.
- CELL3: When an automobile is travelling in the third street cell.
- CELL4: When an automobile is travelling in the fourth street cell.
- WAITING_CELL1: When an automobile is waiting in the first street cell, until the second street cell becomes free again.
- WAITING_CELL2: When an automobile is waiting in the second street cell, until the third street cell becomes free again.
- WAITING_CELL3: When an automobile is waiting in the third street cell, until the fourth street cell becomes free again.

The bibliographic revision of Schreckenberg and Nagel in 1992 suggests a length of 7.5 m for each street cell. Therefore, the physical environment is a street with a total length of 30 m (four street cells).

An automobile travelling inside of the street will occupy one street cell when the model evolves. An automobile can only enter in the first street cell if all the others variables responsible for the first street cell from the other automobiles are equal to zero. This is the information declared in the guard of the first transition, meaning that the first street cell is free (Fig. 7). If this condition is verified, the automobile (in this case, automobile Nr. 4) can enter in the first street cell, which activates channel *approach_cell1*. At this time, the respective control variable for the first street cell of this automobile takes the value 1. With this variable updated, no other automobile can enter the first street cell, because the first street cell has an automobile. To continue to move, the automobile needs to verify if the second street cell is free. If this condition is true, it will move to the third street cell and update the respective control variable of the street cell Nr. 1 for the value zero and cell Nr. 2 for the value one. If the condition to move is not verified, the automobile evolves

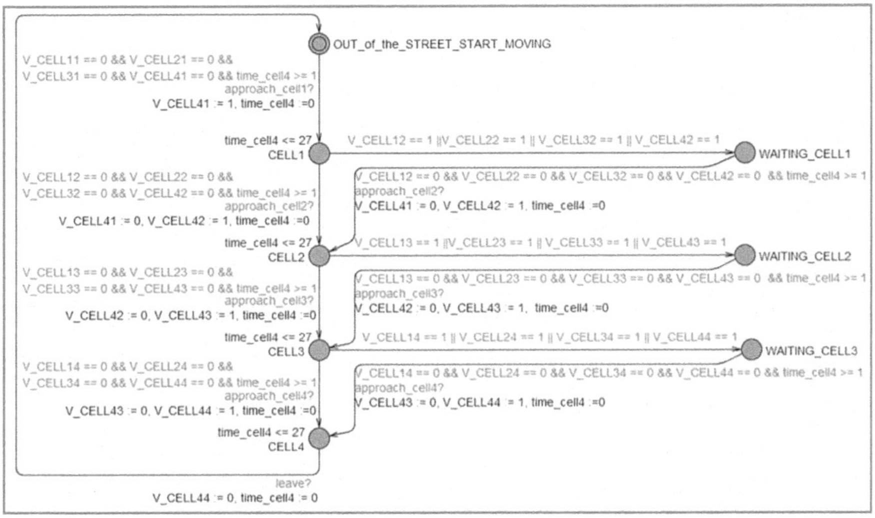

Fig. 8 Automaton automobile for automobile Nr. 4 is presented for the model with one traffic lane with several automobiles, using only variables to control the movement

to the place *waiting_cell1* and only when the cell in front is free can it move. The same procedure is executed for the followings cells.

When the automobile is in the last cell of the street, the channel *leave* is activated and the respective control variable of the last street cell is updated to the value zero, and the automobile leaves the street.

Figure 8 presents the automaton automobile for the model with one street with several automobiles only controlled with variables.

In this model, time conditions were already implemented. The time needed to travel one street cell meter with a range of velocities between 1 km/h and 50 km/h is 27 s and 0.5 s, respectively.

In this model, the physical environment was discretized in street cells with a length of 7.5 m, providing the groundwork for a neighbourhood. The neighbourhood of an automobile is the state of the street's cell in front of it creating interactions between the automobiles. If the cell in front of it is occupied by another vehicle, it will wait until it is free again, and if the frontal street cell is free it can continue to move. The set of rules in this model is creating a queue of automobiles, from the beginning to the end of the street, by the same order that initialized its movements.

This model has the following limitations:

- It is a very simple model, because it considers only one type of vehicle (automobiles).
- To implement in several streets with several automobiles, it is necessary to create each automaton automobile and street and several control variables for each automobile responsible for each street cell; it is humanly impossible to

extend the model to a scenario with such a large number of automobiles and streets.
- If it is necessary to extend this model for intersections, the new upcoming street would be implemented in the automaton automobile, creating strict routes for each automobile, and the level of non-determinism would not be the desirable.
- For questions regarding organization, the features of the street (number of cells and conditions to move further) should be implemented in the automaton street and not in each automaton automobile.

Model Using Functions and Variables to Coordinate the Interactions Between the Automobiles Travelling in a Traffic Lane

In order to improve the limitations of previously presented models, it a new model to solve the same modelling issues was created.

First, we declared in the global declaration the number of cells that the street contains. In this specific example, four street cells with a length of 7.5 m were declared, and this number of cells can be easily amended. Therefore, the physical environment created is a street 30 m long. The number of automobiles was declared using the function *typedef*. This function links and identifies each automobile with a specific integer. In this concrete model, four automobiles were declared. An automobile travelling inside the street occupies one street cell when the model evolves. For the interaction between the automobiles, five channels were created: *approach, leave, stop, stay* and *go*, which are functions of the number of automobiles present in the street. In this model also is assumed, at the beginning of the simulation, that all the automobiles start moving inside the street with some velocity.

Figure 9 presents the explained global declaration of the model one traffic lane with several automobiles controlled with functions and variables.

To simulate the desired behaviour and interactions between automobiles, it was necessary to create a new function and new auxiliary variables in the street declarations. Firstly, the automobiles are allocated for the same order that initialize their movement, in a list the same size as the number of the cells. This means that each element of the list is a street cell correlating to an element on the list with the *ID 0,* the last street's cell (end of the street) and the element with ID equal to the number of cells of the first street's cell (beginning of the street). A new variable length was declared, which is the total number of automobiles inside the street (length of the queue). For this reason, the length will have the maximum value equal to the number of street cells, because one automobile will occupy always one street cell.

The function *enqueue* was created to allocate the automobiles' IDs in the last cell of the street (list ID 0). This function will increase the value of the variable length

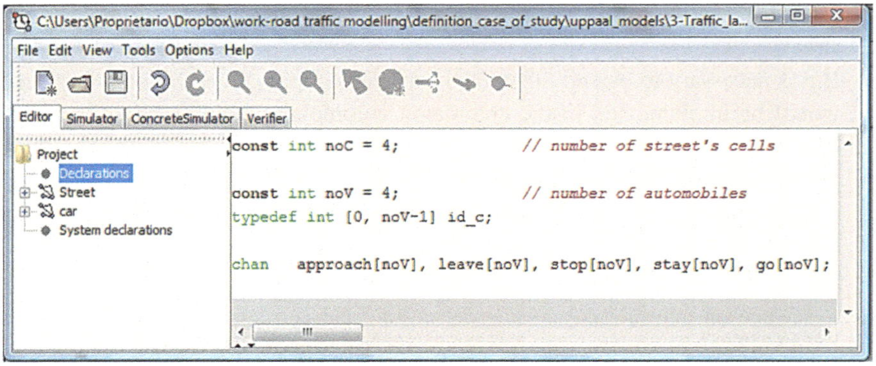

Fig. 9 Global declaration for the model with one traffic lane with several automobiles moving controlled by variables and functions

by one such that the synchronization channel *approach* is activated. If another automobile arrives while the previous is moving inside the street, the new automobile's ID are allocated in the second-to-last cell and the value of the variable length is equal to two until all the street's cells have an automobile inside. Consequently, the function *enqueue* is updated in the same arrow as the synchronization channel *approach*.

The function *dequeue* will release the last street's cell when the respective automobile inside it leaves the street and will decrease the value of the variable length by one. For this reason the function *dequeue* is associated and updated in the same arrow as the synchronization channel *leave*.

Another function created was the function 'front', which after an automobile leaves the street will move the vehicles to the next street's cell. This function is associated with the channel *go*.

The function *tail* was created to allocate an automobile that was forced to stop in the last position of the queue, and for this reason is associated with the synchronized channel *stop*.

Figure 10 presents the street declarations and the functions and new variables created.

The automaton street is more complex and has three places:

- STREET_START: When the sensor Si1 has detected an automobile at the beginning of the street and the street is without an automobile.
- OCCUPIED: When one or more automobiles is inside of the street travelling.
- OUT_of_the_STREET_END: When an automobile was detected by the sensor So1 at the end of the street and it leaves the street.

At the beginning of the simulation, this model assumes that the street is empty, without any automobiles moving inside. For this reason, the evolution from the first place *STREET_START* to the place *OCCUPIED* means that an automobile is travelling inside of the street. The function *enqueue* and the variable length are

```
id_c list[noC];
int[0,noC] length;

void enqueue(id_c element) // Put an automobile at the end of the queue
{
    list[length++] = element;
}
void dequeue()    // Remove the front automobile of the queue
{
    int i = 0;
    length -= 1;
    while (i < length)
    {
        list[i] = list[i + 1];
        i++;
    }
    list[i] = 0;
}
id_c front()    // Returns the front automobile of the queue
{
    return list[0];
}
id_c tail()    // Returns the last automobile of the queue
{
    return list[length - 1];
}
```

Fig. 10 Street's declaration for the model with one traffic lane with several automobiles controlled by variables and functions

updated. In the location *OCCUPIED* the automobile can interact with others. For example, other automobiles can enter into the street and this automobile can be forced to stop and then start again its movement coordinated by the street's condition.

The automaton street only evolves to *STREET_END* when an automobile has been detected by the sensor So1 and is leaving the street. If the street is occupied, it means that the length is greater than zero and the automaton returns to the place *OCCUPIED*. If the street is empty (length equal to zero) the automaton returns to *STREET_START*.

Figure 11 shows the configuration of the automaton street for the model with one traffic lane with several automobiles travelling and coordinated with variables and functions.

The automaton automobile was reconfigured and has five places:

- OUT_of_the_STREET_START_MOVING: When an automobile is moving towards the beginning of the street
- MOVING_INSIDE_the_STREET: When an automobile has been detected by sensor Si1 and is inside of the street travelling.

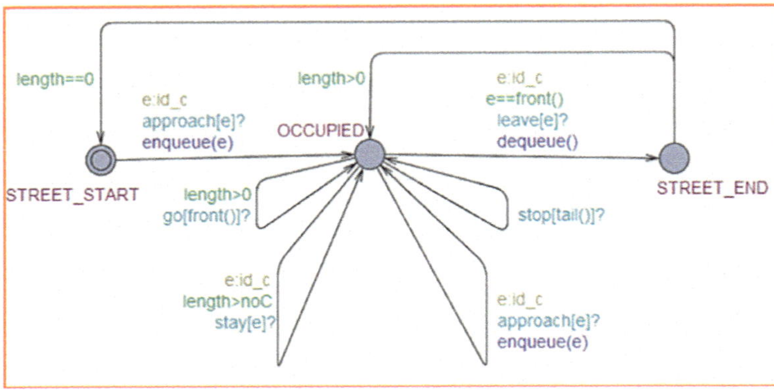

Fig. 11 Street automaton for the model with one traffic lane with several automobiles controlled by variables and functions

- STOPPED_INSIDE_of_the_STREET: When the automobile has stopped inside of the street.
- AGAIN_IN_MOVEMENT: When an automobile starts to move after a stop inside of the street.
- OUT_of_the_STREET_END_MOVING: When an automobile has been detected by sensor S_{01} exiting the street.

The automaton automobile only evolves from the initial location if it has been sensor Si1 detects an automobile at the beginning of the street. From the moment an automobile is inside of the street, it has two possibilities of evolution: cross the entire street without a stop or is forced to stop. If the automobile reaches the end of the street without a stop, the time needed to travel four street cells with velocities between 1 km/h and 50 km/h is 108 s or 2 s, respectively. An automobile is forced to stop if there is another automobile in front of it travelling more slowly than it is or if the time allowed was more or less than the time needed to travel four street cells.

If an automobile is stopped on the street, it will only move if street conditions allow it to go in front. These evolutions in the automobile automaton are made by the synchronization channel *go*. If after that the street is free until the end, the automobile will reach the end of the street, but if it is forced to stop the channel *stay* is activated and the automobile stops again.

Figure 12 shows the configuration of the automaton automobile for the model with one street several automobiles.

In this model, similar to the previously presented model, the physical environment was discretized in street cells with a length of 7.5 m and the concept of neighbourhood is implemented. The neighbourhood of an automobile is the state of the street's cell in front of it creating interactions between the automobiles.

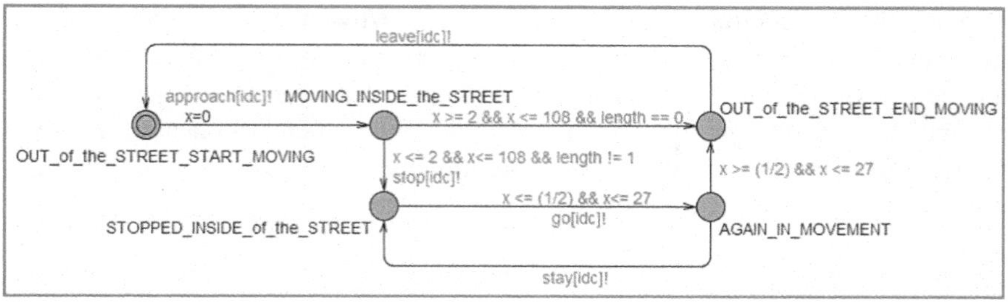

Fig. 12 Automaton automobile for a model with one street several automobiles controlled by variables and functions

The set of rules in this model creates a queue of automobiles at the end of the street if another automobile is travelling in the street or if the street is empty its entire length.

This model has the following limitations and advantages:

- It is a very simple model; it considers only one type of vehicles (automobiles).
- The number of street cells and automobiles can be easily changed; it is only necessary to define the global declaration of those variables.
- To implement it, several streets with several automobiles are created for each street automaton and several variable lengths and a list to allocate the automobiles in each street. It is humanly impossible to extend the model to a scenario with a large number of streets.
- The structure of the automaton street does not allow for the creation of streets with different lengths, because the time needed to travel all the streets is always the same.

The creation of these low complexity models was useful for understanding which modelling technique and structure can be used in order to reach the versatility required. At the end of this subchapter, we present an analysis of the two previous ways of modelling and the following conclusions can be verified:

- The number of vehicles and streets needs to be declared using the function *typedef.*
- One needs to rearrange the structure of the automaton automobile and street.
- To model the physical environment, neither changing the function list nor creating variables are good solutions for coordinating movements. In the case of the function list, the accuracy of the movement is not predictable, and this function it will create problems for calculating the crossing time of each road. It is not possible to extend the model for large scenarios by controlling by variables, because of the number of variables that are needed to implement into the automaton automobile.
- The structure to model the physical environment also needs to be easily changeable and accurate for the introduction of the traffic elements.

With these modelling requirements, we created a matrix of coordinates for modelling the physical environment. This matrix can control accurately the movement of the vehicles, precisely introduce traffic elements and is easily extendable by adding more rows. To model the physical environment of the streets matrices were used in all of the following models.

High Complexity Models

The basic models previously proposed, if applied to models with high levels of complexity or large traffic scenarios, cannot be easily be changed or reused, due the limitations previously explained.

In order to solve these limitations, models based upon the matrices were created to model the traffic environment. A matrix, which is a map of street cells, is a compact data structure that can easily be extended.

The matrices implemented to model the road traffic environment have the following features:

- Each line corresponds to an independent traffic lane of a street.
- The value present in column Nr. 1 corresponds to the number of cells that each street has.
- The following columns are filled with empty street cells or traffic elements.

Due to UPPAAL's limitations, one cannot create vectors with different sizes, and some of the values present in this matrix have no significance.

Traffic Lane with Multiple (1, 2, or 3) Possible Traffic Lanes Travelled by Automobiles in a Free Flow

As the behaviour of a street with several automobiles has already been modelled, the next level of complexity entails extending the map. Thus, at the end of the street upon which the automobiles started its movement, the automobiles will have to choose between one of three streets to continue their trips.

The rule to choose the next street should be based upon statistical data received by sensors implemented at the end of the streets. The data received give the traffic flow's percent for the upcoming streets. If the automobile at the end of the street has two possible next streets; the flow is distributed equally by the two streets. It means that 50% of the automobiles continue travelling on one street and 50% continue their movement on the other next street. When a street ends with three possible next streets, the traffic flow will also be distributed equally. Consequently, each next street will have 33% of the traffic flow circulating on the previous street.

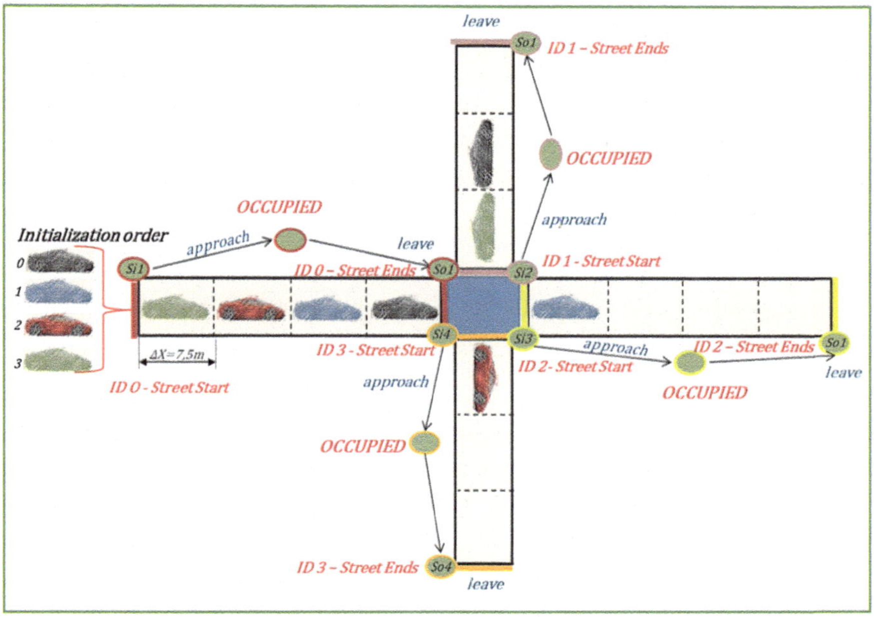

Fig. 13 Schematic diagram of the model with a traffic lane with several automobiles travelling with three possible next streets

Figure 13 presents the model with several automobiles moving in a street, and at the end of the street, another street is chosen. The previous paragraph explained the distribution of the traffic flow with the rules admitted represented in this scheme. The first automobile (black), which arrives in the intersection (blue rectangle), travels on the street ID 1, the second one to reach the intersection (blue) will choose the street ID 2, the third one (red) will choose the street ID 3, and this cycle of choice will continue. This distribution of the traffic flow seems to be reasonable and to create a model more accurate than taking information from each traffic lane that is studied.

The authors have long been concerned with the creation of general models that can be easily extended to or implement in new situations and features. For this reason, the previous definition of the street cell using the function list was erased and in this new model, similar to the automobiles' definition, we used the function *typdef* to create the streets in the global declarations.

To define the physical environment, we created another variable, *maxnoCells*. This variable defines the maximum length of the street considered (maximum number of cells). With this new variable and with the variable number of streets (noS), we created a matrix *idexSC*. This matrix is a map of street cell coordinates. Each line corresponds to a street ID with the first line ID0 and the last line the last street's ID. The column Nr. 1 corresponds to the number of cells that each street has and the following columns are filled with −1, which means a street's cell is empty.

Due to UPPAAL's limitations, it is not possible to create vectors with different sizes, thus many of the −1 s present in the matrix have no significance. The first column is important for limiting the street size. Therefore, it is easy to extend the map and to implement other features in the posterior models.

The choice of the next street in an intersection was implemented with a similar application. First, a new variable, *maxNextStreets,* (maximum next streets) was declared. This variable is equal to three, because at a maximum, an automobile has three possible next streets in this intersection. With this variable and with the variable number of the streets (noS), we created another matrix, *indexMAP*. Each line corresponds to a street ID with the first line the street's ID 0 and the last line the last street's ID. The first column contains the number of possible next streets at the end of the street, and the other columns have the street IDs of the next streets. As presented in the scheme of Fig. 12, the street ID 0 at its end has three possible next streets (street ID 1, 2 or 3), and all the automobiles that are in the last cell of the street IDs 1, 2, or 3 have zero possible next streets. Thus when an automobile leaves the street IDs 1, 2 or 3 are considered to have left the map. The value −1 means that there is no upcoming street, and the automobile will exit the map. The variable b is a control variable, a function of the number of the street. For each value of b an automobile present at the end of the street will have another destination.

The current street for each automobile has been defined, and the value defined in the vector *currentStreet* represents the initial street's ID that an automobile will start the simulation. This vector contains the inputs of the traffic flow in the simulation. The value of *currentStreet* will change every time that an automobile leaves a street and continues to move in another street of the map.

All the channels present are a function of the number of automobiles. The channel approach is activated when a vehicle is detected at the beginning of the street (first street's cell), the channel *travelling* is activated when an automobile has a free cell in front of it of it, and the channel *leave* is activated when an automobile is in the last street's cell.

In Fig. 14 the global variables are presented in the model with one traffic lane and multiple possible choices as upcoming next streets with automobiles moving in a free flow.

The automaton street was simplified, because there are only three channels but continues with three places:

- STREET_STARTS: When a sensor Si1 has detected an automobile at the beginning of a street and there are no automobiles on the street.
- UPDATING_CELLS: When one or more automobiles are travelling on a street.
- STREET_ENDS: When the sensor So1 has detected an automobile at the end of the street and the automobile exits the street.
- This model assumes at the beginning of the simulation, that the entire map of the street is empty, without any automobiles moving inside it. For this reason, the evolution from the first place *STREET_START* to the place *UPDATING_CELLS* means that an automobile is travelling on a street and updates the function *enqueue*. To be safe, verify three rules: First, read the value inside of the vector

```
const int noA = 4;              // Number of AUTOMOBILES
typedef int [0, noA-1] idA;     // Automobiles' ID
//-----------------------------------------------------------------------------------------
const int noS = 4;              // Number of STREETS
typedef int [0,noS-1] idS;     // Streets' ID
//-----------------------------------------------------------------------------------------
const int maxnoCells = 6; // Maximum number of cells (maximum street lenght)

// Description of the matrix number of cells per different street
int indexSC[noS][maxnoCells+1] = {{6,-1,-1,-1,-1,-1,-1},
                                  {5,-1,-1,-1,-1,-1,-1},
                                  {6,-1,-1,-1,-1,-1,-1},
                                  {4,-1,-1,-1,-1,-1,-1}};
//-----------------------------------------------------------------------------------------
// Maximum number of possible streets for a car in the last cell of a street
const int maxNextStreets = 3;

//Description of the matrix street possibilities
int indexMAP[noS][maxNextStreets+1] = {{3, 1, 2, 3},
                                       {0,-1,-1,-1},
                                       {0,-1,-1,-1},
                                       {0,-1,-1,-1}};

// auxiliar variable to control the choice at the street ends
int b[noS] = {0,0,0,0};
//-----------------------------------------------------------------------------------------
// variable which contains the street's ID for each car at the beginning of the simulation
int currentStreetA[noA] = {0, 0, 0, 0};
//-----------------------------------------------------------------------------------------
chan  approach[noA], leave[noA], traveling[noA];
```

Fig. 14 Global declarations of the model traffic lane with several automobiles travelling with three possible next streets

currentStreet, which will define the street where an automobile is arriving. Second the matrix *indexSD* needs to have in its second column the value −1, which means that the first street's cell is empty without any vehicle. Third, the variable *novis* (number the vehicles inside of the street) needs to be smaller or equal to the number of cells, which means at maximum each street can only contain an number of automobiles equal to the number of street cells, because each automobile will occupy one cell.

In the location *UPDATING_CELLS,* the automobiles can interact with each other. For example, another automobile can enter the street with the same rules described previously and another automobile can continue to travelling, if the number of the cells travelled (*nocTA*) is smaller than the size of the street (smaller than the first column of the matrix *indexSC*) and if the cell in front of it is free (the cell needs to have the value −1). It is necessary to read the value present in the *currentStreet* vector to know the street on which an automobile is moving. In this second case, the channel *travelling* and function *moving* are updated.

The automaton street only evolves from the place *UPDATING_CELLS* when the sensor S01 detects an automobile leaving the street. In this case, the variable *nocTA*

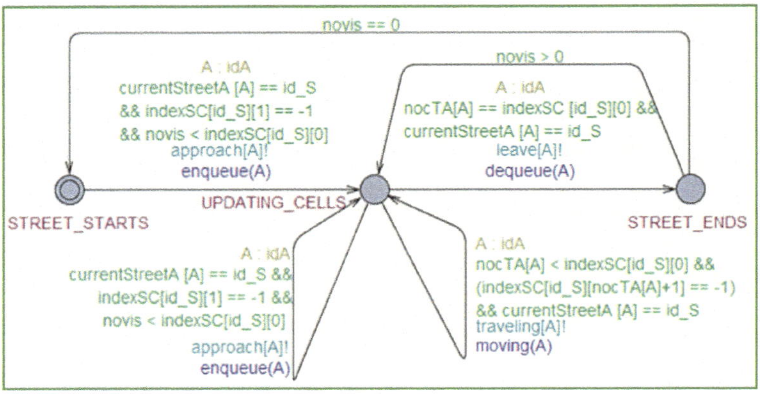

Fig. 15 Street automaton of the model traffic lane with several automobiles travelling with three possible next streets

is equal to the first column of the matrix *indexSC*. To know the street that an automobile is leaving, it is also necessary to read the value present in the vector *currentStreet*. If the variable *novis* is greater than zero, the street automaton returns to the place *UPDATING_CELLS*, because the street is not empty and has automobiles travelling. Another scenario is if the number of the automobiles on the street is equal to zero, the automaton returns to the place *STREET_START* with the street empty.

The automaton street is the "brain" of the simulation having all the information regarding traffic conditions, a set of rules and consequent transition rules.

Figure 15 shows the configuration of the automaton street for the model with one traffic lane and several automobiles travelling with three possible next streets.

In street declarations, three functions were implemented (*enqueue, dequeue* and *moving*), which will contain a set of rules for the correct traffic circulation.

The *enqueue* function defines rules for an automobile that enters a street. In a real scenario an automobile can only enter a street if the street is not full of vehicles and starts its trip from the beginning of the street until the end.

The function moving contains the rules for an automobile continues to move. In the reality an automobile can only move further if there is free space in front of it.

The function *dequeue* comprehends the rules for an automobile leaving a street. The function *dequeue* will verify if the automobile is in the last street's cell, and if this condition is true, the automobile exits the street.

Figure 16 presents the explained street declarations and the created functions and variables. The street declarations are the places where all the transition rules for a street are implemented.

The automaton automobile was reconfigured and has three places:

- OUT_of_the_STREET_START_MOVING: When an automobile is moving towards the beginning of the street.

```
int nocTA[noA] = {1,1,1,1};     // number of cells travelled by an automobile
int novis ;          // number of vehicles inside of the street

void  enqueue(idA automobile)   // Put an element (AUTOMOBILE) in the first cell of the street
{   // if the first cell of the street is free and novis is smaller than first column of indexSC
    if ((indexSC[id_S][nocTA[automobile]] == -1) && (novis < indexSC[id_S][0]))
    {
        indexSC[id_S][nocTA[automobile]] := automobile;     // add an element in the first cell
        novis++;
    }
}
void moving (idA automobile)    // Moving an element (AUTOMOBILE) inside of the street
{ // if the cell in front is free and nocTA is smaller than first column of indexSC
    if ((indexSC[id_S][nocTA[automobile]+1] == -1) && (nocTA[automobile] < indexSC[id_S][0]))
    {
        indexSC[id_S][nocTA[automobile]+1] := automobile;  // add an element in the cell in front
        indexSC[id_S][nocTA[automobile]] := -1;     //release the previous cell
        nocTA[automobile]++;
    }
}
void dequeue(idA automobile) // Remove the front element (AUTOMOBILE) of the street
{ // if nocTA is equal to the first column of indexSC
    if (nocTA[automobile] == indexSC[id_S][0])
    {
        indexSC[id_S][nocTA[automobile]] := -1; //release the last cell of the street
        nocTA[automobile] := 1;   // reset the number of the cells for the initial value
        novis--;
    }
}
```

Fig. 16 Street declarations for the model with a traffic lane with several automobiles travelling with three possible next streets

- MOVING_INSIDE_the_STREET: When the sensor Si1 has detected an automobile, and the automobile is travelling on the street.
- OUT_of_the_MAP: When the sensor So1 detects the automobile, and there are no possible next streets, they are considered out of the map.

Figure 17 shows the configuration of the automaton automobile for the model with one traffic lane and several automobiles travelling with three possible next streets.

This model is the first model with relevant traffic road rules included and a significant level of detail. At the end of this subchapter, we briefly present some features of this model:

- The physical environment is a one-dimensional grid of rectangular cells, all equal in size (7.5 m of length).
- It is a single cell model, because each automobile occupies only one street cell in each time iteration, and the cells can only have two possible states: occupied by an automobile or empty.
- The size of the neighbourhood is the same for each cell. The model is anisotropic, because the automobiles only respond to stimulus in front of it.

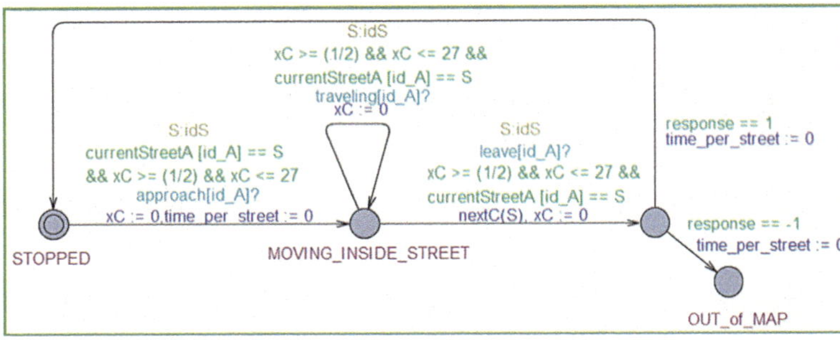

Fig. 17 *Automobile* automaton for the model traffic lane with several automobiles travelling with three possible next streets

- This model is a dynamic system with a closed number of automobiles and streets. The evolution in space and time depending on the same rules. Only if the cell in front is free, can the automobile proceed on the street.
- The time is a stochastic feature and its choice is completely non deterministic. In an instant *t*, the automobile can circulate in a cell at velocity of 50 km/h and in the instant t + 1, it can move at 2 km/h. This model presents heavy breaking and quick accelerations, even if the street is completely free.
- It is a very simplified model; one only considers a type of vehicles (automobiles).
- This model can easily be extended.

In fact, this model has the potential to be extended to new traffic scenarios and be implemented with other types of vehicles and traffic elements. Even in those scenarios, this model provides a solid basis.

In addition, one can consider other approaches and behavioural properties [31, 32] for the whole system.

Conclusions

The cellular automata allow for the observation of different phenomena, dissecting components into individual variables, allowing one to understand how local changes affect the whole grid of cells.

The formalism-timed automata are appropriate for a modular approach due to its elementary structure. The resolution (level of detail) and system size (the network size to be covered) obtained are appropriate for the proposed model.

In the context of urban traffic theory, the cellular automata in microscopic models have the capacity to simulate in detail all the elements presented in this traffic environment. The quantities of traffic elements implemented generate a model containing a large number of evolutionary rules and interactions.

The main goal of this work was the creation of a systematic approach for a complex urban traffic scenario. The structure implemented has the potential to be expanded. This systematic approach can be easily and limitlessly extended, the only limitation being the computational power available.

This model has flexibility in accordance with the environment that it is applied, because it can implement a large group of traffic elements and possible interactions. The stimuli created in each vehicle depend on the traffic elements contained in the street and the traffic conditions, because of the modular approach.

As future work, the authors will implement the distributed control of an illustrated case study considering some issues and solutions used on industrial networks [33–38] and adapted to this field of application. An analysis of performance of the developed distributed controller will also be considered [39, 40].

References

1. Clark J, Daigle G (1997) The importance of simulation techniques in its research and analysis. In: Andradóttir S, Healy KJ, Withers DH, Nelson BL (eds) Proceedings of the 1997 winter simulation conference, USA
2. Lima EB (2007) Modelos microscópicos para simulação do tráfego baseados em autômatos celulares. Universidade Federal Fluminense, Niterói, Brazil, Dissertação de Mestrado em Computação
3. May AD (1997) Traffic flow fundamentals. Prentice Hall
4. Vico FJ, Basagoiti FJ, Platas RG, Lobo D (2005) Modelado y simulación del tráfico en vías urbanas y periurbanas en base a la estimación de tiempos de recorrido. Universidad de Málaga and Tecnologías Viales Aplicadas, TEVA, SL, Malaga, ETSI Informática
5. Nagel K, Schreckenberg M (1992) A cellular automaton model for freeway traffic. J de Phys I, France, 2221–2229
6. Nakanishi K, Itoh K, Igarashi Y, Bando M (1996) Solvable optimal velocity models and asymptotic trajectory. Department of Physics, Kyoto University, Kyoto 606-01, Physics Division, Dept. of Education, Niigata University, Niigata 950-21, and Physics Division, Aichi University, Miyoshi, Aichi 470-02, 1996, Japan
7. Chowdhury D, Santen L, Schadschneider A (2000) Statistical physics of vehicular traffic and some related systems. Phys Rep 329:199–329
8. Chua L (2005) A nonlinear dynamics perspective of wolfram's new kind of science. In: Bernoulli shift to universal computation, inaugural lecture of the international Francqui chair, katholieke Universiteit Leuven, June, 2005
9. Murata T (1989) Petri nets: properties, analysis and applications. In: Proceedings of the IEE, vol 77, no 4. Department of Electrical Engineering and Computer Science, University of Illinois, Chicago, USA Apr 1989
10. Maerivoet S, Moor BD (2005) Transportation planning and traffic flow models. 05–155, Katholieke Universiteit Leuven, Department of Electrical Engineering ESAT-SCD (SISTA), July 2005
11. Maerivoet S, Moor BD (2005) Cellular automata models of road traffic. In: Physics reports (ed) Department of Electrical Engineering ESAT-SCD (SISTA), Katholieke Universiteit Leuven, Kasteelpark Arenberg 10, 3001 Leuven, Belgium, 12 Sept 2005
12. Wolfram S (1983) Statistical mechanics of cellular automata. Rev Mod Phys 55:601–644
13. Wolfram S (2002) A new kind of science. Wolfram Media, Inc., ISBN 1-579-955008-8

14. Wang J Petri nets for dynamic event-driven system modeling. Department of Software Engineering, Monmouth University, West Long Branch, USA
15. Nagel K (1996) Particle hopping models and traffic flow theory. Phys Rev E 53(5):4655–4672
16. Rothery R (1998) Traffic flow theory. Transportation Research Board (TRB). Special Report, p 165
17. Gardner M (1970) Mathematical games—the fantastic combinations of John Conway's new solitaire game "life". Sci Am, 120–123
18. Milner R (1989) Communication and Concurrency. Prentice-Hall International, Englewood Cliffs
19. Alur R, Dill DL (1994) A theory of timed automata. Theor Comput Sci Elsevier 126:183–235
20. Angulo FRF (2014) Autómata celular
21. Eppstein D (2014) Cellular automaton
22. Vaandrager F A first introduction to Uppaal—a job shop example. Institute for Computing and Information Sciences, Radboud University Nijmegen, Heijendaalseweg 135, 6525 AJ Nijmegen, Netherlands
23. Machado J, Seabra E, Campos JC, Soares F, Leão CP (2011) Safe controllers design for in-dustrial automation systems. Comput Ind Eng 60(4):635–653
24. Moreira A (2003) Universality and decidability of number-conserving cellular automata. Theor Comput Sci 292:711–721
25. Crutchfield JP, Kaneko K (1987) Phenomenology of spatiotemporal chaos. In: Directions in chaos. World Scientific, pp 272–353
26. Behrmann G, David A, Larsen KG (2004) A tutorial on UPPAAL. In: Proceedings of the 4th international school on formal methods for the design of computer, communication, and software systems (SFM-RT'04). LNCS 3185
27. Kaneko K (1990) Simulating physics with coupled map lattices. In: Kawasaki K, Onuki A, Suzuki M (eds) Formation, dynamics, and statistics of patterns. World Scientific, pp 1–52
28. Kier LB, Seybold PG, Cheng C-K Cellular automata modeling of chemical systems. Published in Springer (ed), Center for the study of biological complexity. Virginia Commonwealth University, Richmond Virginia, USA
29. Kunz G, Machado J, Perondi E (2015) Using timed automata for modeling, simulating and verifying networked systems controller's specifications. Neural Comput. Appl., 1–11
30. Machado J, Denis B, Lesage J-J (2006) A generic approach to build plant models for des verification purposes. In: Proceedings—eighth international workshop on discrete event systems, WODES 2006, pp 407–412
31. Campos AM (2014) Autómato Celular
32. Campos JC, Machado J, Seabra E (2008) Property patterns for the formal verification of automated production systems. In: IFAC proceedings volumes (IFAC-PapersOnline), 17 (1 Part 1)
33. Barros C, Leao CP, Soares F, Minas G, Machado J (2013) Issues in remote laboratory developments for biomedical engineering education. In: International conference on interactive collaborative learning, ICL 2013, art. no 6644585, pp 290–295
34. Costa J, Carvalho N, Soares F, Machado J (2009) The fins protocol for complex industrial applications: a case study. In: ICINCO 2009—6th international conference on informatics in control, automation and robotics. Proceedings, 2 RA, pp 348–354
35. Leão CP, Soares FO, Machado JM, Seabra E, Rodrigues H (2011) Design and development of an industrial network laboratory. In: Int J Emerg Technol Learn 6(Special Issue 2):21–26
36. Leão CP, Soares F, Rodrigues H, Seabra E, Machado J, Farinha P, Costa S (2012) Web-assisted laboratory for control education: remote and virtual environments. Communications in Computer and Information Science, 282 CCIS, pp 62–72
37. Silva M, Pereira F, Soares F, Leão CP, Machado J, Carvalho V (2015) An overview of industrial communication networks. Mech Mach Sci 24:933–940
38. Silva PCM (2001) Teoria do fluxo de tráfego. Universidade de Brasília, Brasil, Material didático do curso de engenharia de tráfego

39. Ceccarelli M, Carbone G, Ottaviano E (2005) Multi criteria optimum design of manipulators. Bull Pol Acad Sci Tech Sci 53(1):9–18
40. Thomas F, Ottaviano E, Ros L, Ceccarelli M (2005) Performance analysis of a 3-2-1 pose estimation device. IEEE Trans Rob 21(3):288–297

Process Analysis of Cable-Driven Parallel Robots for Automated Construction

Tobias Bruckmann, Arnim J. Spengler, Christian K. Karl, Christopher Reichert and Markus König

Abstract This contribution presents an introduction to cable robots, their properties and their mechatronic design for application in automated construction. Today, most steps involved in the construction process are performed manually. Thus, the integration of automated functions demands a closer look at the production and logistic paradigms, these are the main focus of this paper. Based on case studies, the authors discuss upcoming transformations in shell production by comparing the conventional construction process with proposed processes involving cable-driven parallel robots. The focus is on bricklaying and working methods for the installation of prefabricated elements. Adaptations to site logistics and changes in work organization are also considered. A case study and sensitivity analysis based on system dynamics modeling are introduced, and the conceptual design of an experimental prototype is presented. The results of the investigations show that the use of a cable-driven robot is feasible for onsite construction, enabling automation of processes to save time and cost. The study investigates crucial parameters and the potential for cable robots in the field of construction.

T. Bruckmann (✉) · C. Reichert
University of Duisburg-Essen, Duisburg, Germany
e-mail: tobias.bruckmann@uni-due.de

C. Reichert
e-mail: christopher.reichert@uni-due.de

A. J. Spengler
Institute of Construction Management,
University of Duisburg-Essen, Duisburg, Germany
e-mail: Arnim@arnim-spengler.de

C. K. Karl
Specialized Didactics in Construction Technology,
University of Duisburg-Essen, Duisburg, Germany
e-mail: christian.karl@uni-due.de

M. König
Ruhr University Bochum, Duisburg, Germany
e-mail: koenig@inf.bi.rub.de

E. Ottaviano et al. (eds.), *Mechatronics for Cultural Heritage and Civil Engineering*,
Intelligent Systems, Control and Automation: Science and Engineering 92,
https://doi.org/10.1007/978-3-319-68646-2_3

Introduction and State of the Art

Since the 1980s, the application of automated robotic systems in mechanical engineering has been an enabling strategy. In nearly all fields of production, robots have increased productivity and precision in repetitive processes, thereby enabling reduction of costs. From an economic standpoint, the use of robots constitutes an inexpensive solution for mass production. In the context of Industry 4.0, robot systems become a vital factor for made-to-order production.

In contrast, until now, the successful application of robots toward construction has been limited to special cases. This is mainly due to three factors:

1. In the construction industry, CAD systems traditionally were only used for creating 2D plans, and these plans could not be interpreted by computer systems. 3D models were rarely used, but over the years the employed systems have grown more powerful so that advanced applications like robotized automated production processes can use this data. Recently, the method of Building Information Modeling (BIM) has enabled this process.
2. Conventional robots are characterized by limited size and workspace (workspace diameter is usually well below 8 m), a feature which does not allow the size of most buildings to be covered. The introduction of moving robots is confronted by problems associated with accuracy, performance and workspace. Therefore, this approach increases costs. Additionally, moving robots can only be used for isolated steps, such as bricklaying.
3. Buildings are always unique and complex in their totality. This is the reason why experience in robotic processes can't simply be transferred from the stationary industry to the construction industry.

Accordingly, two of these factors are about to change drastically in the near future. First, the introduction of BIM will extend beyond production planning as it will allow buildings to be constructed digitally [1]. This allows the required production steps needed to feed a robotic system to be defined. Second, the recently developed, so-called cable-driven parallel robots, or simply cable robots, will allow the creation of huge automated systems that can easily span dozens of meters and therefore allow the volume of a building to be covered.

This contribution presents an introduction to cable robots, their properties and their mechatronic design. As most steps in construction are conducted manually, production paradigms in construction need to be examined, and this is a focus of this paper.

Based on case studies, the authors discuss emerging transformations in shell production by comparing the conventional construction process with the prospective process influenced by cable-driven parallel robots. In this context, aspects such as feasible building types and geometries, technology, work tasks and flows, workspace and safety, materials management and site logistics as well as occupational aspects will be considered. The focus is on bricklaying and working paving methods. Since the concept being presented is at a quite early stage close to fundamental research, all development steps need to be accompanied by economic assessments to ensure its feasibility for future applications. Accordingly, the contribution discusses the elementary economy of cable robots for several production steps as well as for several types of buildings.

Conventional Construction Process and Paradigm Changes Through Digitalization and Automation

In spite of prefabricated elements or other methods to increase efficiency, handwork is still an essential feature of the building process. This is due to several factors:

- Buildings are expensive and unique.
- The building process is minted by handicraft.
- The build consists of many small, individual working steps.
- Constantly changing work steps are difficult to map using robots.
- Digital building models (using the BIM method for example) have not yet been enforced on a broad level.

Like other areas of industry and manufacturing, the building and planning process will become more and more digitized. These changes are visible to the whole construction industry. Currently, planners, architects, building owners, investors, building companies or public authorities accept and even demand digital methods. These methods, like BIM, allow for advanced applications that would not have been possible before. The approach presented in this paper uses digital models to provide this information for a cable driven robot. The robot processes and uses this information to carry out the building process. Other projects like the onsite 3D printer BigDelta (cf. http://www.wasproject.it), Kamermaker (cf. http://www.kamermaker.com), Hadrian (cf. http://www.fbr.com.au) or the D-Shape project (cf. http://d-shape.com/) are examples of this new building strategy.

Cable Robots for Shell Production

Robots are flexible machines and—in conventional approaches—typically find use in repeating tasks. Their flexibility is mainly due to their ability to follow easily exchangeable programmed motion patterns. Currently, the generation of motion patterns is coupled with engineering processes based on computer-aided engineering (CAE) which allows these motions to be generated automatically.

Accordingly, robots have enjoyed success in many branches of mechanical manufacturing. Most series products that are produced in high-wage countries derive value from automated processes using robots.

Typical industrial robots use a serial structure i.e. manipulator arms that consist of a chain of joints and links where each element has a unique predecessor. While this is a simple, cheap and adaptable architecture, it is inherently limited in terms of workspace: Since the robotic arm needs to support its own weight, elastic effects accumulate along the arm and manifest in pose errors of the end effector where the payload is attached. Additionally, the arm may be very heavy since the links and joints must support the mass of all subsequent components. Thus, most serial robots have a limited range of three to four meters. While their payload can approach one ton, they may weigh up to 5 tons.

Recently, new approaches for bricklaying based on serial robot manipulators have been presented [2]. The Australian company Fastbrick Robotics is developing a large-scale manipulator called Hadrian, which is large enough to move bricks along the manipulator arm to any desired position within a planned building. In parallel, the American company Construction Robotics has presented the SAM robot, a conventional serial robot mounted on a moving platform, equipped with a wide range of pose sensors. Both approaches are currently under practical evaluation. In the context of innovative fundamental research, Kohler presented approaches based on flying drones and moving industrial robots [3, 4].

While the serial robot is very popular in industrial applications, the parallel robot has been investigated extensively but is used mainly in niche applications today. The parallel robot is characterized by end effectors, which are usually the tool or payload to be moved, that are supported by multiple kinematic chains. While this is advantageous for load distribution, precision and payload-to-weight ratio, its workspace compared to construction space is usually smaller relative to equivalent serial structures due to the limited stroke of the linear actuators used in popular designs like the Stewart-Gough platform. Furthermore, the required joints are more complicated and thereby more expensive.

In 1985, Landsberger and Sheridan [5] proposed a new design for parallel robots using cables: Since the limited workspace of many conventional parallel robot designs is mainly due to the limited stroke of the employed linear actuators, they introduced cables coiled on winches as force transmission elements. This enables virtually unlimited actuator lengths and allows the practical generation of workspaces dozens of meters in diameter. Additionally, there is no need for complicated joints which makes the cable-driven parallel robot (in short, cable robot) a mechanically simple, cheap and intrinsically modular design.

Nevertheless, this approach includes special properties that mainly influence the geometrical design, the layout of the end effector (usually a moving platform) and the control approach. The remainder of this chapter describes the specialization of a cable robot for automated construction, including the development of a mechatronic bricklaying unit as the end effector. Fundamental introductions can also be found in [2, 6].

Modeling and Workspace

As mentioned above, the workspace of conventional robots is limited by geometrical constraints, such as actuator strokes or link lengths. Since a cable robot uses cables to move the end effector, which can be coiled for dozens of meters, this limitation vanishes. Still, there are workspace limitations. In practice, a leading constraint is that cables can pull but cannot push. Accordingly, only tensile forces can be applied by the cables.

Assuming an end effector with n degrees-of-freedom (usually six for a platform moving in space), at least m cables are needed to fix the platform i.e.

$$m \geq n + 1 \tag{1}$$

Under this condition, a *fully tensed* cable-driven parallel robot can be built: For a certain typical workspace, where some winches are "above" the workspace and others are "below". This enables the creation of inner tension within the system without the need for external wrenches. This actuation redundancy $r = m - n$ even allows active variation of the system's tension: If for a given pose one or more (to be precise, up to r) cables increase the tension, the others can react with an appropriate force and re-establish equilibrium. This allows full constraint of the payload, thereby leading to effective vibration suppression and enabling variation in the stiffness of the system [7].

It should be noted that there are drawbacks to the fully tensed approach. For instance, cables must come from "below" the platform; therefore, collisions between the cables or between cables and the environment are possible (see Fig. 1).

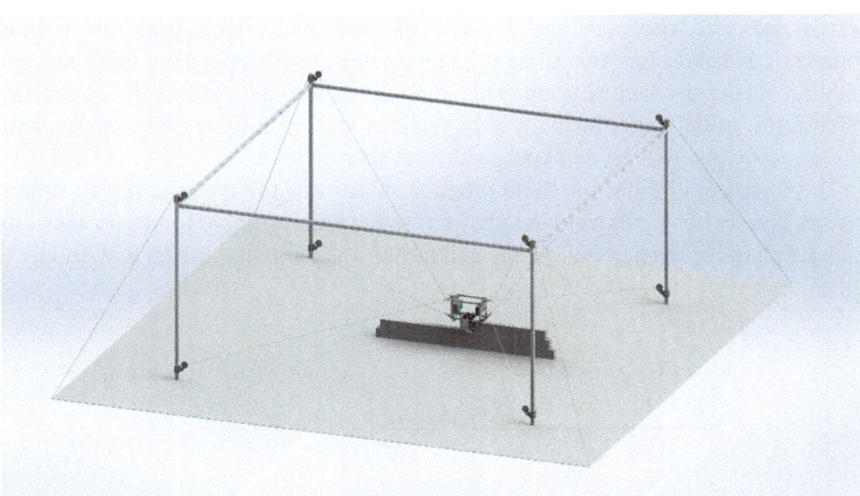

Fig. 1 Fully tensed cable-driven parallel robot. Source [12]

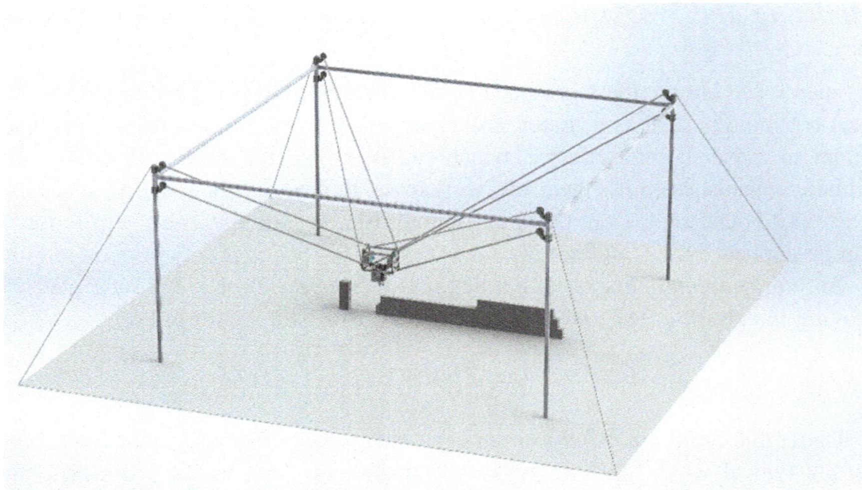

Fig. 2 Suspended cable-driven parallel robot. Source [12]

As an alternative approach, *suspended* cable-driven parallel robots (see Fig. 2) can be employed that only use cables from "above" the platform. This scenario resembles a combination of collaborating cranes. It is easy to see that, again, more than n cables might be employed, which allows for variation of the inner tension, albeit by a very small margin. Suspended robots may be prone to low stiffness effects, such as oscillations and reduced accuracy. This is especially the case when less than n cables are used and the payload is able to perform unconstrained motions.

Moveable pulleys can be used to change from a fully tensed to a suspended design and vice versa. At the expense of additional components and a more complex control design, this strategy incorporates the advantages of both concepts within a single mechanical setup. This approach is also advised for the application at hand [6]. Initial experiments will be realized using a fully tensed system, while future prototypes will be equipped with movable pulleys.

The force equilibrium for cable robots is defined as follows, see [8]: Assuming vectors l_μ; $\mu = 1, \ldots, m$ from the platform to the pulleys, cable forces (or tensions) f_μ and a platform wrench (i.e. forces and torques) w including inertia and gravity, it holds:

$$\underbrace{\begin{bmatrix} \nu_1 & \cdots & \nu_m \\ \boldsymbol{p}_1 \times \nu_1 & \cdots & \boldsymbol{p}_m \times \nu_m \end{bmatrix}}_{A^T} \underbrace{\begin{bmatrix} f_1 \\ \vdots \\ f_m \end{bmatrix}}_{f} + \underbrace{\begin{bmatrix} \boldsymbol{f}_p \\ \tau_p \end{bmatrix}}_{w} = 0, \tag{2}$$

where

$$\nu_\mu = \frac{l_\mu}{\|l_\mu\|_2} \tag{3}$$

As this is an underdetermined system of equations (m unknowns, n equations and $m \geq n + 1$), the variable tension levels are reflected in the model. The determination of appropriate cable tensions in real-time is a non-trivial task since amongst all forces f that are solutions to Eq. (2), only the ones fulfilling two constraints are technically feasible:

1. Cables need to have a certain minimum tension to avoid slackness, i.e.

$$f \geq f_{\min} \geq 0 \tag{4}$$

   where $f_{\min}$ needs to be set reasonably while accounting for cable mass, expected cable lengths and desired (passive) stiffness of the platform
2. Cables have a breaking load and the actuators have a limited maximum torque, i.e.

$$f_{\max} \geq f \tag{5}$$

In total, this results in an allowable force range

$$f_{\max} \geq f \geq f_{\min} \geq 0 \tag{6}$$

The solution to this problem has been subject to extensive research [9, 10].

End-Effector Design for Bricking Experiments

The design of an end effector for a given task is a critical objective. On one hand, its geometry and mass have a major influence on the force equilibrium described in the last section and therefore contribute to the determination of the robot's workspace. On the other hand, it must provide several functions that allow the application task to be realized.

To facilitate a feasibility study aimed at erecting a single brick wall in a laboratory environment, a cable robot was engineered (see Fig. 3) that includes a basic low-cost end effector (see Fig. 4). Initial experiments included the following minimal set of features:

- Detection and pose measurement of bricks as delivered onsite
- Gripping of bricks (where pose of fixed brick must be exactly known relative to the end effector)
- Transport of bricks
- Detection of wall

Fig. 3 Laboratory prototype for bricking a single wall. Source [12]

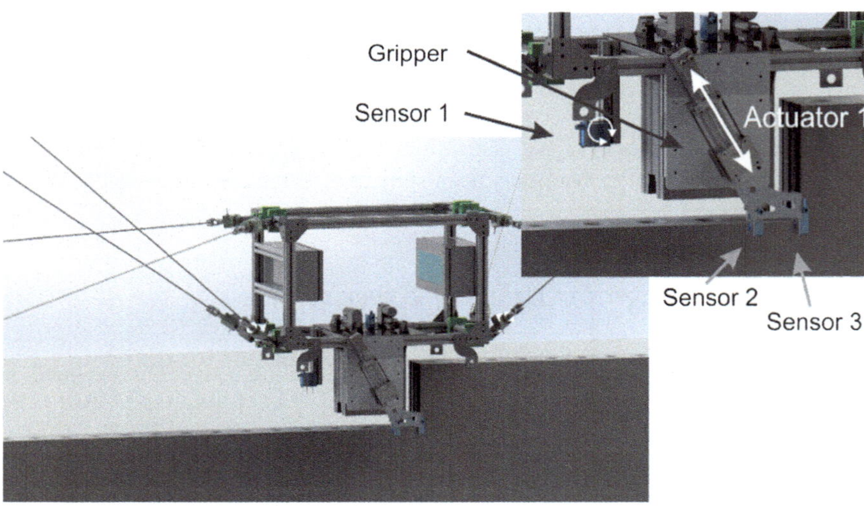

Fig. 4 Bricking using a cable-driven parallel robot. Based on [12]

- Pose measurement of end effector relative to intended dropping location
- Dropping of brick with defined pressure (for dropping on mortar)

Accordingly, the realized end effector is equipped with a set of mechatronic components that enable bricklaying according to the following process.

The end effector is equipped with a gripper that can fix a brick by pressing steel plates, arranged along the wall surface, against its flanks. Rotation of the brick is not considered by the gripper at this time.

First, sensor 1 (that is capable of pivoting) is looking horizontally towards a brick that is to be gripped while actuator 1 is retracted. This enables measurement of the pose variables of the brick relative to the end effector. Once the gripper has closed, the brick pose is known. The robot starts to move to the goal position of the brick, and then sensor 1 is rotated to look downwards while actuator 1 is deployed. Accordingly, Sensor 1 is able to provide the relative height coordinate when approaching the wall. In parallel, once the distance is small enough, sensors 2 and 3 continuously measure the alignment of the end effector relative to the wall. Once the wall is visible for sensors 2 and 3, the back and forth motion of actuator 1 further enables measurement of the air gap between the intended contact surfaces of

Table 1 Cable robot geometry

Winch no.	Platform connection point coordinates (w.r.t. platform center) [m]	Pulley coordinates (w.r.t. platform center) [m]
1	[0.25, 0.25, 0.25]'	[15, −15, 20]'
2	[0.25, −0.25, 0.25]'	[−15, −15, 20]'
3	[−0.25, −0.25, −0.25]'	[15, −15, 20]'
4	[−0.25, 0.25, −0.25]'	[−15, −15, 20]'
5	[−0.25, 0.25, 0.25]'	[15, 15, 20]'
6	[−0.25, −0.25, 0.25]'	[−15, 15, 20]'
7	[0.25, −0.25, −0.25]'	[15, 15, 20]'
8	[0.25, 0.25, −0.25]'	[−15, 15, 20]'

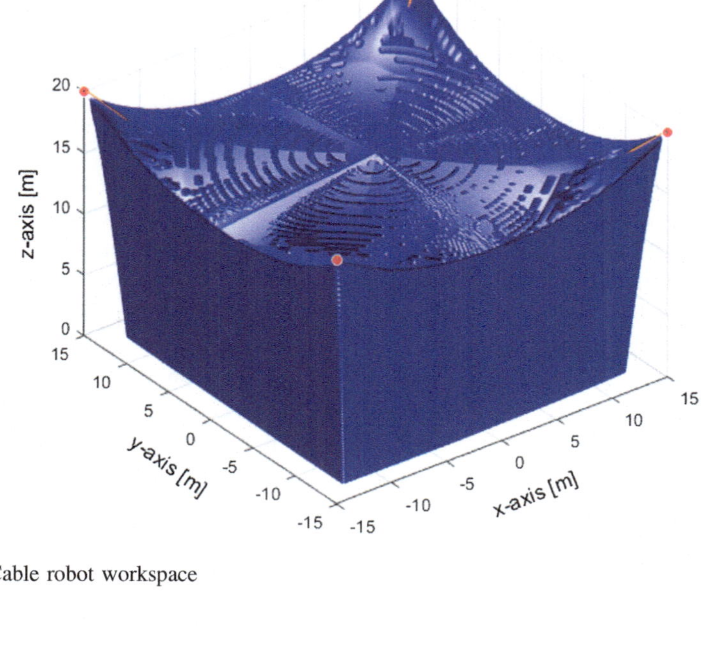

Fig. 5 Cable robot workspace

brick and wall until the brick reaches its targeted position. As the gripper may perform translational motion in the horizontal direction of the wall, supported by a spring, sensor 1 can be used in horizontal direction as well to ensure a good alignment of the bricks.

If mortar is applied (which still must be done manually at this stage of development), the cable force sensors report the pressure that is applied while dropping the brick. Currently, a dedicated force sensor is being developed for the gripper that also enables precise measurement of this pressure.

A preliminary design for the cable robot has been set up to perform initial investigations (see Table 1). The resulting workspace is shown in Fig. 5.

The specific cable stiffness has been set to 25000 N/m, the end effector total mass is 250 kg. The minimum cable force was 25 N to avoid slackness while the maximum force provided by the winches is 2500 N.

Transformations in Shell Production Through Cable Robots

The use of cable robots results in changes to the site equipment as a whole. Today, the logistics processes of construction sites are dominated by tower cranes and various material storage methods. The materials are stored on the ground and lifted by cranes to the appropriate location when needed. Material deliveries can be made on different times, which are largely independent of the time of installation. Just-in-time deliveries are used for filigree, heavy components or concrete. The preparation of the construction site is marked by initial groundwork.

In contrast to current protocols, construction by way of a cable driven robot will impact the construction process as well as the logistic setup of the whole construction site. Although the use of cable driven robots on construction sites will have significant impact on other processes, tasks, economic relations and even business models outside of the site, the authors focus only the construction site, including material delivery in the so-called access zone. Details regarding the different zones will be presented in the corresponding chapter of this paper.

A general inspection of the shell production process makes it obvious that the classical hoist device, the tower crane, can be omitted. Due to the use of presorted bricks and prefabricated concrete elements, there is no longer need for work places and storages for steel works, carpentry and formworks. By using high-strength adhesive in the bricklaying process, it is even possible to replace mortar, thus saving the space for both material and preparation. On the other hand, more magazines (usually steel storage containers) are needed to store different end effector modules and robot equipment. As real time data exchange will be needed during the construction process, the importance of data processing increases. This results in higher demands on the data networking onsite. Moreover, to realize the maximum benefit of digital construction, the requirements of the BIM models employed will be expanded to address specific attributes needed by the cable robot.

In terms of auxiliary devices, the cable robot construction site requires fork-lifters and lifters during the whole shell process to support the robot by providing material such as bricks etc.

Automated Bricking Process and Materials Management

In this study, bricklaying and pave out are considered. Other steps such as finishing are not included in this paper. Site logistics are changing as a result of the use of cable robots.

Figure 6 shows the site logistics for the basic case study of the actual research activities. In this case, an access zone is arranged between the cable robot and the area for unloading trucks. This zone is not mandatory, but it reduces the waiting

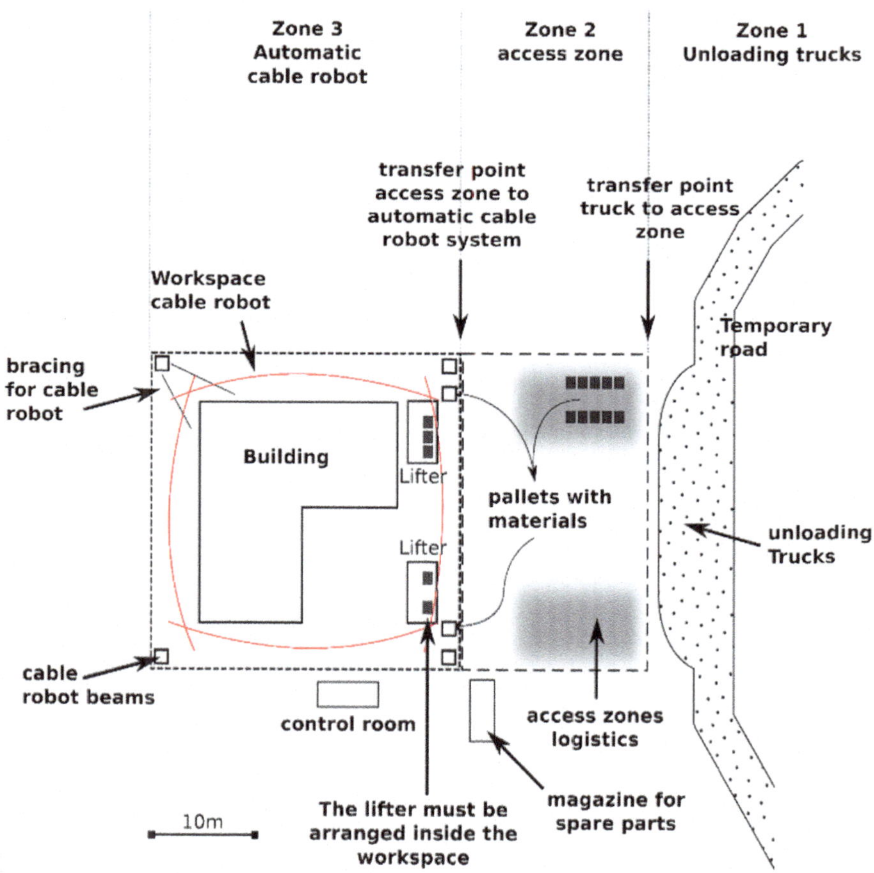

Fig. 6 Site logistics with cable robot

period for trucks, and it helps to minimize the impact of inactive lifters and incorrect material deliveries.

The three site logistic zones have the following functions that will be detailed later:

- **Zone 1** is for unloading the trucks. The pallets with material are transported by trucks and unloaded by fork-lift trucks.
- **Zone 2** is the storage and access zone for the cable robot. Fork-lift trucks move the material to the storage location.
- **Zone 3** is for the automatic cable robot system. The fork-lifters move the pallets with material onto the lifter. The lift raises the material to the installation height (working layer) and the cable robot withdraws the materials needed for the actual working setup. This is shown in Fig. 7.

Note that the cable robot can only work in the actual working layer and the space above.

Consequently, collisions become an issue as the building grows during the course of construction. This mandates that the building and its walls are erected layer-wise. For walls, this means that one brick layer of all walls must be completed

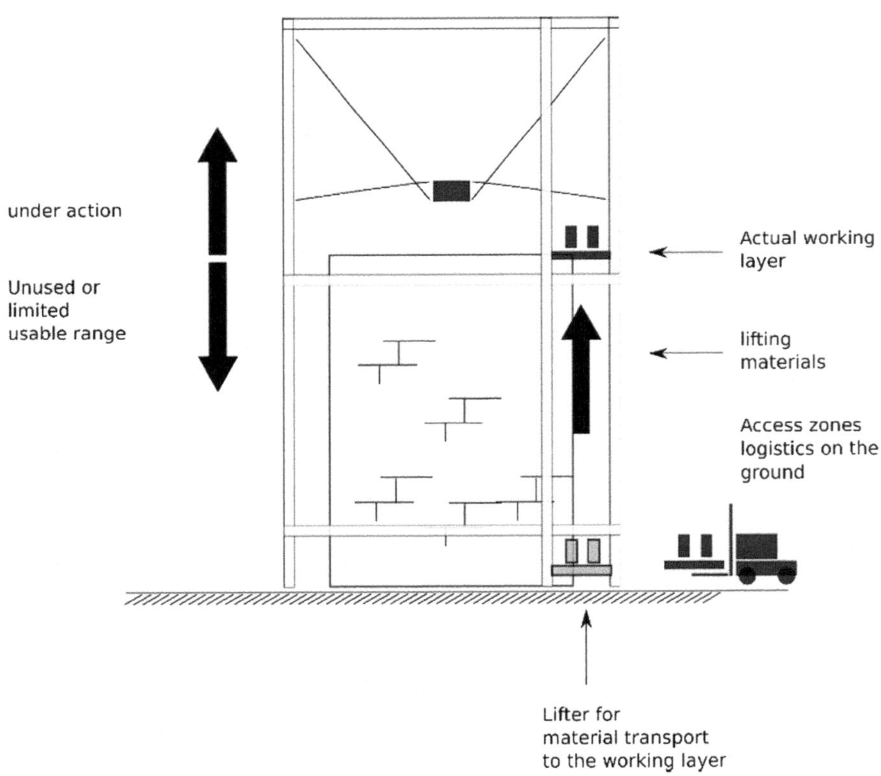

Fig. 7 Site view of the site with cable robot, fork lifter, lifter and building

before the next layer is started (note that conventionally, single walls are completed). Therefore, the robot must climb i.e. its pulleys must move upward (see Fig. 7). At the same time, materials have to be lifted to the actual working layer. These lifted materials must be supplied in a precise order (sequential storing). Deviation from the sequential storing method will lead to a failure that must be fixed before robot operations can restart by unloading at ground-level and refilling the lifters with the correct materials in the correct sequence.

In this chapter, the calculations performed later use a simplification of the trajectory calculation. Here, the paths to be followed are simplified as straight lines in space. A more sophisticated approach for complex numerical simulations can be found in [5]. Consequently, similar to a classical crane, the elements of the total trajectory time need to be composed.

$$t_{\text{workingcycle}} = t_{\text{picking}} + t_{\text{running}} + t_{\text{fitting}} \left[\frac{m}{s}\right] \tag{7}$$

$$t_{\text{picking}} = const \left[\frac{m}{s}\right] \tag{8}$$

$$t_{\text{running}} = \frac{p}{t_{\text{speed}}} \left[\frac{m}{s}\right] \tag{9}$$

$$p = |\vec{p}(x, y, z)| = \left| \begin{Bmatrix} p_x \\ p_y \\ p_z \end{Bmatrix} \right| = \sqrt{p_x^2 + p_y^2 + p_z^2} \tag{10}$$

$$t_{\text{speed}} = \text{const} \left[\frac{m}{s}\right] \tag{11}$$

$$t_{\text{fitting}} = \text{const} \left[\frac{m}{s}\right] \tag{12}$$

In terms of further considerations and optimizations, the center of mass of the building (center of production) should be considered as a benchmark for the determination of average distance and speed values.

While until now, only bricklaying was highlighted, both in the end-effector design of the robot in the last section as well as in this section, other working steps like the transfer of pre-fabricated elements need to be included. From an economic standpoint, it is reasonable to cover as many processes involved in the construction of a building as possible. Figure 8 shows an exemplary cross-section where bricks and pre-fabricated elements are used to construct walls and ceilings.

For continuous construction progress and optimal utilization, it is necessary to restart the bricklaying process after the pre-fabricated elements are placed. In future research, the composition of sequential working steps involving different materials and parts will be a point of focus, as it is important for continuous progress of construction, economic building, and minimization of downtimes.

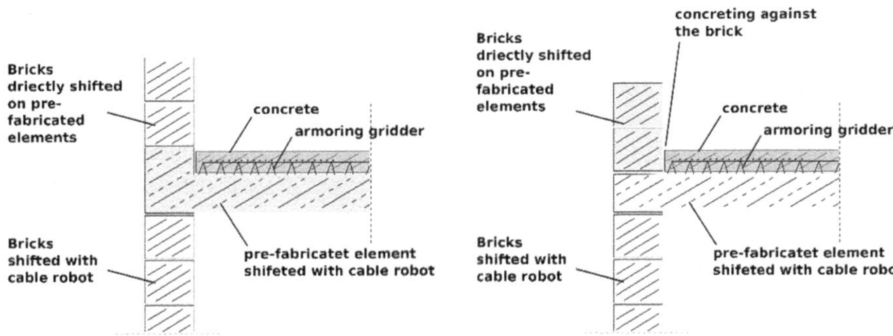

Fig. 8 Exemplary cross-section for pre-fabricated elements

Site Logistics in Detail

In this study, site logistics is divided into three zones, which are described in more detail below.

Zone 1 is for unloading all needed materials. Its dimensions are the same as those of a conventional loading zone. The unloading process can be done with the crane installed on the truck or the fork lift. The following list shows the differences between a conventional loading zone and a zone in a cable robot context:

- In the ideal case, a site with a cable robot does not necessarily need a tower crane. The tower crane is very slow in comparison with the robot and therefore slows the logistic processes on the site. If both are used, there exists a risk of collision between the robot and crane. This must be prevented under all conditions.
- The consumption of necessary building materials depends on the speed of the robot. If possible, the capacity of the access zone is tuned to the building speed and the turnover of the fastest delivery with the aim of minimizing the risk of standstill resulting from a lack of material. Just-in-time delivery carries the risk of stagnating the working processes. Because the robot requires specific material for a given building process, it can only build step by step. In contrast, a human can work on other parts of the building.
- When the site stagnates without finishing an actual working step, it is very expensive, and it is detrimental to the effectiveness of the overall system. From an economic standpoint, this should be avoided.

Zone 2 is for buffering the materials temporarily in an access zone. As specified in the previous section, the access zone has an important function. Using simulation methods, the optimal dimensions can be found. If this area too small, the logistics process can be suboptimal. This is a risk for the site. In contrast, when the area is too large, space is wasted and the area is not used optimally.

- Materials that will be placed by the robot during the next process steps must be stored in the access zone. If a component is absent, the site stands still until it is delivered. The first priority of the access zone is to minimize the risk of standstill.
- The building process is a sequential process, and all materials for building must be arranged in the right order. The access zone can be used to perform this serialization.
- For a continuous building process, the lifters must be loaded in rotation and in the right order.

Zone 3 contains the robot and the building. The workspace of the robot needs to be larger than the volume of the building, thus the robot geometry must be designed accordingly. As mentioned, in analogy to some tower cranes, the capacity of a robot to climb is optional. In zone 3, the actual creation of value occurs. The working zone is the subject of the case study in the next section.

Operational Transformation

Through the use of a cable robot the whole work organization, especially within the socio-technical area (working system), has to be adapted.

The elementary factors of operational production are affected as follows:

- Executive work: The object-related construction processes are carried out in zone 3 by the cable robot. This should lead to an increase in the execution performance, which in turn increases productivity. Besides the reduction of the physical workload, which will lead to an increase in physiological motivation in the workforce, the altered working conditions and safer working environment will increase the psychological motivation as well.
- Equipment, utilities and material: This is one of the most critical aspects. These resources must be present in a sufficient amount and at the right time. Depending on the available space for onsite storage of spare parts, construction material etc. resources should be available to ensure a continuous production process at all times. In addition to the elementary factors, the dispositional factors (planning, management and organization) of the operational production are also affected.

- Organization: Rethinking the entire site organization, at both the operational and management level, is reasonable. Long-term impacts on corporate strategies and business models are very likely.
- Operational management: Due to changed logistics and work processes, the redesign of tasks, responsibilities and other considerations is necessary.
- Planning: Continuous use of BIM will be inherent.
- Monitoring: In addition to traditional onsite construction supervision, the site can be monitored and coordinated via continuous data connection from a control

room. In addition, a head control room can monitor multiple sites and allocate resources, facilitating management of multiple projects.

When considering prospective changes in the context of operational production factors, it is clear that there will be a shift from executive work towards dispositive work. The changed roles and responsibilities impose the acquirement of additional competencies and skills.

Vocational education and training concepts relevant to several professions must adjust in response to the drive toward the use of cable robots. In particular, curricula for the operational construction workforce, including bricklayers, concrete construction workers and construction equipment operators, should be redesigned. With the goal of imparting a thorough understanding of the technology and new working methods, depending on the characteristics of specific professions, addenda to curricula should include the following: mechatronics, computer networks, communication systems, real-time systems, programming and algorithms, databases, modeling and simulation, and data security. Additionally, there will be a need to educate foremen, site managers, area managers and company management.

A similar development has already occurred in the automotive industry. As an example, the profession of car mechanic has changed over time to include new professions in response to technological developments, such new roles include the car electrician and the car mechatronic technician (all as apprenticeships). A similar development is taking place in the construction industry; apprenticeships for professions such as agricultural and construction mechatronics technician were founded as early as 2014. This profession is already engaged in 3D printing, however, the current focus is construction equipment and machinery (especially designing and printing of spare parts) and not the building itself.

The execution of construction processes is the subject of the following case study.

Economic Case Study

The case study models simulate the operative level. The aim is to have a construction project model that can represent connections and dependencies, and, more importantly, predict tendencies and dynamics within the construction process. Additionally, elementary supply chains will be included in the project model to allow examination of their influence on the project. The present case study and subsequent dynamic simulations should provide answers to the following questions:

- Which parameters affect the economy of the robot?
- How do the costs change with the passing of time?
- Which features can be identified that deviate from the normal building process?
- What risks arise from automation of the system?

Traditionally, the construction of a building shell depends on the formworks of the processes, reinforcing, concreting and brickworking [3]. As discussed in this paper, making use of a cable robot will fundamentally change the aforementioned process. For instance, there will be no need for formworks and reinforcing, and concreting will be minimized. Brickworking will be, from the process perspective, unchanged. Therefore, a comparison of the brickworking of the outer wall and the inner wall is the subject of this case study. Furthermore, only the working steps for completion of the building shell are considered in detail. Processes considered herein include brickwork and precast elements of shell construction. Earthworks and foundations are assumed to be completed. Resources like materials, tools and personnel are included in the calculations. The tasks neglected by this study are considered to be services contributed by subcontractors or third-parties. The current results are based on the design of the actual prototype, initial simulations and case studies. The conclusions show the actual state of research. These research activities focus on bricklaying and the identification of construction methods that work well with the cable robot. The identified construction methods will be evaluated in practice in the near future.

To investigate the questions posed, a fictitious building is considered. This case study can only show a small part of the theoretical options and this example is not optimized on the basis of economic parameters, but it illustrates the differences between the conventional building process and processes employing a cable robot. A system dynamics model based on data from the prototype and estimated values is used. The parameters were chosen so that fundamental statements can be taken. Figure 9 shows the 1st floor with inner walls and parts of the ceiling. In total, the building has three floors and a float roof.

Other input parameters include the following:

- The building area is 350 m^2. This area is also supposed to be covered by the prefabricated elements for the ceiling.
- The area for bricklaying the outside wall is 260 m^2 per each floor, including openings.
- The area for bricking the inner wall is 220 m^2 per floor, including openings.

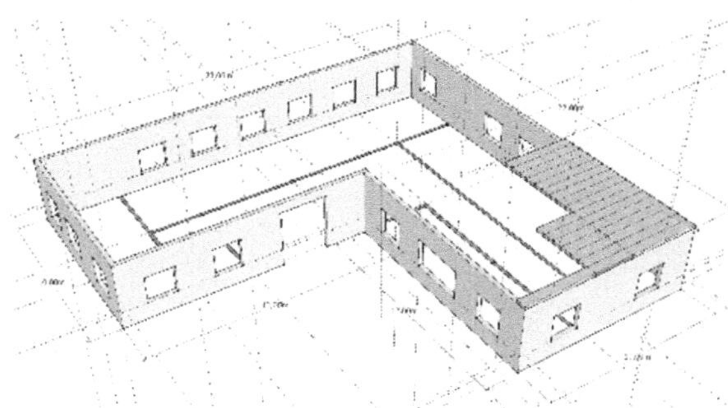

Fig. 9 Principle sketch of the 1st floor

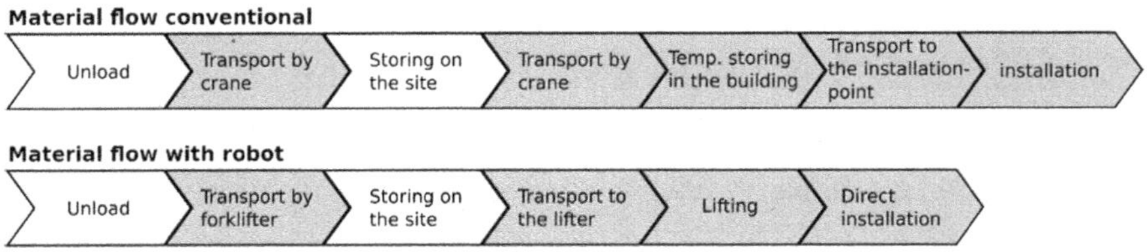

Fig. 10 Material flows of bricking with and without cable robot

- The building process for the floating roof is the same as for the ceilings.
- Only the shell construction is supported by the robot. All further construction processes are done conventionally.

The onsite flows with and without a robot are different. As an example, Fig. 10 illustrates the differences in the material flow for install bricks.

A comparison of the process chains shows that the process chain with a robot encompasses fewer steps. In contrast to the setup of a conventional construction site, which is primarily composed of parts containers, tower cranes and storage space [11], a cable robot obviates the need for tower cranes and interim storage directly at the installation point. Moreover, it will be shown that the total time from unloading to installation using a cable robot is significantly reduced relative to the conventional bricklaying process.

The unloading, transport and storing processes are analogous to bricklaying without a robot. Major distinctions emerge with respect to transport inside the

Table 2 Initializing parameters for the case study

Parameter		Description
Material parameters		
Average material transport with cable robot*	12 m	
Bricksize (outer walls)	L/W/H = 1.00/0.24/0.5 m	
Bricksize (inner walls)	L/W/H = 1.00/0.24/0.5 m	
Size of prefabricated elements	L/W/H = 3/0.45/0.15 m = 0.5 t	
Cable robot parameters		
Platform speed*	1 m/s	Parameters to be verified on prototype
Grip material*	0.3 min	
Settling material*	0.3 min	
Limit of material weight*	0.5 t	
Total system cost*	1,500,000 €	Parameter is estimated
Deprecation of robot	46,000 €/month (6.5% interest)	
Cost assembly and dismantling for robot*	15,000 €/per use	
Operating cost robot*	9,000 €/month	
Personal*	20,000 €/month	2 skilled workers

*Estimated parameters were recognized unfavorable

created building. From this stage, large differences in the cycle times of the procedures become obvious. For example, transport by tower cranes, miniature cranes or pushcarts is much slower than transport by a lifter and a robot.

In the following discussion, the time required for shell construction will be investigated. At the current stage of research, precise economic forecasts would be premature, but the simulation results based on the parameters in Table 2 suggest the automated approach provides time and cost benefits (vide infra).

Figure 11 shows the essential results of the case study. A very interesting result is that loading the pre-fabricated elements requires nearly the same time as the conventional method using a tower crane (5 h for loading). The robot is much faster per element, but this advantage is nullified by the low payload that mandates the use of smaller and more lightweight pre-fabricated elements. This low payload means that nearly 260 pieces of elements must be moved (versus 20 elements when loading with a tower crane).

With respect to the bricklaying processes, the low payload plays no role. Here, the higher speed of the robot generates a major economic advantage!

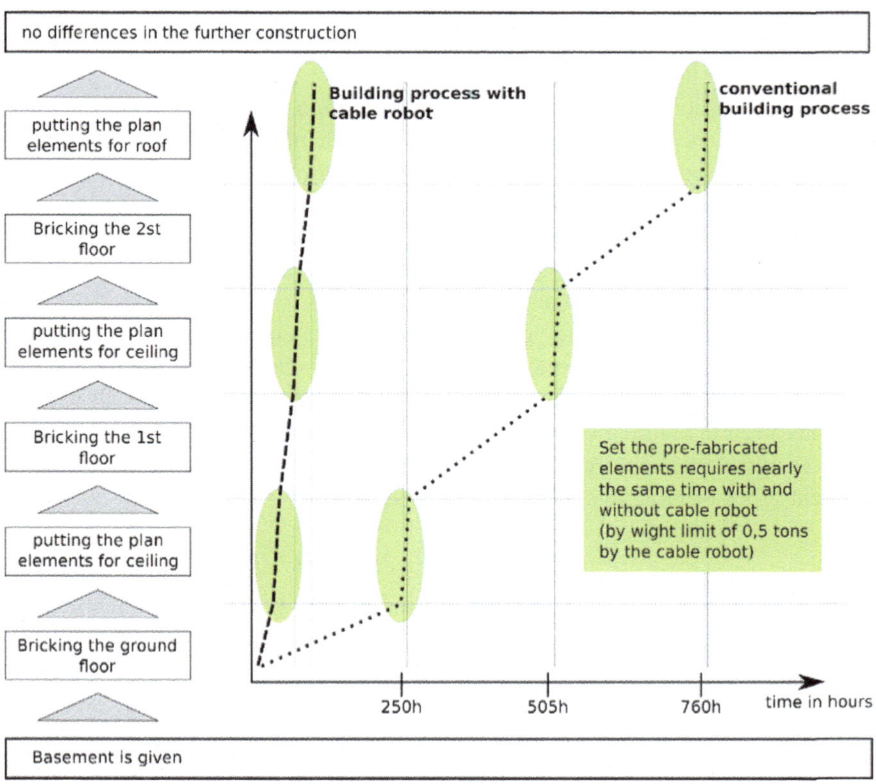

Fig. 11 Building process with and without cable robot

In conclusion, this first case study shows the potential associated with the use of a cable robot for shell construction. The results show that some parameters are vitally important when using a cable robot, these parameters include (in order):

1. Total working hours per day
2. Moving speed of the platform
3. Depreciation of the robot
4. Total price of the complete system

The most influential factor is the number of working hours per day. In this case study, the system had to work for a minimum of 11 h a day to ensure an economic building process.

Discussion and Conclusion

This article presented an interdisciplinary research approach that links the construction industry and modern mechatronic systems. Specifically, the application of a cable robot for the automated construction of a building was presented on a conceptual level. In addition to discussing paradigm changes, a detailed examination of automated processes was presented within the framework of bricklaying. The design of a prototype was also presented.

Initial discussions show that automated construction techniques, which can make use of components like bricks and prefabricated elements, can be applied advantageously to all building types, including those with multiple floors. A concluding economic feasibility analysis indicates that this concept holds significant potential for time and cost savings. Clearly, a robot requires novel specific site logistics. A standstill must be avoided. For a detailed analysis, BIM models with dimensions, materials and processes are needed.

In the near future, a prototype will be realized and experiments will be performed, providing insight into the potential accuracy and practical performance of the strategy outlined herein. In parallel, a more detailed simulation of all working steps involved in building construction need to be modelled to enable more detailed analyses of technical processes as well as predictions of the economic feasibility for different types of buildings under various conditions.

References

1. Epstein E (2012) Implementing successful building information modeling. Artech House, Norwood, Britain
2. Bruckmann TH, Mattern H Spengler A, Malkwitz A, König M (2016). Automated construction of masonry buildings using cable-driven parallel robots. In: Proceedings of the ISARC 2016

3. Willmann J (2014) Gramazio F, Kohler M, Roboter und Megastrukturen: Neue Maßstäbe in der digitalen Fabrikation. in Technik in Bayern 04
4. Willmann J, Gramazio F, Kohler M (2015) If robots conquer airspace: the architecture of the vertical city. In: Springer optimization and its applications, pp 1–11, Switzerland
5. Landsberger SE, Sheridan TB (1985). A new design for parallel link manipulator. In: Proceedings systems man and cybernetics conference, pp 812–814
6. Mattern H, Bruckmann T, Spengler A, König M (2016). Simulation of automated construction using wire robots. In: Proceedings of the 2016 winter simulation conference
7. Reichert C et al (2015) Dynamische Rekonfiguration eines seilbasierten Manipulators zur Verbesserung der mechanischen Steifigkeit. Tagung VDI Mechatronik, Dortmund, Deutschland
8. Bruckmann T et al (2008) Wire robots part I—kinematics, analysis & design. Parallel manipulators, New Developments. Ryu J-H Vienna, Austria, I-Tech Education pp 109–132
9. Gouttefarde M et al (2015). A versatile tension distribution algorithm for n-DOF parallel robots driven by n + 2 cables. Robotics 31(6): 1444–1457
10. Müller K et al (2015) Analysis of a real-time capable cable force computation method. Cable-driven parallel robots. Ed. Pott A, Bruckmann T Springer International Publishing pp 227–238
11. Karl CK, Ibbs W (2016) Developing modular-oriented simulation models using system dynamics libraries. Springer International Publishing, Cham
12. Meik M (2016) Entwurf eines Endeffektors und Implementierung einer Regelung für einen Seilroboter, Master Thesis, University of Duisburg-Essen

Design of Reconfigurable Cable-Driven Parallel Robots

Lorenzo Gagliardini, Marc Gouttefarde and Stéphane Caro

Abstract This chapter is dedicated to the design of Reconfigurable Cable-Driven Parallel Robots (RCDPRs) where the locations of the cable exit points on the base frame can be selected from a finite set of possible values. A task-based design strategy for discrete RCDPRs is formulated. By taking into account the working environment, the designer divides the prescribed workspace or trajectory into parts. Each part shall be covered by one configuration of the RCDPR. Placing the cable exit points on a grid of possible locations, numerous CDPR configurations can be generated. All the possible configurations are analysed with respect to a set of constraints in order to determine the parts of the prescribed workspace or trajectory that can be covered. The considered constraints account for cable interferences, cable collisions, and wrench feasibility. The configurations satisfying the constraints are then compared in order to find the combinations of configurations that accomplish the required task while optimising one or several objective function(s). A case study comprising the design of a RCDPR for sandblasting and painting of a three-dimensional tubular structure is finally presented. Cable exit points are reconfigured, switching from one side of the tubular structure to another, until three external sides of the structure are covered. The optimisation includes the minimisation of the number of cable attachment/detachment operations required to switch from one configuration to another one, minimisation of the size of the RCDPR, and the maximisation of the RCDPR stiffness.

L. Gagliardini
IRT Jules Verne, 44340 Bouguenais, France
e-mail: lorenzo.gagliardini.at.work@gmail.com

M. Gouttefarde (✉)
LIRMM-CNRS-UM, 161, Rue Ada, 34095 Montpellier Cedex 5, France
e-mail: marc.gouttefarde@lirmm.fr

S. Caro
CNRS–LS2N, École Centrale de Nantes, 1, Rue de la Noë, 44321 Nantes Cedex 3, France
e-mail: Stephane.Caro@ls2n.fr; Stephane.Caro@irccyn.ec-nantes.fr

E. Ottaviano et al. (eds.), *Mechatronics for Cultural Heritage and Civil Engineering*, Intelligent Systems, Control and Automation: Science and Engineering 92,
https://doi.org/10.1007/978-3-319-68646-2_4

Introduction

Cable-Driven Parallel Robots (CDPRs) form a particular class of parallel robots whose moving platform is connected to a fixed base frame by cables. Hereafter, the connection points between the cables and the base frame will be referred to as exit points. The cables are coiled on motorised winches. Passive pulleys may guide the cables from the winches to the exit points. A central control system coordinates the motors actuating the winches. Thereby, the pose and the motion of the moving platform are controlled by modifying the cable lengths. An example of CDPR is shown in Fig. 1.

CDPRs have several advantages such as a relatively low mass of moving parts, a potentially very large workspace due to size scalability, and reconfiguration capabilities. Therefore, they can be used in several applications, e.g. heavy payload handling and airplane painting [1], cargo handling [15], warehouse applications [14], large-scale assembly and handling operations [28, 33], and fast pick-and-place operations [18, 21, 29]. Other possible applications include the broadcasting of sporting events, haptic devices [8, 10, 31], support structures for giant telescopes [34, 35], and search and rescue deployable platforms [23, 24]. Recent studies have been performed within the framework of an ANR Project CoGiRo [2] where an efficient cable layout has been proposed [11] and used on a large CDPR prototype called CoGiRo.

CDPRs can be used successfully if the tasks to be fulfilled are simple and the working environment is not cluttered. When these conditions are not satisfied, Reconfigurable Cable-Driven Parallel Robots (RCDPRs) may be required to achieve the prescribed goal. In general, several parameters can be reconfigured, as described in section "Classes of RCDPRs". Moreover, these reconfiguration parameters can be selected in a discrete or a continuous set of possible values.

Preliminary studies on RCDPRs were performed in the context of the *NIST RoboCrane* project [5]. Izard et al. [16] also studied a family of RCDPRs for in-

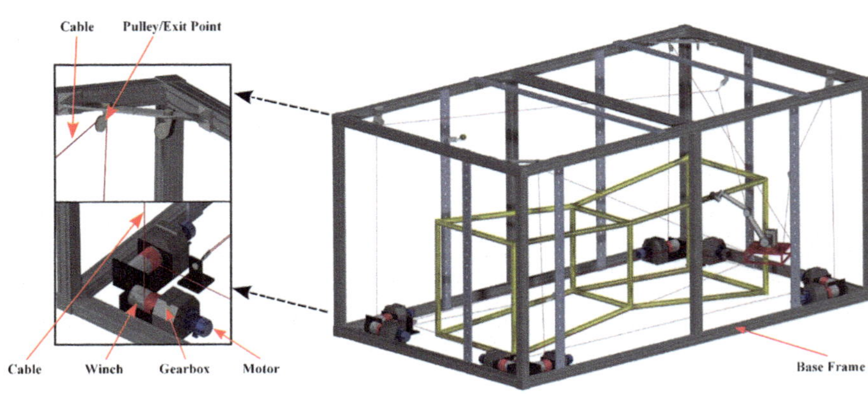

Fig. 1 Architecture of a CDPR developed in the framework of the IRT Jules Verne CAROCA project

Fig. 2 CAROCA prototype: a reconfigurable cable-driven parallel robot working in a cluttered environment (Courtesy of IRT Jules Verne and STX France)

dustrial applications. Rosati et al. [32, 36] and Zhou et al. [38, 39] focused their work on planar RCDPRs. Recently, Nguyen et al. [26, 27] proposed reconfiguration strategies for large-dimension suspended CDPRs mounted on overhead bridge cranes. Contrary to these antecedent studies, this chapter considers discrete reconfigurations where the locations of the cable exit points are selected from a finite set (grid) of possible values. Hereafter, reconfigurations are limited to the cable exit point locations and the class of RCDPRs whose exit points can be placed on a grid of positions is defined as *discrete RCDPRs*.

Figure 2 shows the prototype of a reconfigurable cable-driven parallel robot developed at IRT Jules Verne within the framework of CAROCA project. This prototype is reconfigurable for the purpose of being used for industrial operations in a cluttered environment. Indeed, its pulleys can be displaced onto the robot frame faces such that the collisions between the cables and the environment can be avoided during operation. The prototype has eight cables, can work in both suspended and fully constrained configurations and can carry up to 400 kg payloads. It contains eight motor-geardhead-winch sets. The nominal torque and velocity of each motor are equal to 15.34 Nm and 2200 rpm, respectively. The ratio of the twp-stage gearheads is equal to 40. The diameter of the Huchez™ industrial winches is equal to 120 mm. The CAROCA prototype is also equipped with 6 mm non-rotating steel cables and a B&R control board using Ethernet Powerlink™ communication.

To the best of our knowledge, no design strategy has been formulated in the literature for discrete RCDPRs. Hence, section "Design Strategy for RCDPRs" presents a novel task-based design strategy for discrete RCDPRs. By taking into account the working environment, the designer divides the prescribed workspace or trajectory into n_t parts. Each part will be covered by one and only one configuration of the

RCDPR. Then, for each configuration, the designer selects a cable layout, parametrising the position of the cable exit points. The grid of locations where the cable exit points can be located is defined by the designer as well. Placing the exit points on the provided set of possible locations, it is possible to generate many CDPR configurations. All the possible configurations are analysed with respect to a set of constraints in order to verify which parts of the prescribed workspace or trajectory can be covered. The configurations satisfying the constraints are compared in order to find the combinations of n_t configurations that accomplish the required task and optimise at the same time one or several objective function(s). A set of objective functions, dedicated to RCDPRs, is provided in section "Global Objective Functions". These objective functions aim at maximising the productivity (production cycle time) and reducing the reconfiguration time of the cable exit points. Let us note that if the design strategy introduced in section "Design Strategy for RCDPRs" does not produce satisfactory results, the more advanced but complex method recently introduced by the authors in [9] can be considered.

In order to analyse the advantages and limitations of the proposed design strategy, a case study is presented in section "Case Study: Design of a RCDPRs for Sandblasting and Painting of a Large Tubular Structure". It involves the design of an RCDPR for sandblasting and painting of a three-dimensional tubular structure. The tools performing these operations are embarked on the RCDPR moving platform, which follows the profile of the tubular structure. Each side of the tubular structure is associated to a single configuration. Cable exit points are reconfigured switching from one side of the tubular structure to another, until three external sides of the structure are sandblasted and painted. The cable exit point locations of the three configurations to be designed are optimised so that the number of cable attachment/detachment operations required to switch from a configuration to another is minimised. The size of the RCDPR is also minimised while its stiffness is maximised along the trajectories to be followed.

Classes of RCDPRs

CDPRs usually consist of several standard components: A fixed base, a moving platform, a set of m cables connecting the moving platform to the fixed base through a set of pulleys, a set of m winches, gearboxes and actuators, and a set of internal and external sensors. These components are usually dimensioned in such a way that the geometry of the CDPR does not vary during the task. However, by modifying the CDPR geometry, the capabilities of CDPRs can be improved. RCDPRs are then defined as CDPRs whose geometry can be adapted by reconfiguring part of their components. RCDPRs can then be classified according to the components, which are reconfigured and the nature of the reconfigurations.

Fig. 3 CableBot designs with cable exit points fixed to a grid (left) and with cable exit points sliding on rails (right). Courtesy of the European FP7 project CableBot

Reconfigurable Elements and Technological Solutions

Part of the components of an RCDPR may be reconfigured in order to improve its performances. The geometry of the RCDPRs is mostly dependent on the locations of the cable exit points, the locations of the cable attachment points on the moving platform, and the number of cables.

The locations of the cable exit points $A_i, i = 1, \ldots, m$ have to be reconfigured to avoid cable collisions when the environment is strongly cluttered. Indeed, modifying the cable exit point locations can increase the RCDPR workspace size. Furthermore, the reconfiguration of cable exit points provides the possibility to modify the layout of the cables and improve the performance of the RCDPR (such as its stiffness). From a technological point of view, the cable exit points A_i are displaced by moving the pulleys orienting the cables and guiding them to the moving platform. Pulleys are connected on the base of the RCDPR. They can be displaced by sliding them on linear guides or fixing them on a grid of locations, as proposed in the concepts of Fig. 3. These concepts have been developed in the framework of the European FP7 Project CableBot [4, 7, 25]. Alternatively, pulleys can be connected to several terrestrial or aerial unmanned vehicles, as proposed in [17, 22, 37].

The geometry of the RCDPR and the cable layout can be modified as well by displacing the cable anchor points on the moving platform, $B_i, i = 1, \ldots, m$. Changing the locations of points B_i allows the stiffness of the RCPDR as well as its wrench (forces and moments) capabilities to be improved. A modification of the cable anchor points may also result in an increase of the workspace dimensions. The reconfiguration of points B_i can be performed by attaching and detaching the cables at different locations on the moving platform.

The number m of cables has a major influence on performance of the RCDPR. Using more cables than DOFs can enlarge the workspace of suspended CDPRs [11] or yields fully constrained CDPRs where internal forces can reduce vibrations, e.g. [18]. However, the larger the number of cables, the higher the risk of collisions. In this case, the reconfiguration can be performed by attaching or detaching one or several cable(s) to/from the moving platform and possibly to/from a new set of exit points. Furthermore, by attaching and detaching one or several cable(s), the architecture

Table 1 CDPR reconfigurable parameter classification

Reconfigurable parameter	Discrete domain	Continuous domain
Exit point locations	Yes	Yes
Platform anchor point locations	Yes	Yes
Cable number	Yes	No

of the RCDPRs can be modified, permitting both suspended and fully constrained CDPR configurations.

Discrete and Continuous Reconfigurations

According to the reconfigured components and the associated technology, reconfiguration parameters can be selected over a continuous or discrete domain of values, as summarised in Table 1. Reconfigurations performed over a *discrete domain* consist of selecting the reconfigurable parameters within a finite set of values. Modifying the number of cables is a typical example of a discrete reconfiguration. Discrete reconfigurations also apply to cable anchor points, when the cables can be installed on the moving platform at a (discrete) number of specific locations, e.g. its corners. Another example of discrete RCDPR is represented in Fig. 3 (left). In this concept, developed in the framework of the European FP7 Project CableBot, cable exit points are installed on a predefined grid of locations on the ceiling. Discrete reconfigurations are performed off-line, interrupting the task the RCDPR is executing. For this reason, the set up time for these RCDPRs can be relative long. On the contrary, RCDPRs with discrete reconfigurations can use the typical control schemes already developed for CDPRs. Furthermore, they do not require to motorise the cable exit points, thereby avoiding a large increase of the CDPR cost.

Reconfigurations performed over a *continuous domain* provide the possibility of selecting the geometric parameters over a continuous set of values delimited by upper and lower bounds. A typical example of continuous RCDPR is represented in Fig. 3 (right), which illustrates another concept developed in the framework of the European FP7 Project CableBot. In this example, the cable exit points slide on rails fixed on the ceiling. Reconfigurations can be performed on-line, by continuously modifying the reconfigurable parameters during the task execution. The main advantages of continuous reconfigurations are the reduced set-up time and the local optimisation of the RCDPR properties. However, modifying the locations of the exit points in real time may require the design of a complex control scheme. Furthermore, the cost of RCDPRs with continuous reconfigurations is significantly higher than the cost of discrete RCDPRs when the movable pulleys are actuated.

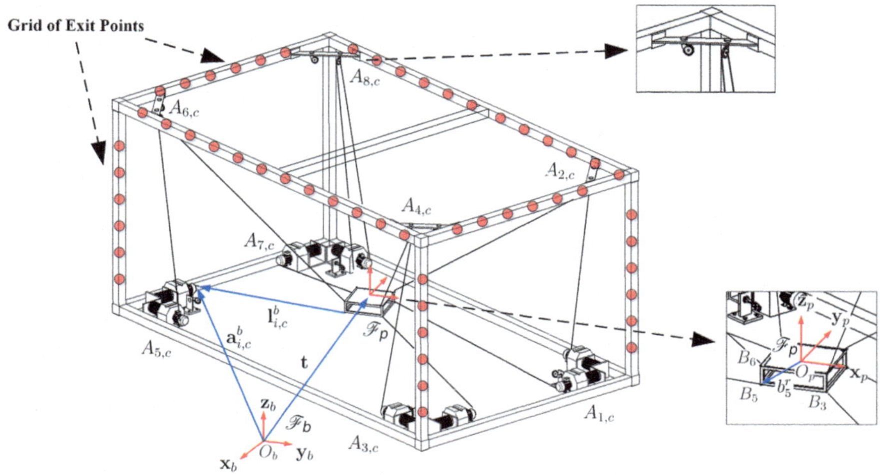

Fig. 4 Schematic of a RCDPR. The red points represent the possible locations of the cable exit points, where the pulleys can be fixed

Nomenclature for RCDPRs

Similarly to CDPRs, an RCDPR is mainly composed of a moving platform connected to the base through a set of cables, as illustrated in Fig. 4. The moving platform is driven by m cables, which are actuated by winches fixed on the base frame of the robot. The cables are routed by means of pulleys to exit points from which they extend toward the moving platform. The main difference between this chapter and previous works on CDPRs is the possibility to displace the cable exit points on a grid of possible locations.

As illustrated in Fig. 4, $\mathscr{F}_b$, of origin O_b and axes $\mathbf{x}_b$, $\mathbf{y}_b$, $\mathbf{z}_b$, denotes a fixed reference frame while $\mathscr{F}_p$ of origin O_p and axes $\mathbf{x}_p$, $\mathbf{y}_p$ and $\mathbf{z}_p$, is fixed to the moving platform and thus called the moving platform frame. The anchor points of the i-th cable on the platform are denoted as $B_{i,c}$, where c represents the configuration number. For the c-th configuration, the exit point of the i-th cable is denoted as $A_{i,c}$, $i = 1, \ldots, m$. The Cartesian coordinates of each point $A_{i,c}$, with respect to $\mathscr{F}_b$, are given by the vector $\mathbf{a}^b_{i,c}$ while $\mathbf{b}^b_{i,c}$ is the position vector of point $B_{i,c}$ expressed in $\mathscr{F}_b$. Neglecting the cable mass, the vector $\mathbf{l}^b_{i,c}$ directed along the i-th cable from point $B_{i,c}$ to point $A_{i,c}$ can be written as:

$$\mathbf{l}^b_{i,c} = \mathbf{a}^b_{i,c} - \mathbf{t} - \mathbf{R}\mathbf{b}^p_{i,c} \qquad i = 1, \ldots, m \tag{1}$$

where $\mathbf{t}$ is the moving platform position, i.e. the position vector of O_p in $\mathscr{F}_b$, and $\mathbf{R}$ is the rotation matrix defining the orientation of the moving platform, i.e. the orientation of $\mathscr{F}_p$ with respect to $\mathscr{F}_b$. The length of the i-th cable is then defined by the 2-norm of the cable vector $\mathbf{l}^b_{i,c}$, namely, $l_{i,c} = \|\mathbf{l}^b_{i,c}\|_2$, $i = 1, \ldots, m$.

In order to balance an external wrench (combination of a force and a moment), each cable generates on the moving platform a wrench proportional to its tension $\tau_i = 1, \ldots, m$. The cables balance the external wrench $\mathbf{w}_e$, according to the following equation [30]:

$$\mathbf{W}\boldsymbol{\tau} + \mathbf{w}_e = 0 \tag{2}$$

The cable tensions are collected into the vector $\boldsymbol{\tau} = \left[\tau_1, \ldots, \tau_m\right]$ and multiplied by the wrench matrix $\mathbf{W}$ whose columns are composed of the unit wrenches $\mathbf{w}_i$ exerted by the cables on the platform:

$$\mathbf{W} = \begin{bmatrix} \mathbf{d}^b_{1,c} & \mathbf{d}^b_{2,c} & \ldots & \mathbf{d}^b_{m,c} \\ \mathbf{R}\mathbf{b}^p_{1,c} \times \mathbf{d}^b_{1,c} & \mathbf{R}\mathbf{b}^p_{2,c} \times \mathbf{d}^b_{2,c} & \ldots & \mathbf{R}\mathbf{b}^p_{m,c} \times \mathbf{d}^b_{m,c} \end{bmatrix} \tag{3}$$

where $\mathbf{d}^b_{i,c}$, $i = 1, \ldots, m$ are the unit cable vectors associated with the cth configuration:

$$\mathbf{d}^b_{i,c} = \frac{\mathbf{l}^b_{i,c}}{l_{i,c}}, \qquad i = 1, \ldots, m \tag{4}$$

Design Strategy for RCDPRs

Similarly to CDPRs, the design of RCDPRs requires the dimensioning of all its components. In this chapter, the design of RCDPRs focuses on the selection of the cable exit point locations. The other components of the RCDPR are required to be chosen in advance.

Design Problem Formulation

The RCDPR design strategy proposed in this section and illustrated in Fig. 5 consists of ten steps. The design can be formulated as a mono-objective or hierarchical multi-objective optimisation problem. The designer defines a prescribed workspace or moving platform trajectory and divides it into n_t parts. Each part should be covered by one and only one configuration. The design variables are the locations of the cable exit points for the n_t configurations covering the n_t parts of the prescribed workspace or trajectory. The *global objective functions* investigated in this chapter (section "Global Objective Functions") aim to reduce the overall complexity of the RCDPR and the reconfiguration time. The optimisation is performed while verifying a set of user-defined constraints such as those presented in section "Constraint Functions ".

Step I. *Task and Environment*. The designer describes the task to be performed. He/She specifies the nature of the problem, defining if the motion of

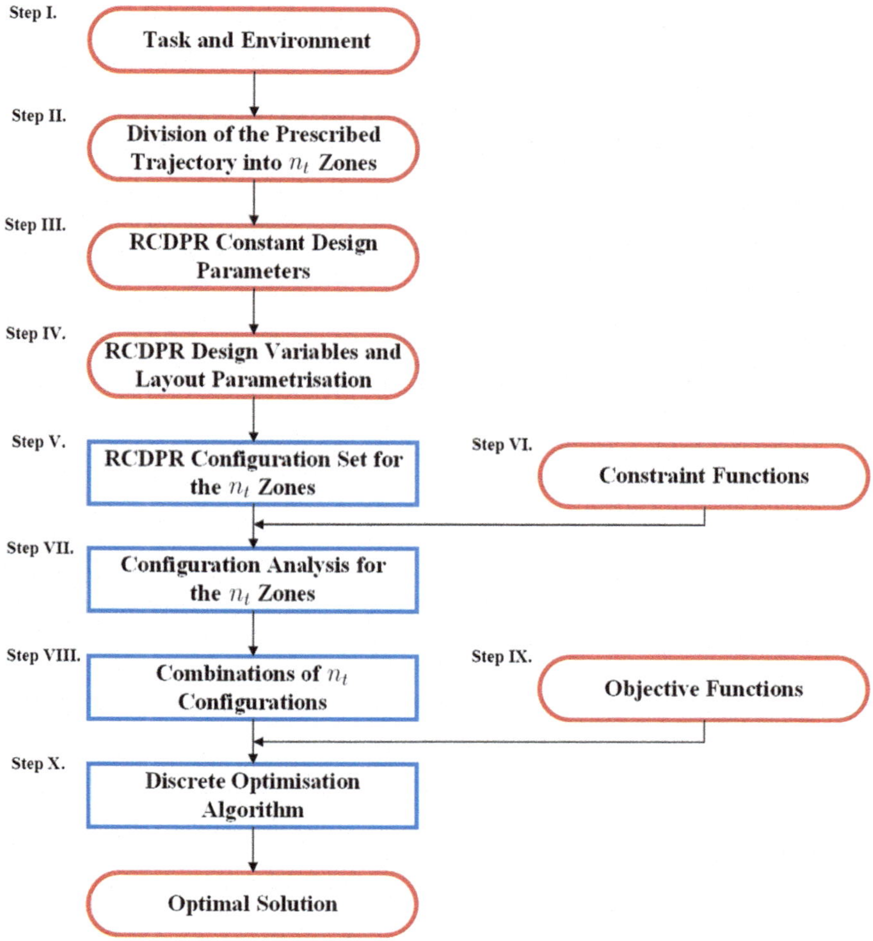

Fig. 5 Design strategy for RCDPRs

the moving platform is static, quasi-static or dynamic. According to the nature of the problem, the designer defines the external wrenches applied to the moving platform and, possibly, the required moving platform twist and accelerations. The prescribed workspace or trajectory of the moving platform is given. A description of the environment is provided as well, including the possible obstacles encountered during the task execution.

Step II. *Division of the Prescribed Trajectory*. Given the prescribed workspace or moving platform trajectory, the designer divides it into n_t parts, assuming that each of them is accessible by one and only one configuration of the RCDPR. The division may be performed by trying to predict the possible collisions of the cables and the working environment.

Step III. *Constant Design Parameters*. The designer defines a set of constant design parameters and their values. The parameters are collected in the constant design parameter vector $\mathbf{q}$.

Step IV. *Design Variables and Layout Parametrisation*. For each part of the prescribed workspace or moving platform trajectory, the designer defines the cable layout of the associated configuration. The cable layout associated with the t-th part of the prescribed workspace or trajectory defines the locations of the cable exit points, parametrised with respect to a set of $n_{t,v}$ design variables, $u_{t,v}$, $v = 1, \dots, n_{t,v}$. The design variables are defined as a discrete set of $\varepsilon_{t,v}$ values, $[u]_{t,v}$, $v = 1, \dots, n_{t,v}$.

Step V. *RCDPR Configuration Set*. For each part of the prescribed trajectory, the possible configurations, which can be generated combining the values $[u]_{t,v}$, $v = 1, \dots, n_{t,v}$ of the $n_{t,v}$ design variables, are computed. Therefore, $n_{t,\mathscr{C}} = \prod_{v=1}^{n_{t,v}} \varepsilon_{t,v}$ possible configurations are generated for the tth part of the prescribed workspace or trajectory.

Step VI. *Constraint Functions*. The user defines a set of n_ϕ constraint functions, ϕ_k, $k = 1, \dots, n_\phi$. These functions are applied to all possible configurations associated to the n_t parts of the prescribed workspace or trajectory.

Step VII. *Configuration Analysis*. For each portion of the prescribed workspace or trajectory, all the possible configurations generated at Step V with respect to the n_ϕ user-defined constraint functions are tested. The $n_{f,t}$ configurations satisfying the constraints all over the tth part of the prescribed workspace or trajectory are defined hereafter as feasible configurations.

Step VIII. *Feasible Configuration Combination*. The set of n_t configurations that lead to the achievement of the prescribed task are computed. Each set is composed by selecting one of the $n_{f,t}$ feasible configurations for each part of the prescribed workspace or trajectory. The number of feasible configuration sets generated during this step is equal to $n_{\mathscr{C}}$.

Step IX. *Objective Functions*. The designer defines one or more global objective function(s), $\mathscr{V}_t, t =, 1, \dots, n_{\mathscr{V}}$, where $n_{\mathscr{V}}$ is equal to the number of global objective functions taken into account. The global objective functions associated with RCDPRs do not focus solely on a single configuration. They analyse the properties of the combination of n_t configurations comprising the RCDPR. If several global objective functions are to be solved simultaneously, the optimisation problem can be classically reduced to a mono-objective optimisation according to:

$$\mathscr{V} = \sum_{t=1}^{n_{\mathscr{V}}} \mu_t \mathscr{V}_t, \quad \mu_t \in [0, 1], \quad \sum_{t=1}^{n_{\mathscr{V}}} \mu_t = 1 \tag{5}$$

The weighting factors $\mu_t, t = 1, \dots, n_{\mathscr{V}}$, are defined according to the priority assigned to each objective function $\mathscr{V}_t$, the latter lying between

0 and 1. If several global optimisation functions have to be solved hierarchically, the designer will define those functions according to their order of priority, $t = 1, \ldots, n_{\mathscr{V}}$, where $\mathscr{V}_1$ has the highest priority and $\mathscr{V}_{n_{\mathscr{V}}}$ the lowest one.

Step X. *Discrete Optimisation Algorithm.* The design problem is formulated as an optimisation problem and solved by analysing all the $n_{\mathscr{C}}$ set of feasible configurations. The analysis is performed with respect to the global objective functions defined at Step IX. The sets of n_t configurations with the best global objective function value are determined. If a hierarchical multi-objective optimisation is required, the following procedure is applied:

a. The algorithm analyses the $n_{\mathscr{C}}$ sets of feasible configurations with respect to the global objective function which currently has the highest priority, $\mathscr{V}_t$ (the procedure is initialised with $t = 1$).
b. If only one set of configuration optimises $\mathscr{V}_t$, this solution is considered as the optimum. On the contrary, if $n_{\mathscr{C},t}$ multiple solutions optimise $\mathscr{V}_t$, the algorithm proceeds to the following step.
c. The algorithm analyses the $n_{\mathscr{C},t}$ sets of optimal solutions with respect to the global objective function with lower priority, $\mathscr{V}_{t+1}$. Then, $t = t + 1$ and the procedure moves back to Step b.

Global Objective Functions

The design strategy proposed in the previous section aims to optimise the characteristics of the RCDPR. The optimisation may be performed with respect to one or several *global objective functions*. The objective functions used in this chapter are described hereafter.

RCDPR Size

The design optimisation problem may aim to minimise the size of the robot, defined as the convex hull of the cable exit points. The Cartesian coordinates of exit point $A_{i,c}$ are defined as $\mathbf{a}^b_{i,c} = [a^x_{i,c}, a^y_{i,c}, a^z_{i,c}]^T$. The variables $\underline{s}_x$, $\underline{s}_y$ and $\underline{s}_z$ denote the lower bounds on the Cartesian coordinates of the cable exit points along the axes $\mathbf{x}_b$, $\mathbf{y}_b$ and $\mathbf{z}_b$, respectively:

$$\underline{s}_x = \min a^x_{i,c}, \ \forall i = 1, ..., m, \ c = 1, ..., n_t \tag{6}$$

$$\underline{s}_y = \min a^y_{i,c}, \ \forall i = 1, ..., m, \ c = 1, ..., n_t \tag{7}$$

$$\underline{s}_z = \min a^z_{i,c}, \ \forall i = 1, ..., m, \ c = 1, ..., n_t \tag{8}$$

The upper bounds on the Cartesian coordinates of the RCDPR cable exit points, along the axes $\mathbf{x}_b$, $\mathbf{y}_b$, $\mathbf{z}_b$, are denoted by $\bar{s}_x$, $\bar{s}_y$ and $\bar{s}_z$, respectively.

$$\bar{s}_x = \max a^x_{i,c}, \ \forall i = 1, ..., m, \ c = 1, ..., n_t \tag{9}$$

$$\bar{s}_y = \max a^y_{i,c}, \ \forall i = 1, ..., m, \ c = 1, ..., n_t \tag{10}$$

$$\bar{s}_z = \max a^z_{i,c}, \ \forall i = 1, ..., m, \ c = 1, ..., n_t \tag{11}$$

Hence, the objective function related to the size of the robot is expressed as follows:

$$\mathscr{V} = (\bar{s}_x - \underline{s}_x)(\bar{s}_y - \underline{s}_y)(\bar{s}_z - \underline{s}_z) \tag{12}$$

Number of Cable Reconfigurations

According to the reconfiguration strategy proposed in this chapter, reconfiguration operations require the displacement of the cable exit points, and consequently attaching/detaching operations of the cables. These operations are time consuming. Hence, an objective can be to minimise the number of reconfigurations, n_r, defined as the number of exit point changes to be performed in order to switch from configuration $\mathscr{C}_i$ to configuration $\mathscr{C}_j$. By reducing the number of cable attaching/detaching operations, the RCDPR set up time could be significantly reduced.

Number of Configuration Changes

During the reconfiguration of the exit points, the task executed by the RCDPR has to be interrupted. These interruptions impact the task execution time. Therefore, it may be necessary to minimise the number of interruptions, n_i, in order to improve the effectiveness of the RCDPR. The objective function $\mathscr{V} = n_i$ associated with this goal measures the number of configuration changes, n_i, to be performed during a prescribed task.

RCDPR Complexity

The higher the number of configuration sets $n_{\mathscr{C}}$ allowing to cover the prescribed workspace or trajectory, the more complex the RCDPR. When the RCDPR requires a large number of configurations, the base frame of the CDPR may become complex. In order to minimise the complexity of the RCDPR, an objective can be to minimise the overall number of exit point locations, $\mathscr{V} = n_e$, required by the $n_{\mathscr{C}}$ configuration sets. Therefore, the optimisation aims to maximise the number of exit point locations shared among two or more configurations.

Constraint Functions

Any CDPR optimisation problem has to take into account some constraints. Those constraints represent the technical limits or requirements that need to be satisfied. The constraints used in this chapter are described hereafter.

Wrench Feasibility

Since cables can only pull on the platform, the tensions in the cables must always be non-negative. Moreover, cable tensions must be lower than an upper bound, τ_{max}, which corresponds either to the maximum tension τ_{max1} the cables (or other mechanical parts) can bear, or to the maximum tension τ_{max2} the motors can provide.

The cable tension bounds can thus be written as:

$$0 \leq \tau_i \leq \tau_{max}, \quad \forall i = 1, \ldots, m \tag{13}$$

where $\tau_{max} = \min \{\tau_{max1}, \tau_{max2}\}$.

Due to the cable tension bounds, RCDPRs can balance only a bounded set of external wrenches. In this chapter, the set of external wrenches applied to the platform and that the cables have to balance is called the required external wrench set and is denoted $[\mathbf{w}_e]_r$. Moreover, the set of of admissible cable tensions is defined as:

$$[\boldsymbol{\tau}] = \{\tau_i \mid 0 \leq \tau_i \leq \tau_{max}, \ i = 1, \ldots, m\} \tag{14}$$

A pose (position and orientation) of the moving platform is then said to be wrench feasible if the following constraint holds:

$$\forall \mathbf{w}_e \in [\mathbf{w}_e]_r, \ \exists \boldsymbol{\tau} \in [\boldsymbol{\tau}] \quad \text{such that} \quad \mathbf{W}\boldsymbol{\tau} + \mathbf{w}_e = 0 \tag{15}$$

Eq. (15) can be rewritten as follows:

$$\mathbf{C}\mathbf{w}_e \leq \mathbf{d}, \quad \forall \mathbf{w}_e \in [\mathbf{w}_e]_r \tag{16}$$

Methods to compute matrix $\mathbf{C}$ and vector $\mathbf{d}$ are presented in [6, 12].

Cable Lengths

Due to technological reasons, cable lengths are bounded between a minimum cable length, l_{min}, and a maximum cable length, l_{max}:

$$l_{min} \leq l_{i,c} \leq l_{max}, \ \forall i = 1, \ldots, m \tag{17}$$

The minimum cable lengths are defined so that the RCDPR moving platform is not too close to the base frame. The maximum cable lengths depend on the properties of the winch drums that store the cables, in particular their lengths and their diameters.

Cable Interferences

A second constraint is related to the possible collisions between cables. If two or more cables collide, the geometric and static models of the CDPR are not valid anymore and the cables can be damaged or their lifetime severely reduced.

In order to verify that cables do not interfere, it is sufficient to determine the distances between them. Modeling the cables as linear segments, the distance $d^{cc}_{i,j}$ between the ith cable and the jth cable can be computed, e.g. by means of the method presented in [20]. There is no interference if the distance is larger than the diameter of the cables, ϕ_c:

$$d^{cc}_{i,j} \geq \phi_c \quad \forall i,j = 1,\ldots,m, \quad i \neq j \tag{18}$$

The number of possible cable interferences to be verified is equal to $C^m_2 = \frac{m!}{2!(m-2)!}$. Note that, depending on the way the cables are routed from the winches to the moving platform, possible interferences of the cable segments between the winches and the pulleys may have to be considered.

Collisions Between the Cables and the Environment

Industrial environments may be cluttered. Collisions between the environment and the cables of the CDPR should be avoided. In general, for fast collision detection, the environment objects (obstacles) are enclosed in bounding volumes such as spheres and cylinders. When more complex shapes have to be considered, their surfaces are approximated with polygonal meshes. Thus, collision analysis can be performed by computing the distances between the edges of those polygons and the cables, e.g. by using [20]. Many other methods may be used, e.g., those described in [4].

In the case study presented in section "Case Study: Design of a RCDPRs for Sand blasting and Painting of a Large Tubular Structure", a tubular structure is considered. The ith cable and the kth structure tube will not collide if the distance between the cable and the axis (straight line segment) of the structure tube is larger than the sum of the cable radius $\phi_c/2$ and the tube radius $\phi_s/2$, i.e.:

$$d^{cs}_{i,k} \geq \frac{(\phi_c + \phi_s)}{2} \quad \forall i = 1,\ldots,m, \forall k = 1,\ldots,n_{st} \tag{19}$$

where n_{st} denotes the number of tubes composing the structure.

Pose Infinitesimal Displacement Due to the Cable Elasticity

Cables are not perfectly rigid body. Under load, they are notably subjected to elongations that may induce some moving platform displacements. In order to quantify the stiffness of the CDPR, an elasto-static model may be used:

$$\delta \mathbf{w_e} = \mathbf{K}\delta \mathbf{p} = \mathbf{K}\begin{bmatrix}\delta \mathbf{t}\\ \delta \mathbf{r}\end{bmatrix} \tag{20}$$

where $\delta \mathbf{w_e}$ is the infinitesimal change in the external wrench applied to the platform, $\delta \mathbf{p}$ is the infinitesimal displacement screw of the moving platform and $\mathbf{K}$ is the stiffness matrix whose computation is explained in [3]. $\delta \mathbf{t} = \left[\delta t_x, \delta t_y, \delta t_z\right]^{\mathrm{T}}$ is the variation in the moving platform position and $\delta \mathbf{r} = [\delta r_x, \delta r_y, \delta r_z]^{\mathrm{T}}$ is the vector of the infinitesimal (sufficiently small) rotations of the moving platform around the axes $\mathbf{x}_b$, $\mathbf{y}_b$ and $\mathbf{z}_b$.

The pose variation should be bounded by the positioning error threshold vector, $\delta \mathbf{t} = \left[\delta t_{x,c}, \delta t_{y,c}, \delta t_{z,c}\right]$, where $\delta t_{x,c}$, $\delta t_{y,c}$ and $\delta t_{z,c}$ are the bounds on the positioning errors along the axes $\mathbf{x}_b$, $\mathbf{y}_b$ and $\mathbf{x}_b$, and the orientation error threshold vector, $\delta \boldsymbol{\phi} = \left[\delta \gamma_c, \delta \beta_c, \delta \alpha_c\right]$, where $\delta \gamma_c$, $\delta \beta_c$ and $\delta \alpha_c$ are the bounds on the platform orientation errors about the axes $\mathbf{x}_b$, $\mathbf{y}_b$ and $\mathbf{z}_b$, i.e.:

$$-\left[\delta t_{x,c}, \delta t_{y,c}, \delta t_{z,c}\right] \le \left[\delta t_x, \delta t_y, \delta t_z\right] \le \left[\delta t_{x,c}, \delta t_{y,c}, \delta t_{z,c}\right] \tag{21}$$
$$-\left[\delta \gamma_c, \delta \beta_c, \delta \alpha_c\right] \le \left[\delta \gamma, \delta \beta, \delta \alpha\right] \le \left[\delta \gamma_c, \delta \beta_c, \delta \alpha_c\right] \tag{22}$$

Case Study: Design of a RCDPRs for Sandblasting and Painting of a Large Tubular Structure

Problem Description

The necessity to improve the production rate of large tubular structures has incited companies to investigate new technologies. These technologies should be able to reduce manufacturing time associated with the assembly of the structure parts or the treatment of their surfaces. Painting and sandblasting operations over wide tubular structures can be realised by means of RCDPRs, as illustrated in the present case study.

Task and Environment

The tubular structure selected for the given case study is 20 m long, with a cross section of 10 m x 10 m. The number of tubes to be painted is equal to twenty. Their

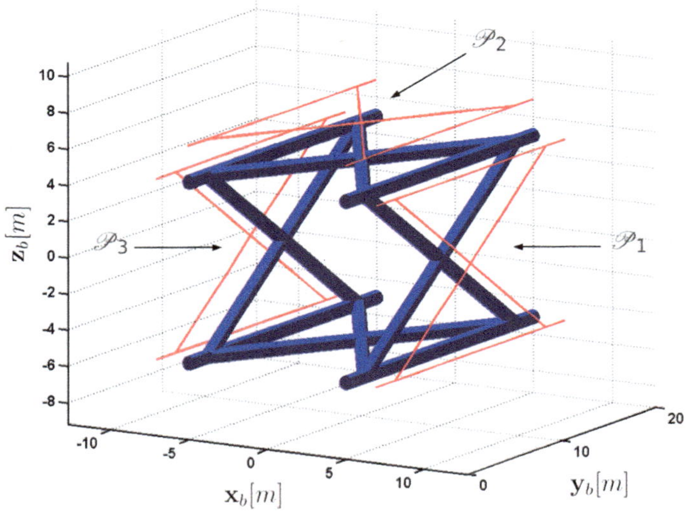

Fig. 6 Case study model and prescribed paths $\mathscr{P}_1$, $\mathscr{P}_2$ and $\mathscr{P}_3$ of the moving platform CoM

diameter, ϕ_s, is equal to 0.8 m. The sandblasting and painting operations are realised indoor. The structure lies horizontally in order to reduce the dimensions of the painting workshop. The whole system can be described with respect to a fixed reference frame, $\mathscr{F}_b$, of origin O_b and axes $\mathbf{x}_b$, $\mathbf{y}_b$, $\mathbf{z}_b$, as illustrated in Fig. 6.

Sandblasting and painting tools are embarked on the RCDPR moving platform. The Center of Mass (CoM) of the platform follows the profile of the structure tubes and the tools perform the required operations. The paths to be followed, $\mathscr{P}_1$, $\mathscr{P}_2$ and $\mathscr{P}_3$, are represented in Fig. 6. Note that each path $\mathscr{P}_i, i = 1, \ldots, 3$ is discretised into 38 points $\mathrm{P}_{j,i}, j = 1, \ldots, 38 \quad i = 1, \ldots, 3$ and that n_p denotes the corresponding total number of points. The offset between paths $\mathscr{P}_i, i = 1, \ldots, 3$ and the structure tubes is equal to 2 m. No path will be assigned to the lower external side of the structure, since it is sandblasted and painted from the ground.

Division of the Prescribed Workspace

In order to avoid collisions between the cables and structure, reconfigurations of the cable exit points are necessary. Each external side of the structure should be painted by only one robot configuration. Three configurations are necessary to work on the outer part of the structure, configuration $\mathscr{C}_i$ being associated to path $\mathscr{P}_i$, $i = 1$, 2 and 3, in order not to interrupt the painting and sandblasting operations during their execution. Passing from one configuration to another, one or more cables are disconnected from their exit points and connected to other exit points located elsewhere. For each configuration, the locations of the cable exit points are defined as variables of the design problem. In the present case study, the dimensions of the platform as well as the position of the cable anchor points on the platform are fixed.

Constant Design parameters

The number of cables, $m = 8$, the cable properties, and the dimensions of the platform are given. Those parameters are the same for the three configurations. The moving platform of the RCDPR analysed in this case study is driven by steel cables. The maximum allowed tension in the cables, τ_{max}, is equal to 34950 N and we have:

$$0 < \tau_i \leq \tau_{max}, \quad \forall i = 1, \ldots, 8 \tag{23}$$

Moreover, l_p, w_p and h_p denote the length, width and height of the platform, respectively: $l_p = 30$ cm, $w_p = 30$ cm and $h_p = 60$ cm. The mass of the moving platform is $m_{MP} = 60$ kg. The design (constant) parameter vector $\mathbf{q}$ is expressed as:

$$\mathbf{q} = \left[m, \phi_c, k_s, \tau_{max}, l_p, w_p, h_p, m_{MP}\right]^{\mathrm{T}} \tag{24}$$

Constraint Functions and Configuration Analysis

The design problem aims to identify the locations of points $A_{i,c}$ for the configurations $\mathscr{C}_1$, $\mathscr{C}_2$ and $\mathscr{C}_3$. At first, in order to identify the set of feasible locations for the exit points $A_{i,c}$, the three robot configurations are parameterised and analysed separately in the following paragraphs. A set of exit points is feasible if the design constraints are satisfied along the whole path to be followed by the moving platform CoM. The analysed constraints are: wrench feasibility, cable interferences, cable collisions with the structure, and the maximum moving platform infinitesimal displacement due to the cable elasticity.

Both suspended and fully constrained eight-cable CDPR architectures are used. In the suspended architecture, gravity plays the role of an additional cable pulling the moving platform downward, thereby keeping the cables under tension. The suspended architecture considered in this work is inspired by the CoGiRo CDPR prototype [11, 19]. For the non-suspended configuration, note that eight cables is the smallest possible even number of cables that can be used for the platform to be fully constrained by the cables.

Collisions between the cables as well as collisions between the cables and structure tubes should be avoided. Since sandblasting and painting operations are performed at low speed, the motion of the CDPR platform can be considered quasi-static. Hence, only the static equilibrium of the robot moving platform will be considered. The wrench feasibility constraints presented in section “Constraint Functions” are considered such that the required external wrench set $[\mathbf{w}_e]_r$ is an hyperrectangle defined as:

$$-50\,\text{N} \leq f_x, f_y, f_z \leq 50\,\text{N} \tag{25}$$

$$-7.5\,\text{Nm} \leq m_x, m_y, m_z \leq 7.5\,\text{Nm} \tag{26}$$

where $\mathbf{w}_e = \left[f_x, f_y, f_z, m_x, m_y, m_z\right]^T$, f_x, f_y and f_z being the force components of $\mathbf{w}_e$ and m_x, m_y, and m_z being its moment components. Besides, the moving platform infinitesimal displacements, due to the elasticity of the cables, are constrained by:

$$-5\,\text{cm} \leq \delta t_x, \delta t_y, \delta t_z \leq 5\,\text{cm} \tag{27}$$

$$-0.1\,\text{rad} \leq \delta r_x, \delta r_y, \delta r_z \leq 0.1\,\text{rad} \tag{28}$$

Configuration $\mathscr{C}_1$

A fully-constrained configuration has been assigned to configuration $\mathscr{C}_1$. The exit points $A_{i,1}$ have been arranged in a parallelepiped layout. The edges of the parallelepiped are aligned with the axes of frame $\mathscr{F}_b$. This layout can be fully described by means of five variables: u_1, u_2 and u_3 define the Cartesian coordinates of the parallelepiped center, while u_4 and u_5 denote the half-lengths of the parallelepiped along the axes $\mathbf{x}_b$ and $\mathbf{y}_b$, respectively. Therefore, the Cartesian coordinates of the exit points $A_{i,1}$ are expressed as follows:

$$\mathbf{a}^b_{1,1} = \left[u_1 + u_4,\ u_2 + u_5,\ u_3\right]^{\mathrm{T}} \quad \mathbf{a}^b_{2,1} = \left[u_1 + u_4,\ u_2 + u_5,\ -u_3\right]^{\mathrm{T}} \tag{29}$$

$$\mathbf{a}^b_{3,1} = \left[u_1 - u_4,\ u_2 + u_5,\ u_3\right]^{\mathrm{T}} \quad \mathbf{a}^b_{4,1} = \left[u_1 - u_4,\ u_2 + u_5,\ -u_3\right]^{\mathrm{T}} \tag{30}$$

$$\mathbf{a}^b_{5,1} = \left[u_1 - u_4,\ u_2 - u_5,\ u_3\right]^{\mathrm{T}} \quad \mathbf{a}^b_{6,1} = \left[u_1 - u_4,\ u_2 - u_5,\ -u_3\right]^{\mathrm{T}} \tag{31}$$

$$\mathbf{a}^b_{7,1} = \left[u_1 + u_4,\ u_2 - u_5,\ u_3\right]^{\mathrm{T}} \quad \mathbf{a}^b_{8,1} = \left[u_1 + u_4,\ u_2 - u_5,\ -u_3\right]^{\mathrm{T}} \tag{32}$$

The layout of the first robot configuration is described in Fig. 7. The corresponding design variables are collected into the vector $\mathbf{x}_1$:

$$\mathbf{x}_1 = \left[u_1, u_2, u_3, u_4, u_5\right]^{\mathrm{T}} \tag{33}$$

The Cartesian coordinates of the anchor points $B_{i,1}$ of the cables on the platform are expressed as:

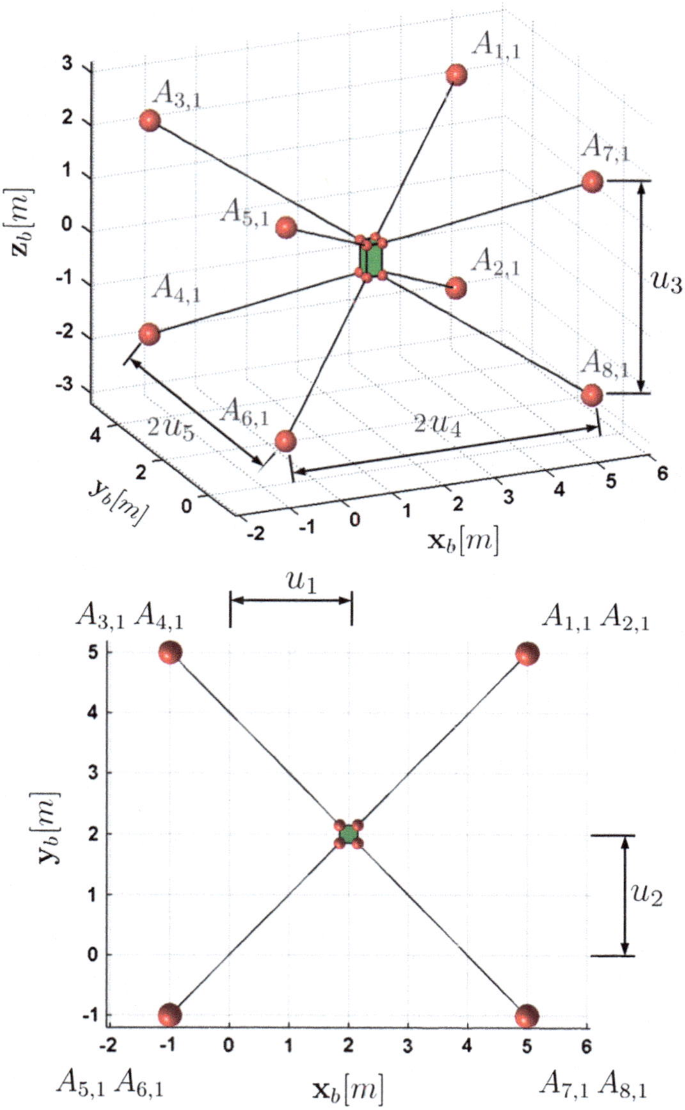

Fig. 7 Design variables parametrising the configuration $\mathscr{C}_1$

$$\mathbf{b}^b_{1,1} = \frac{1}{2}\left[l_p, w_p, h_p\right]^{\mathrm{T}}, \qquad \mathbf{b}^b_{2,1} = \frac{1}{2}\left[l_p, w_p, -h_p\right]^{\mathrm{T}} \tag{34}$$

$$\mathbf{b}^b_{3,1} = \frac{1}{2}\left[-l_p, w_p, h_p\right]^{\mathrm{T}}, \qquad \mathbf{b}^b_{4,1} = \frac{1}{2}\left[-l_p, w_p, -h_p\right]^{\mathrm{T}} \tag{35}$$

$$\mathbf{b}^b_{5,1} = \frac{1}{2}\left[-l_p, -w_p, h_p\right]^{\mathrm{T}}, \qquad \mathbf{b}^b_{6,1} = \frac{1}{2}\left[-l_p, -w_p, -h_p\right]^{\mathrm{T}} \tag{36}$$

$$\mathbf{b}^b_{7,1} = \frac{1}{2}\left[l_p, -w_p, h_p\right]^{\mathrm{T}}, \qquad \mathbf{b}^b_{8,1} = \frac{1}{2}\left[l_p, -w_p, -h_p\right]^{\mathrm{T}} \tag{37}$$

A discretised set of design variables have been considered. The lower and upper bounds as well as the number of values for each variable are given in Table 2. 18225 robot configurations have been generated with those values. It turns out that 4576 configurations satisfy the design constraints along the 38 discretised points of path $\mathscr{P}_1$.

Configuration $\mathscr{C}_2$

A suspended redundantly actuated eight-cable CDPR architecture has been attributed to the configuration $\mathscr{C}_2$ in order to avoid collisions between the cables and the tubular structure. The selected configuration is based on CoGiRo, a suspended CDPR designed and built in the framework of the ANR CoGiRo project [11, 19]. An advantage of this configuration is a large workspace to footprint ratio. The exit points $A_{i,2}$ have been arranged in a parallelepiped layout. The Cartesian coordinates $\mathbf{a}_{i,c}$ are defined as follows:

$$\mathbf{a}^b_{1,2} = \mathbf{a}^b_{2,2} = \left[v_1 - v_4,\ v_2 - v_5,\ v_3\right]^{\mathrm{T}} \tag{38}$$

$$\mathbf{a}^b_{3,2} = \mathbf{a}^b_{4,2} = \left[v_1 - v_4,\ v_2 + v_5,\ v_3\right]^{\mathrm{T}} \tag{39}$$

$$\mathbf{a}^b_{5,2} = \mathbf{a}^b_{6,2} = \left[v_1 + v_4,\ v_2 + v_5,\ v_3\right]^{\mathrm{T}} \tag{40}$$

$$\mathbf{a}^b_{7,2} = \mathbf{a}^b_{8,2} = \left[v_1 + v_4,\ v_2 - v_5,\ v_3\right]^{\mathrm{T}} \tag{41}$$

Variables $v_i, i = 1, \ldots, 5$ are equivalent for configuration $\mathscr{C}_2$ to variables $u_i, i = 1, \ldots, 5$, describing configuration $\mathscr{C}_1$. The layout of this configuration is illustrated in Fig. 8. The design variables of configuration $\mathscr{C}_2$ are collected into the vector $\mathbf{x}_2$:

$$\mathbf{x}_2 = \left[v_1, v_2, v_3, v_4, v_5\right]^{\mathrm{T}} \tag{42}$$

Note that this configuration is composed of couples of exit points theoretically connected to the same locations: $\{A_{1,2}, A_{2,2}\}$, $\{A_{3,2}, A_{4,2}\}$, $\{A_{5,2}, A_{6,2}\}$, and $\{A_{7,2}, A_{8,2}\}$. From a technical point of view, in order to avoid any cable interference, the coupled exit points should be separated by a certain distance. For the design problem at hand, this distance has been fixed to $v_0 = 5$ mm.

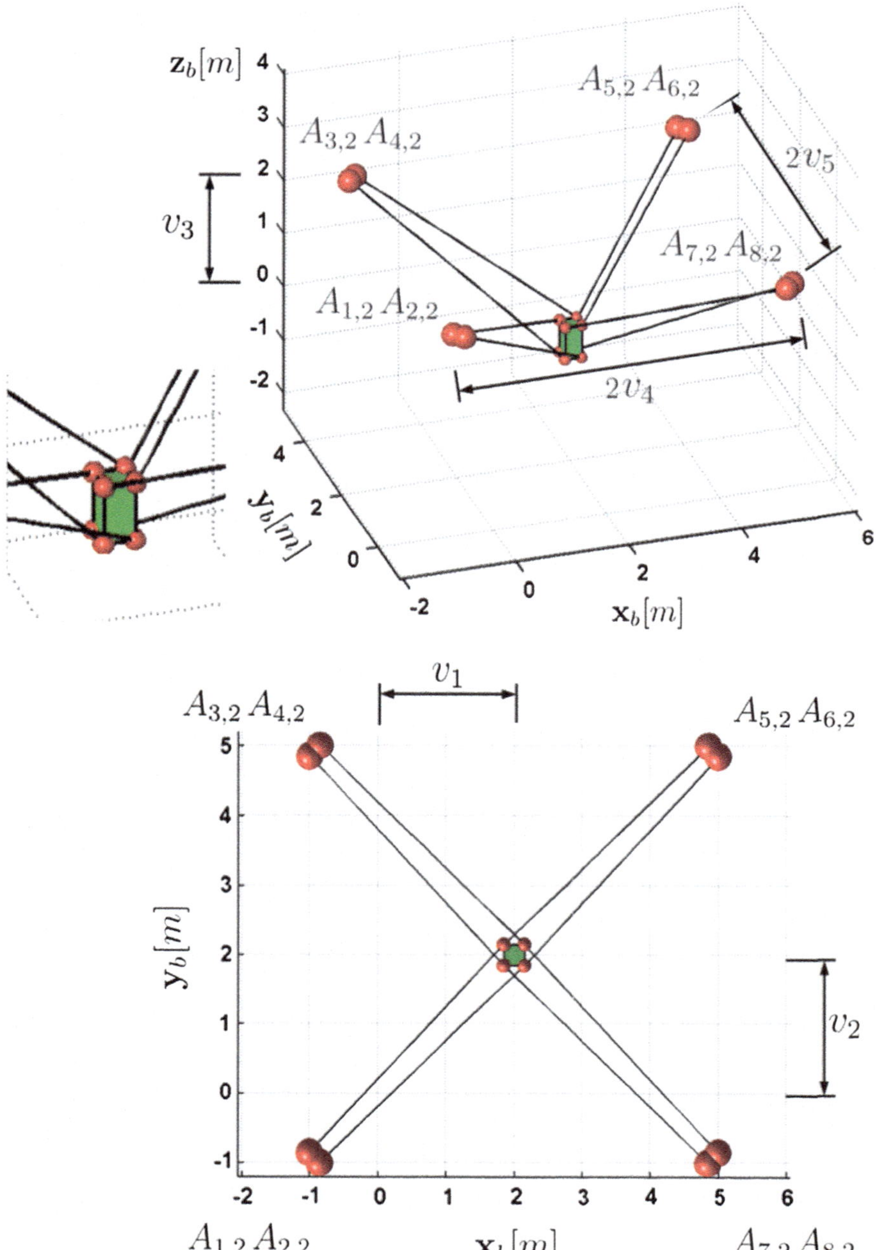

Fig. 8 Design variables parametrising the configuration $\mathscr{C}_2$

$$\mathbf{a}_{1,2}^b = \left[v_1 - v_4',\ v_2 - v_5,\ v_3\right]^{\mathrm{T}} \tag{43}$$

$$\mathbf{a}_{2,2}^b = \left[v_1 - v_4,\ v_2 - v_5',\ v_3\right]^{\mathrm{T}} \tag{44}$$

$$\mathbf{a}_{3,2}^b = \left[v_1 - v_4,\ v_2 + v_5',\ v_3\right]^{\mathrm{T}} \tag{45}$$

$$\mathbf{a}_{4,2}^b = \left[v_1 - v_4',\ v_2 + v_5,\ v_3\right]^{\mathrm{T}} \tag{46}$$

$$\mathbf{a}_{5,2}^b = \left[v_1 + v_4',\ v_2 + v_5,\ v_3\right]^{\mathrm{T}} \tag{47}$$

$$\mathbf{a}_{6,2}^b = \left[v_1 + v_4,\ v_2 + v_5',\ v_3\right]^{\mathrm{T}} \tag{48}$$

$$\mathbf{a}_{7,2}^b = \left[v_1 + v_4,\ v_2 - v_5',\ v_3\right]^{\mathrm{T}} \tag{49}$$

$$\mathbf{a}_{8,2}^b = \left[v_1 + v_4',\ v_2 - v_5,\ v_3\right]^{\mathrm{T}} \tag{50}$$

where $v_4' = v_4 - v_0$ and $v_5' = v_5 - v_0$.

The Cartesian coordinates of points $B_{i,2}$ are defined as:

$$\mathbf{b}_{1,2}^b = \frac{1}{2}\left[l_p, -w_p, h_p\right]^{\mathrm{T}}, \qquad \mathbf{b}_{2,2}^b = \frac{1}{2}\left[-l_p, w_p, -h_p\right]^{\mathrm{T}} \tag{51}$$

$$\mathbf{b}_{3,2}^b = \frac{1}{2}\left[-l_p, -w_p, h_p\right]^{\mathrm{T}}, \qquad \mathbf{b}_{4,2}^b = \frac{1}{2}\left[l_p, w_p, -h_p\right]^{\mathrm{T}} \tag{52}$$

$$\mathbf{b}_{5,2}^b = \frac{1}{2}\left[-l_p, w_p, h_p\right]^{\mathrm{T}}, \qquad \mathbf{b}_{6,2}^b = \frac{1}{2}\left[l_p, -w_p, -h_p\right]^{\mathrm{T}} \tag{53}$$

$$\mathbf{b}_{7,2}^b = \frac{1}{2}\left[l_p, w_p, h_p\right]^{\mathrm{T}}, \qquad \mathbf{b}_{8,2}^b = \frac{1}{2}\left[-l_p, -w_p, -h_p\right]^{\mathrm{T}} \tag{54}$$

Table 2 describes the lower and upper bounds as well as the number of values considered for the configuration $\mathscr{C}_2$. Combining these values, 22275 configurations have been generated. Among these configurations, only 5579 configurations are feasible.

Configuration $\mathscr{C}_3$

The configuration $\mathscr{C}_3$ follows the path $\mathscr{P}_3$. This path is symmetric to the path $\mathscr{P}_1$ with respect to the plane $\mathbf{y}_b O_b \mathbf{z}_b$. Considering the symmetry of the tubular structure, configuration $\mathscr{C}_3$ is thus selected as being the same as configuration $\mathscr{C}_1$. The discretised set of design variables chosen for the configuration $\mathscr{C}_3$ is described in Table 2. The design variables for the configuration $\mathscr{C}_3$ are collected into the vector $\mathbf{x}_3$:

$$\mathbf{x}_3 = \left[w_1, w_2, w_3, w_4, w_5\right]^{\mathrm{T}} \tag{55}$$

where the variables $w_i, i = 1, \ldots, 5$ amount to the variables $u_i, i = 1, \ldots, 5$, describing configuration $\mathscr{C}_1$. Therefore, the Cartesian coordinates of the exit points $A_{i,3}$ are expressed as follows:

Table 2 Design variables associated with configurations $\mathscr{C}_1$, $\mathscr{C}_2$ and $\mathscr{C}_3$

	Variables	Lower bounds	Upper bounds	Number of values
$\mathscr{C}_1$	u_1	5.5	7.5	9
	u_2	8.0	12.0	9
	u_3	6	10	5
	u_4	0.5	2.5	9
	u_5	10	14	5
$\mathscr{C}_2$	v_1	−1	1	9
	v_2	8.0	12.0	5
	v_3	7	11	9
	v_4	5	7.5	11
	v_5	10	14	5
$\mathscr{C}_3$	w_1	−7.5	−5.5	9
	w_2	8.0	12.0	9
	w_3	6	10	5
	w_4	0.5	2.5	9
	w_5	10	14	5

$$\mathbf{a}^b_{1,3} = \left[w_1 + w_4,\ w_2 + w_5,\ -w_3\right]^{\mathrm{T}} \qquad \mathbf{a}^b_{2,3} = \left[w_1 + w_4,\ w_2 + w_5,\ w_3\right]^{\mathrm{T}} \tag{56}$$

$$\mathbf{a}^b_{3,3} = \left[w_1 - w_4,\ w_2 + w_5,\ -w_3\right]^{\mathrm{T}} \qquad \mathbf{a}^b_{4,3} = \left[w_1 - w_4,\ w_2 + w_5,\ w_3\right]^{\mathrm{T}} \tag{57}$$

$$\mathbf{a}^b_{5,3} = \left[w_1 - w_4,\ w_2 - w_5,\ -w_3\right]^{\mathrm{T}} \qquad \mathbf{a}^b_{6,3} = \left[w_1 - w_4,\ w_2 - w_5,\ w_3\right]^{\mathrm{T}} \tag{58}$$

$$\mathbf{a}^b_{7,3} = \left[w_1 + w_4,\ w_2 - w_5,\ -w_3\right]^{\mathrm{T}} \qquad \mathbf{a}^b_{8,3} = \left[w_1 + w_4,\ w_2 - w_5,\ w_3\right]^{\mathrm{T}} \tag{59}$$

Objective Functions and Design Problem Formulation

The RCDPR should be as simple as possible so that the minimisation of the total number of cable exit point locations, $\mathscr{V}_1 = n_e$, is required. Consequently, the number of exit point locations shared by two or more configurations should be maximised. The size of the robot is also minimised to reduce the size of the sandblasting and painting workshop. Finally, the mean of the moving platform infinitesimal displacement due to cable deformations is minimised. The optimisations are performed hierarchically, by means of the procedure described in section "Design Problem For mulation" and the objective functions collected in section "Global Objective Func tions".

Hence, the design problem of the CDPR is formulated as follows:

$$
\begin{aligned}
&\text{minimise} \quad \begin{cases} \mathscr{V}_1 = n_e \\ \mathscr{V}_2 = (\bar{s}_x - \underline{s}_x)(\bar{s}_y - \underline{s}_y)(\bar{s}_z - \underline{s}_z) \\ \mathscr{V}_3 = \dfrac{\|\delta \mathbf{t}\|_2}{n_p} \end{cases} \\
&\text{over} \quad \mathbf{x}_1, \mathbf{x}_2, \mathbf{x}_3 \\
&\text{subject to:} \\
&\forall \mathrm{P}_{m,n},\ m = 1, \dots, 38 \\
&\qquad n = 1, \dots, 3 \quad \begin{cases} \mathbf{Cw} \leq \mathbf{d}, \quad \forall \mathbf{w} \in [\mathbf{w}_e]_r \\ d_{i,j}^{cc} \geq \phi_c \quad \forall i,j = 1, \dots, 8, \quad i \neq j \\ d_{i,k}^{cs} \geq \dfrac{(\phi_c + \phi_s)}{2} \quad \forall i = 1, \dots, 8, \forall k = 1, \dots, 20 \\ -5\,\mathrm{cm} \leq \delta t_x, \delta t_y, \delta t_z \leq 5\,\mathrm{cm} \\ -0.1\,\mathrm{rad} \leq \delta r_x, \delta r_y, \delta r_z \leq 0.1\,\mathrm{rad} \end{cases}
\end{aligned}
\tag{60}
$$

Once the set of feasible solutions have been obtained for each path $\mathscr{P}_i$, a list of RCDPRs with a minimum number of exit points, n_c, is extracted from the list of feasible RCDPRs. Finally, the most compact and stiff RCDPRs from the list of RCDPRs with a minimum number of exit points are the desired optimal solutions.

Optimisation Results

The feasible robot configurations associated with paths $\mathscr{P}_1$, $\mathscr{P}_2$ and $\mathscr{P}_3$ have been identified. For each path, a configuration is selected, aiming to minimise the total number of exit points required by the RCDPR to complete the task. These optimal solutions have been computed in two phases. At first, the 4576 feasible robot configurations for path $\mathscr{P}_1$ are compared with the 5579 feasible robot configurations for path $\mathscr{P}_2$ looking for the couple of configurations having the minimum total number of exit points. The resulting couple of configurations is then compared to the feasible robot configurations for path $\mathscr{P}_3$, and the sets of robot configurations that minimise the overall number n_e of exit points along the three paths are retained. According to the discrete optimisation analysis, 16516 triplets of configurations minimise this overall number of exit points.

A generic CDPR composed of eight cables requires eight exit points $A_i = 1, \dots, 8$ on the base. It is the case for the fully constrained configurations $\mathscr{C}_1$ and $\mathscr{C}_3$. The suspended CDPR presents four coincident couples of exit points. Hence, in the present case study, the maximum overall number of exit points of the RCDPR is equal to 20. The best results provide a reduction of four points. Regarding the configurations $\mathscr{C}_1$ and $\mathscr{C}_2$, points $A_{5,2}$ and $A_{7,2}$ can be coincident with points $A_{3,1}$ and $A_{5,1}$, respectively. Alternatively, points $A_{5,2}$ and $A_{7,2}$ can be coincident with points $A_{1,1}$ and $A_{7,1}$.

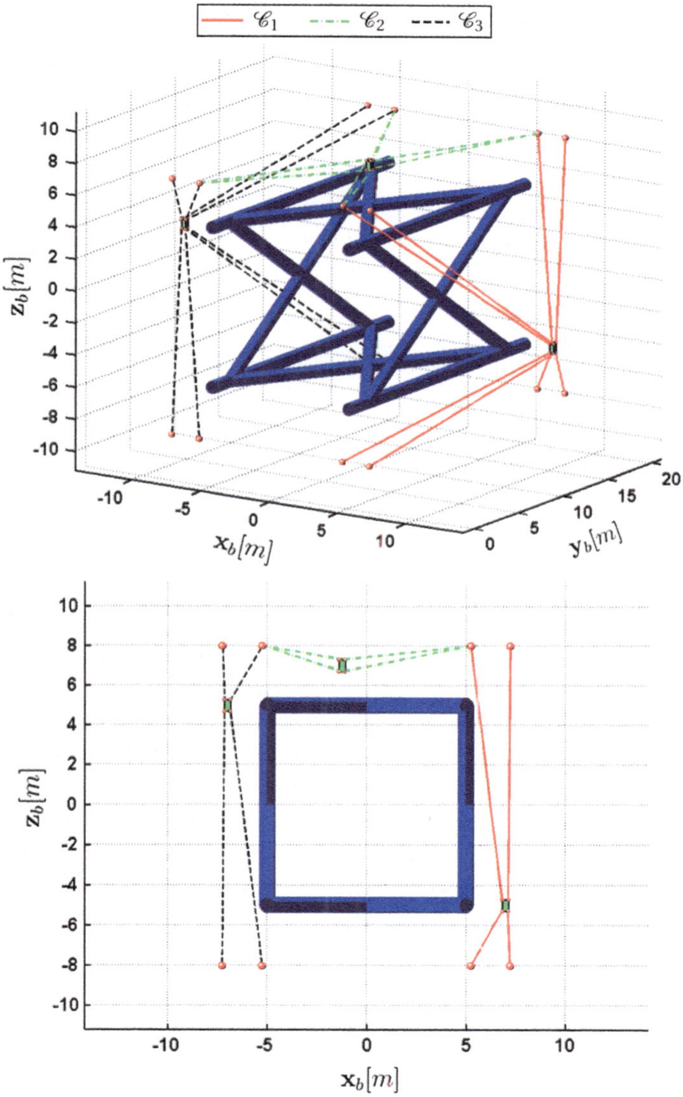

Fig. 9 Optimal reconfigurable cable-driven parallel robot

Table 3 Design parameters of the selected optimum RCDPR

Conf.	var. 1	var. 2	var.3	var.4	var.5
$\mathbf{x}_1$	6.25	10.0	8.0	1.0	11.0
$\mathbf{x}_3$	0	10.0	8.0	5.25	11.0
$\mathbf{x}_3$	−6.25	10.0	8.0	1.0	11.0

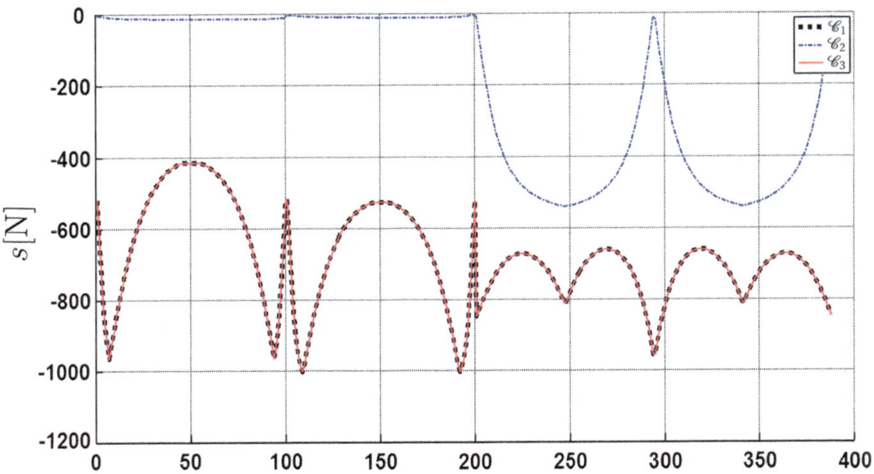

Fig. 10 Minimum degree of constraint satisfaction [13]. The analysis has been performed by discretising the paths $\mathscr{P}_1$, $\mathscr{P}_2$, and $\mathscr{P}_3$ into 388 points

As far as configurations $\mathscr{C}_2$ and $\mathscr{C}_3$ are concerned, points $A_{1,2}$ and $A_{3,2}$ can be coincident with points $A_{8,3}$ and $A_{2,3}$, respectively. Likewise, points $A_{1,2}$ and $A_{3,2}$ can be coincident with points $A_{4,3}$ and $A_{6,3}$, respectively.

The total volume of the robot has been computed for the 16516 triplets of configurations minimising the overall number of exit points. 96 RCDPRs amongst the 16516 triplets of configurations have the smallest size, this minimum size being equal to 5104 m^3. Selection of the best solutions has been promoted through the third optimisation criterion based on the robot stiffness. Twenty solutions provided a minimum mean of the moving platform displacement equal to 1.392 mm. An optimal solution is illustrated in Fig. 9. The corresponding optimal design parameters are given in Table 3.

Figure 10 illustrates the minimum degree of constraint satisfaction s introduced in [13] and computed thereafter along the paths $\mathscr{P}_1$, $\mathscr{P}_2$, and $\mathscr{P}_3$, which were discretised into 388 points. It turns out that the moving platform is in a feasible static equilibrium along all the paths because the minimum degree of constraint satisfaction remains negative. Referring to [13], the minimum degree of constraint satisfaction can be used to test wrench feasibility since it is negative when a platform pose is wrench feasible. Configurations $\mathscr{C}_1$ and $\mathscr{C}_3$ maintain their degree of satisfaction lower than −400 N. On the contrary, configuration $\mathscr{C}_2$ is often close to 0. The poses where s vanishes are such that two cables of the suspended CDPR of configuration $\mathscr{C}_2$ are slack.

The proposed RCDPR design strategy yielded good solutions, but it is time consuming. The whole procedure, performed on an Intel® Core™ i7-3630QM 2.40 GHz, required 19 h of computations, on Matlab® 2013a. Therefore, the development of more efficient strategies for the design of RCDPRs will be part of our future work. Moreover, the mass of the cables may have to be taken into account.

Conclusions

When the task to be accomplished is complicated, and the working environment is extremely cluttered, CDPRs may not succeed in the task execution. The problem can be solved by means of RCDPRs. This chapter focused on RCDPRs whose cable exit points on the base frame can be located on a predefined grid of possible positions. A design strategy for such discrete RCDPRs was introduced. This design strategy assumes that the number of configurations needed to complete the task is defined by the designer according to its experience. The designer divides the prescribed trajectory or workspace into a set of partitions. Each partition has to be entirely covered by one configuration. The position of the cable exit points, for all the configurations, is computed by means of an optimisation algorithm. The algorithm optimises one or more global objective function(s) while satisfying a set of user-defined constraints. Examples of possible global objective functions include the RCDPR size, the overall number of exit points, and the number of cable reconfiguration. A case study was presented in order to validate the RCDPR design strategy. The RCDPR has to paint and sandblast three of the four external sides of a tubular structure. Each of these three sides is covered by one configuration. The design strategy provided several optimal solutions to the case study, minimising hierarchically the overall number of cable exit points, the size of the RCDPR, and the moving platform displacements due to the elasticity of the cables. The computation of the optimal solution required nineteen hours of computation. More complicated tasks may thus require higher computation times. An improvement of the proposed RCDPR design strategy should be investigated in order to reduce this computational effort.

Acknowledgements This research work is part of the CAROCA project managed by IRT Jules Verne (French Institute in Research and Technology in Advanced Manufacturing Technologies for Composite, Metallic and Hybrid Structures). The authors wish to associate the industrial and academic partners of this project, namely, STX, Naval Group, AIRBUS and CNRS.

References

1. Albus J, Bostelman R, Dagalakis N (1992) The NIST spider, a robot crane. J Res Nat Inst Stand Technol 97(3), 373–385
2. ANR Project CoGiRo. http://www.lirmm.fr/cogiro/
3. Behzadipour S, Khajepour A (2005) Stiffness of cable-based parallel manipulators with application to stability analysis. ASME J Mech Design 128(1):303–310
4. Blanchet L (2015) Contribution à la modélisation de robots à câbles pour leur commande et leur conception. Ph.D. thesis, Université Nice Sophia-Antipolis, France
5. Bostelman R, Jacoff A, Proctor F, Kramer T, Wavering A (2000) Cable-based reconfigurable machines for large scale manufacturing. In: Proceedings of the 2000 Japan-USA symposium on flexible automation. Ann Arbor, MI
6. Bouchard S, Gosselin CM, Moore B (2009) On the ability of a cable-driven robot to generate a prescribed set of wrenches. ASME J Mech Robot 2(1):1–10
7. CableBOT. http://www.cablebot.eu/en/

8. Fortin-Coté A, Cardou P, Gosselin C (2014) An admittance control scheme for haptic interfaces based on cable-driven parallel mechanisms. In: Proceedings of the IEEE international conference on robotics and automation (ICRA 2014), pp 819–925. Hong Kong
9. Gagliardini L, Caro S, Gouttefarde M, Girin A (2016) Discrete reconfiguration planning for cable-driven parallel robots. Mech Mach Theory 100:313–337
10. Gallina P, Rosati G, Rossi A (2001) 3-DOF wire driven planar haptic interface. J Intel Robot Syst 32(1):23–36
11. Gouttefarde M, Collard JF, Riehl N, Baradat C (2015) Geometry selection of a redundantly actuated cable-suspended parallel robot. IEEE Trans Robot 31(2):501–510
12. Gouttefarde M, Krut S (2010) Characterization of parallel manipulator available wrench set facets. In: Advances in robot kinematics, pp 475–484. Springer
13. Guay F, Cardou P, Cruz A, Caro S (2014) Measuring how well a structure supports varying external wrenches. In: New advances in mechanisms, transmissions and applications. Mechanisms and machine science, vol 17, pp 385–392. Springer
14. Hassan M, Khajepour A (2009) Analysis of large-workspace cable-actuated manipulator for warehousing applications. In: Proceedings of the ASME international design engineering and technology conference and computer and information in engineering conference (IDETC/CIE 2009), pp 45–53. San Diego, CA
15. Holland C, Cannon D (2004) Cable array robot for material handling
16. Izard JB, Gouttefarde M, Michelin M, Tempier O, Baradat C (2013) A reconfigurable robot for cable-driven parallel robotic research and industrial scenario proofing. In: Cable-driven parallel robots. Mechanisms and machine science, vol 12, pp 135–148. Springer
17. Jiang Q, Kumar V (2013) The inverse kinematics of cooperative transport with multiple aerial robots. IEEE Trans Robot 29(1):136–145
18. Kamawura S, Kino H, Won C (2000) High-speed manipulation by using parallel wire-driven robots. Robotica 18(1):13–21
19. Lamaury J, Gouttefarde M, Chemori A, Herve PE (2013) Dual-space adaptive control of redundantly actuated cable-driven parallel robots. In: Proceedings of the IEEE international conference on robotics and automation (ICRA 2013), pp 4879–4886. Tokyo, Japan
20. Lumelsky V (1985) On fast computation of distance between line segments. Inform Process Lett 21(3):55–61
21. Maeda K, Tadokoro S, Takamori T, Hiller M, Verhoeven R (1999) On design of a redundant wire-driven parallel robot WARP manipulator. In: Proceedings of the IEEE international conference on robotics and automation (ICRA 1999), vol 2, pp 895–900. Detroit, MI
22. Manubens M, Devaurs D, Ros L, Cortés J (2013) Motion planning for 6D manipulation with aerial towed-cable systems. In: Proceedings of robotics: science and systems. Berlin, Germany
23. Merlet JP (2008) Kinematics of the wire-driven parallel robot MARIONET using linear actuators. In: Proceedings of the IEEE international conference on robotics and automation (ICRA 2008), pp 3857–3862. Passadena, CA
24. Merlet JP, Daney D (2010) A portable, modular parallel wire crane for rescue operations. In: Proceedings of the IEEE international conference on robotics and automation (ICRA 2010), pp 2834–2839. Anchorage, AK
25. Nguyen DQ (2014) On the study of large-dimension reconfigurable cable-driven parallel robots. Ph.D. thesis, Université Montpellier 2, Montpellier, France
26. Nguyen DQ, Gouttefarde M (2014) Study of reconfigurable suspended cable-driven parallel robots for airplane maintenance. In: Proceedings of the IEEE international conference on intelligent robots and systems (IROS 2014), pp 1682–1689. Chicago, IL
27. Nguyen DQ, Gouttefarde M, Company O, Pierrot F (2014) On the analysis of large-dimension reconfigurable suspended cable-driven parallel robots. In: Proceedings of the IEEE international conference on robotics and automation (ICRA 2014), pp 5728–5735. Hong Kong
28. Pott A, Meyer C, Verl A (2010) Large-scale assembly of solar power plants with parallel cable robots. In: Proceedings of the international symposium on robotics and 6th German conference on robotics (ISR/ROBOTIK 2010), pp 1–6. Munich, Germany

29. Pott A, Mtherich H, Kraus W, Schmidt V, Miermeister P, Verl A (2013) IPAnema: a family of cable-driven parallel robots for industrial applications. In: Cable-driven parallel robots. Mechanisms and machine science, vol 12, pp 119–134. Springer
30. Roberts R, Graham T, Lippitt T (1998) On the inverse kinematics, statics, and fault tolerance of cable-suspended robots. J Robot Syst 15(10):581–597
31. Rosati G, Gallina P, Masiero S (2007) Design, implementation and clinical test of a wire-based robot for neurorehabilitation. IEEE Trans Neural Syst Rehabil Eng 15(4), 560–569
32. Rosati G, Zanotto D, Agrawal SK (2011) On the design of adaptive cable-driven systems. ASME J Mech Robot 3(2)
33. Williams R, Xin M, Bosscher P (2008) Contour-crafting-cartesian-cable robot system concepts: workspace and stiffness comparisons. In: Proceedings of the ASME international design engineering and technology conference and computer and informatics in engineering conference (IDETC/CIE 2008), vol 2, pp 31–38. Brooklyn, NY
34. Yao R, Li H, Zhang X (2013) A modeling method of the cable driven parallel manipulator for FAST. In: Cable-driven parallel robots. Mechanisms and machine science, vol 12, pp 423–436. Springer
35. Yao R, Tang X, Wang J, Huang P (2010) Dimensional optimization design for the four-cable driven parallel manipulator in FAST. IEEE/ASME Trans Mechatr 15(6):932–941
36. Zanotto D, Rosati G, Minto S, Rossi A (2014) Sophia-3: a semiadaptive cable-driven rehabilitation device with a tilting working plane. IEEE Trans Robot 30(4):974–979
37. Zhou A, Tang CP, Krovi V (2012) Analysis framework for cooperating mobile cable robots. In: Proceedings of the IEEE international conference on robotics and automation (ICRA 2012), pp 3128–3133. Saint Paul, MN
38. Zhou X, Jun S, Krovi V (2013) Tension distribution shaping via reconfigurable attachment in planar mobile cable robots. Robotica 32(2):245–256
39. Zhou X, Jun S, Krovi V (2014) Stiffness modulation exploiting configuration redundancy in mobile cable robots. In: Proceedings of the IEEE international conference on robotics and automation (ICRA 2014), pp 5934–5939. Hong Kong

Distance Geometry in Active Structures

Josep M. Porta, Nicolás Rojas and Federico Thomas

Abstract Distance constraints are an emerging formulation that offers intuitive geometrical interpretation of otherwise complex problems. The formulation can be applied in problems such as position and singularity analysis and path planning of mechanisms and structures. This paper reviews the recent advances in distance geometry, providing a unified view of these apparently disparate problems. This survey reviews algebraic and numerical techniques, and is, to the best of our knowledge, the first attempt to summarize the different approaches relating to distance-based formulations.

Introduction

A structure can be seen as a complex multibody system (see Fig. 1). While rigid structures have been widely used in construction, passive mobile structures are commonly used, for instance, as shock absorbers. The advent of automation, however, opened the possibility for building active structures [1], i.e., structures which can actively vary their geometry as needed. Such structures are mechanisms, since their motions are typically achieved by means of actuated elements such as revolute joints or variable-length bars. Due to their shape versatility, variable geometry structures have myriad potential applications including robot arms [2], hyper-redundant manipulators [3], flight simulators [4], payload vibration reduction systems [5], the manip-

J. M. Porta (✉) · F. Thomas
Institut de Robòtica i Informàtica Industrial, CSIC-UPC, Barcelona, Spain
e-mail: porta@iri.upc.edu

F. Thomas
e-mail: fthomas@iri.upc.edu

N. Rojas
Faculty of Engineering, Dyson School of Design Engineering,
Imperial College, London, England
e-mail: n.rojas@imperial.ac.uk

E. Ottaviano et al. (eds.), *Mechatronics for Cultural Heritage and Civil Engineering*,
Intelligent Systems, Control and Automation: Science and Engineering 92,
https://doi.org/10.1007/978-3-319-68646-2_5

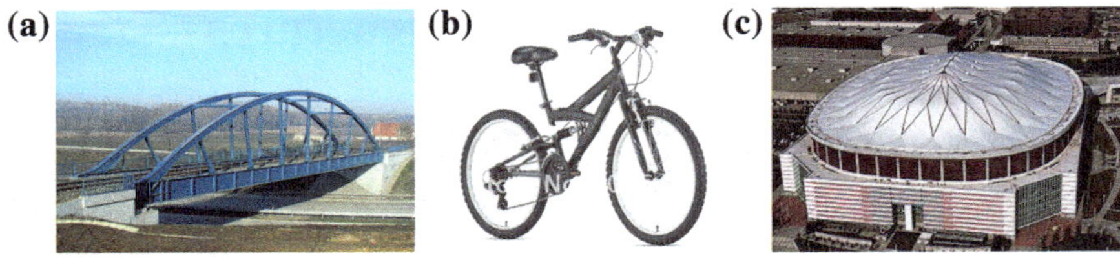

Fig. 1 Structures are typically composed by interconnected rigid bodies and they can have different types of mobility. **a** A fixed bridge is an example of rigid structure. **b** The shock absorbers of a bike are passive mobile structures. **c** Retractable roofs are active structures

ulation of large payloads [6], morphing wings [7], space applications [8], and civil engineering structures [9].

The design of novel variable geometry structures rely on having a complete characterization of their valid configurations. Such configurations are defined by a system of equations encoding the assembly, task, or contact constraints intervening in the problem, and the goal is to analyze the motion capabilities by studying the solutions and properties of such system. The equations can be encoded with different formulations, and the analysis can be significantly simplified if the right formulation is chosen.

The dominant formulation is based on homogeneous transforms using the parameters proposed by Denavit and Hartenberg [10]. This formulation encodes the relative relation between the reference frames associated with the bodies connected by a given joint. Althouth the motion simulation using such parameters is straightforward, the motion analysis using them is challenging, because the resulting equations typically involve complex trigonometric expressions.

Over the past few years, several works have shown that deviating from this classical approach and formalizing the motion analysis problems using distance constraints can be very advantageous. Distance constraints provide intuitive geometric insights on aspects of the motion analysis problem which are difficult to discern otherwise. Moreover, these geometric insights allow the derivation of solutions common to a group of problems that must otherwise be treated on a case-by-case basis [11–13].

In some mechanisms, the configuration space is composed of isolated points. This is what happens, for example, when solving the position analysis problem of serial or parallel manipulators. In other situations, the valid configurations form a variety, and the problem is to analyze relevant subsets of this variety. This is the case when analyzing the singularity loci. In systems in which the dimension of the variety is very high, global analysis of the configuration space is unfeasible, and the main problem is to find collision and/or singularity free paths connecting any two given configurations. Next, we review the existing distance geometry approaches to these three fundamental problems.

Position Analysis

The position analysis problem consists in computing all the valid configurations of a constrained multibody system. This problem appears, for instance, in the inverse kinematic of serial manipulators [14], as illustrated in Fig. 2. It also appears when solving the forward kinematics of parallel structures [15], when planning the motion of deployable structures [16], in robot grasping [17], in constraint-based object positioning [18], or in simultaneous localization and map-building [19]. The problem appears in other domains as well, such as in the dynamical simulation of multibody systems [20], in parametric computer-aided design (CAD) [21], or in the conformational analysis of biomolecules [22].

Traditional approaches translate the original geometric problem into a system of kinematic equations derived from the independent kinematic loops in the problem. Existing techniques for solving such systems of equations can be classified into algebraic and numerical methods. The use of independent loop equations has seldom been questioned, yet the resulting expressions are not particularly well suited for either the algebraic or numerical methods: first, because arbitrary reference frames

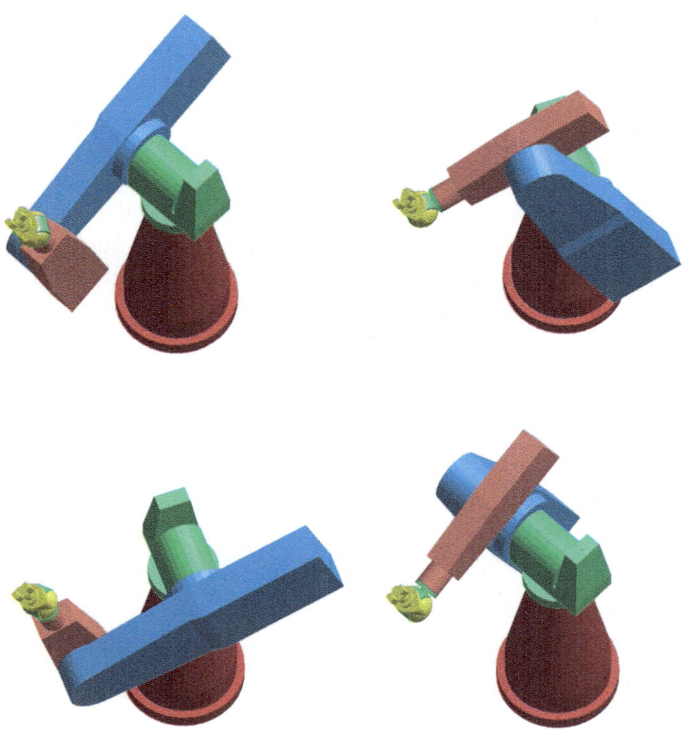

Fig. 2 The inverse kinematic problem of a serial robot consists in finding the robot configurations that position the end effector in a given pose. For the manipulator in the figure, this problem has eight different solutions. Four of them are shown here

have to be included, and second, because all formulas involve translations and rotations simultaneously, thus leading to complex trigonometric equations. The distance-based formulation offers an alternative that avoids these two problems and has been shown to provide novel insights in both algebraic and numerical methods.

Algebraic Methods

Algebraic methods use variable elimination to reduce the initial system to a univariate polynomial whose roots, once back-substituted into other equations, yield all solutions of the original system [23].

To apply the variable elimination methods to equations resulting from the kinematic loops, the trigonometric expressions must be replaced to obtain a system of polynomial equations. Typically the tangent half-angle substitution is used, but it misses possible roots at $\pm\pi$. Moreover, the resulting expressions are complicated, limiting the scalability of this approach. Some of the successful results obtained with this approach can be attributed to clever manual manipulation of the expressions, which are difficult to generalize [24]. Using a distance-based approach, the original geometric problem is translated into a graph where the nodes are points on the structure and the edges are distance, area, or volume constraints involving these points. Relying on this formulation, Rojas and Thomas [27] proposed a procedure for solving the position analysis of complex planar mechanisms (see Fig. 3). The procedure analyzes the two possible assemblies of a triangle, given the distances between its vertices. This basic problem is elegantly formulated using the so-called bilateration matrices. More complex mechanisms are then analyzed, decomposing them in strips of triangles and chaining bilateration matrices for triangles sharing one edge. This directly produces a scalar algebraic resultant in most cases, which can be transformed into a polynomial by clearing radicals.

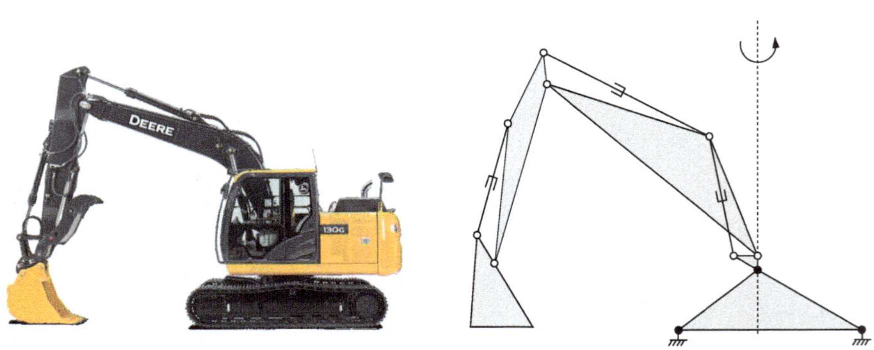

Fig. 3 An excavator (left) can be modeled as a planar mechanism (right) with a global rotation about a vertical axis. White and black dots represent movable and fixed revolute joints, respectively

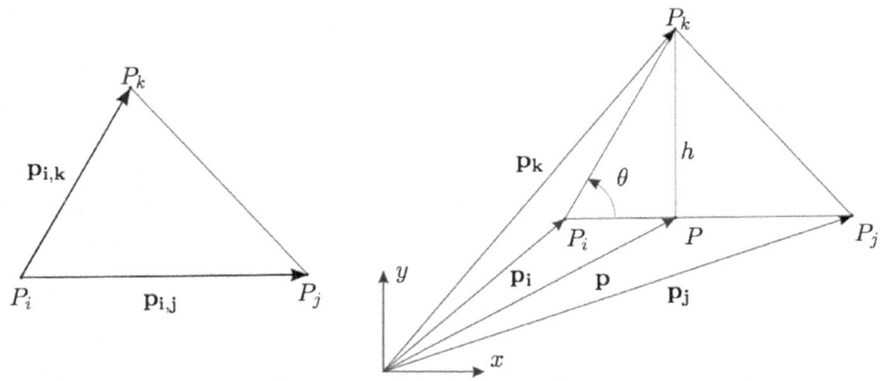

Fig. 4 The bilateration problem (left) and the associated notation (right)

Let us review this procedure in more detail. According to the notation in Fig. 4, the objective of the bilateration operation is to determine the location of point P_k given known locations for points P_i and P_j. The position vector of the orthogonal projection of P_k onto $\overline{P_iP_j}$ can be expressed as

$$\mathbf{p} = \mathbf{p}_i \sqrt{\frac{D(i,k)}{D(i,j)}} \cos\theta \, \mathbf{p}_{i,j} = \mathbf{p}_i + \frac{D(i,j;i,k)}{D(i,j)} \mathbf{p}_{i,j} \tag{1}$$

where

$$D(i_1,\dots,i_n;j_1,\dots,j_n) = \begin{vmatrix} 0 & 1 & \dots & 1 \\ 1 & s_{i_1j_1} & \dots & s_{i_1j_n} \\ 1 & \vdots & \ddots & \vdots \\ 1 & s_{i_nj_1} & \dots & s_{i_nj_n} \end{vmatrix} \tag{2}$$

is the Cayley-Menger determinant of n points, with $s_{i,j}$ the square of $d_{i,j}$, the distance between P_i and P_j, and $D(i_1,\dots,i_n) = D(i_1,\dots,i_n;i_1,\dots,i_n)$ is the Cayley-Menger determinant.

Using $\mathbf{p}$, the position of P_k can be expressed as

$$\mathbf{p}_k = \mathbf{p} \pm \frac{\sqrt{D(i,j,k)}}{D(i,j)} \mathbf{S} \, \mathbf{p}_{i,j}, \tag{3}$$

where $\mathbf{S} = \begin{pmatrix} 0 & -1 \\ 1 & 0 \end{pmatrix}$, and the $\pm$ sign accounts for the two mirror-symmetric locations of P_k with respect to the line defined by $\overline{P_iP_j}$. Substituting (1) in (3), we get

$$\mathbf{p}_{i,k} = \frac{D(i,j;i,k)}{D(i,j)} \mathbf{p}_{i,j} + \frac{D(i,j,k)}{D(i,j)} \mathbf{S} \, \mathbf{p}_{i,j}, \tag{4}$$

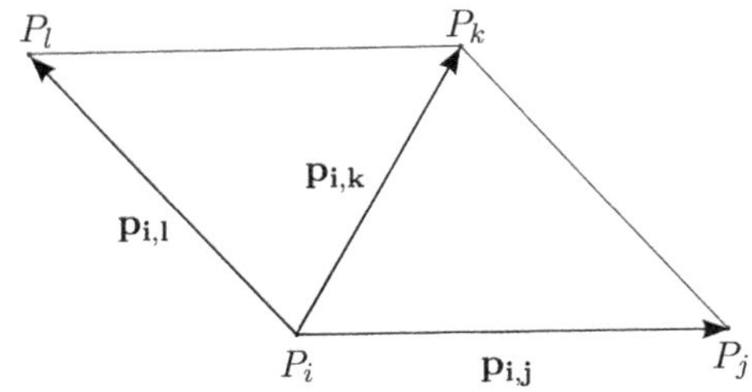

Fig. 5 Two triangles sharing one edge

which can be expressed in a more compact form as

$$\mathbf{p}_{i,k} = \mathbf{Z}_{i,j,k}\, \mathbf{p}_{i,j} \tag{5}$$

where

$$\mathbf{Z}_{i,j,k} = \frac{1}{D(i,j)} \begin{pmatrix} D(i,j;i,k) & \pm\sqrt{D(i,j,k)} \\ \pm\sqrt{D(i,j,k)} & D(i,j;i,k). \end{pmatrix} \tag{6}$$

Note that (5) expresses $\mathbf{p}_{i,k}$ as a function of $\mathbf{p}_{i,j}$. When we have a strip of triangles, i.e., a sequence of triangles each sharing one edge with the previous one in the sequence, we can chain the bilateration process, i.e., multiply bilateration matrices, and express a vector in the final triangle as a function of a vector in the initial one. In the same way, we can derive expressions for vectors between points whose relative distance is initially unknown. For instance, in the situation in Fig. 5, we have that

$$\mathbf{p}_{j,l} = \mathbf{p}_{i,l} - \mathbf{p}_{i,j} = (\mathbf{Z}_{j,k,l}\, \mathbf{Z}_{i,j,k} - \mathbf{I})\, \mathbf{p}_{i,j}. \tag{7}$$

It can then be shown that the squared distance between P_j and P_l is given by

$$s_{j,l} = \det(\mathbf{Z}_{j,k,l}\, \mathbf{Z}_{i,j,k} - \mathbf{I})\, s_{i,j}, \tag{8}$$

which expresses $s_{j,l}$ as a function of known edge lengths.

For larger problems, the same procedure can be applied, identifying strips of triangles with known edge lengths fully covering the mechanism. In most mechanisms with mobility zero, this process typically requires the introduction of unknown distances. Actually, the number of unknown distances to introduce coincides with the *coupling number* of the mechanism. Thus, the method can be applied to linkages with a coupling number higher than one. In the final expression, one of the distances in the triangle at the end of the strip is expressed as a function of an unknown distance in the first triangle. This expression, therefore, is directly a scalar algebraic equation which can be converted into a polynomial by clearing radicals. This direct procedure for deriving resultant polynomials is much simpler than those in the literature, and makes it possible to solve problems such as that in Fig. 6, which contains six independent kinematic loops, a number that has not been attained with elimination methods [26].

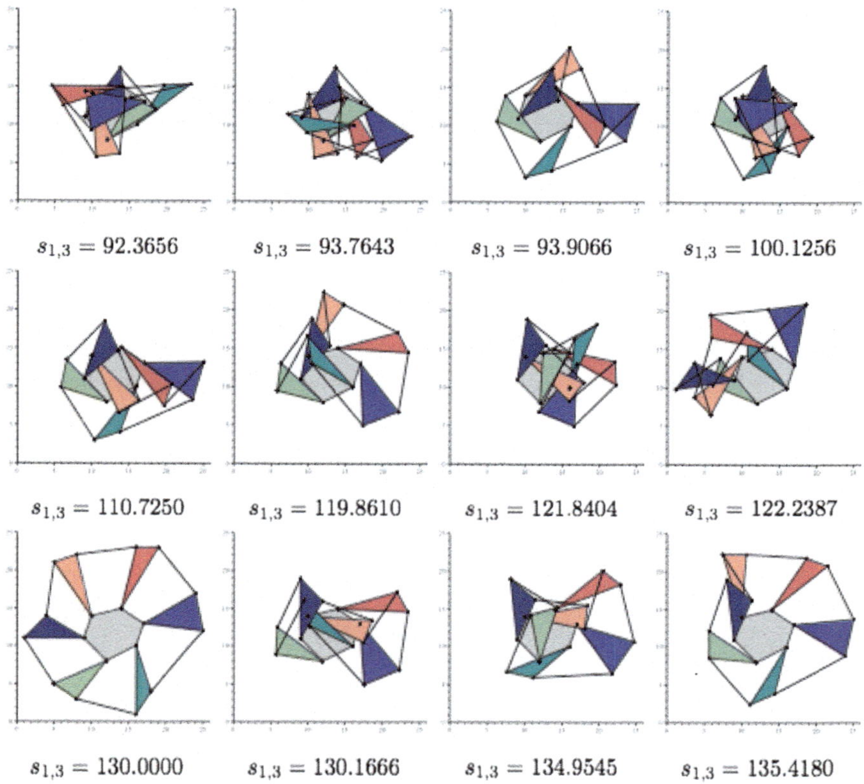

Fig. 6 Some of the valid configurations of a 13-link Watt-Baranov truss. The position analysis of this mechanism using algebraic methods has been shown to be feasible when formulating the problem in terms of distance constraints [25]

For systems with mobility one, we obtain a polynomial depending on two unknown distances which can be used for path tracking: one of the unknowns is fixed, and the system is solved for the remaining variable. The advantage in using a distance formulation in this case is that path-crossing conditions can be readily identified, since they correspond to alignments of particular points in the problem. Thus, the distance-based formulations simplify the path-tracking procedures, as compared with previous approaches [27].

This distance-based algebraic approach can be generalized to 3D using fans of tetrahedra instead of strips of triangles. This enables solving in closed form the position analysis of variable geometry trusses much more complex than those solvable with traditional formulations [28].

Numerical Methods

Although the distance-based algebraic methods have proved quite efficient in fairly non-trivial position analysis problems, this technique explodes in complexity with the size of the problem. Thus, to address larger problems, one must resort to numerical techniques. Several distance-based alternatives are available in the literature.

Trilateration Methods

The position analysis problem can be seen as that of determining all the possible values for the unknown distances in the graph of distance constraints encoding the problem. If the graph is represented in the form of an adjacency matrix, the problem boils down to completing the matrix from the distances initially fixed. Once the matrix is completed, standard linear algebraic methods can be used to give coordinates to the points in the problem [29].

In some problems, the matrix completion process can be performed in an incremental constructive way. In $\mathbb{E}^3$, if all but one of the relative distances between five points are known, the unknown distance can be readily determined by trilateration. Using the notation in Fig. 7, we have that

$$s_{4,5} = \frac{2}{D(1,2,3)} \left(D(1,2,3,4;1,2,3,5)\Big|_{s_{4,5}=0} \pm \sqrt{D(1,2,3,4)\,D(1,2,3,5)} \right), \quad (9)$$

where $D(1,2,3,4;1,2,3,5)\Big|_{s_{4,5}=0}$ denotes the corresponding Cayley-Menger bideterminant with $s_{4,5}$ set to 0. Observe that no point coordinates appear in the result, only inter-point distances, and that two solutions are possible, corresponding to the two possible signs for the square root in the expression.

A problem is trilaterable if it is possible to determine a trilateration sequence to compute the initially unknown distances in the problem. This sequence can be

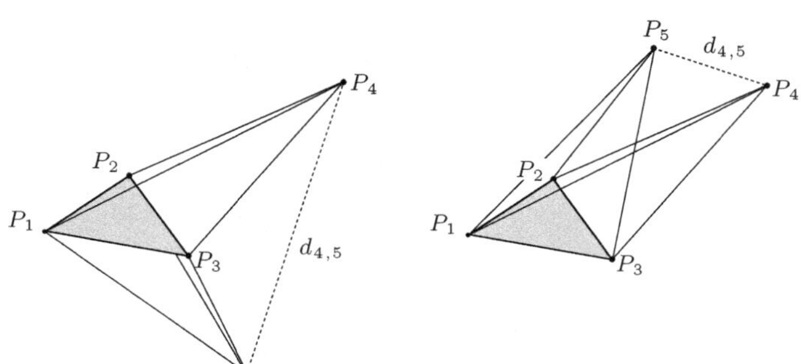

Fig. 7 Trilateration can be used to compute the distance between P_4 and P_5 from their distances to P_1, P_2, and P_3, which form a fixed triangle. Two solutions are possible, depending on the location of P_4 and P_5 with respect to the plane defined from points P_1, P_2, and P_3

readily determined by subgraph matching, i.e., finding parts of the original graph which match with the trilateration subgraph shown in Fig. 7. Porta et al. [11] showed that the inverse/direct kinematics of the most usual serial/parallel robots are trilaterable problems, which greatly simplifies their resolution. Moreover, [30] showed that if the searched subgraph includes six points with only one unknown distance between them, this unknown distance is linear with respect to the rest of the distances, and thus it has only one possible solution. This avoids the generation of distance completions that have to be discarded in a post-processing stage, since they would include tetrahedra with orientations incompatible with the original problem. For greater detail on the role of orientations in distance geometry, see Sect. Bound Smoothing with Orientation Constraints.

Note that the trilateration process is closely related to the algebraic approach described in Sect. Algebraic Methods. The main difference is that in the trilateration, only one new unknown distance appears at each step, whereas in the algebraic method, the first step involves two unknown distances. This is why the former method is purely numerical while the latter generates a symbolic expression in one variable distance.

To the best of our knowledge, the combination of the trilateration step and the procedure to determine a trilateration sequence was first introduced by Porta et al. [11, 30], and was later independently proposed by Lavor et al. [31] in the context of structural biology, but in Cartesian space, i.e., relying on the coordinates of the points and assigning coordinates to the trilaterated point at each step. More recently, the same authors proposed an equivalent algorithm in distance space [32].

Branch-and-Prune Methods

Unfortunately, not all problems admit a trilateration sequence, and thus general solvers must be devised to determine the valid distance matrices from the initially known distances. Porta et al. present alternative general solvers relying on a branch-and-prune technique [33–35]. These solvers iteratively eliminate regions of the space of distances where the distance constraints are not satisfied. When the distance space cannot be further reduced, it is split, and the reduction and split procedure is recursively applied to the two resulting sub-spaces. This process can be seen as an extension of the classic bound-smoothing techniques [36], and isolates the valid solutions for the input problem in the form of interval matrices at the desired resolution. Since splitting the search space is trivial, the key operation of the branch-and-prune methods is the procedure used to shrink the boxes in distance space.

The method presented in [34, 35] introduces variable substitutions to convert the quadratic expression resulting from the Cayley-Menger determinants into multilinear equations. The graph of a multilinear function defined on an axis-aligned box is included in the convex hull of the evaluation of the function in the corners of the domain [37]. The solution of $f(\mathbf{x}) = 0$ can then be bounded to the intersection of this hull with the plane $f(\mathbf{x}) = 0$. Since the computation of this intersection can be difficult, the method projects the hull onto each coordinate plane, as depicted in Fig. 8

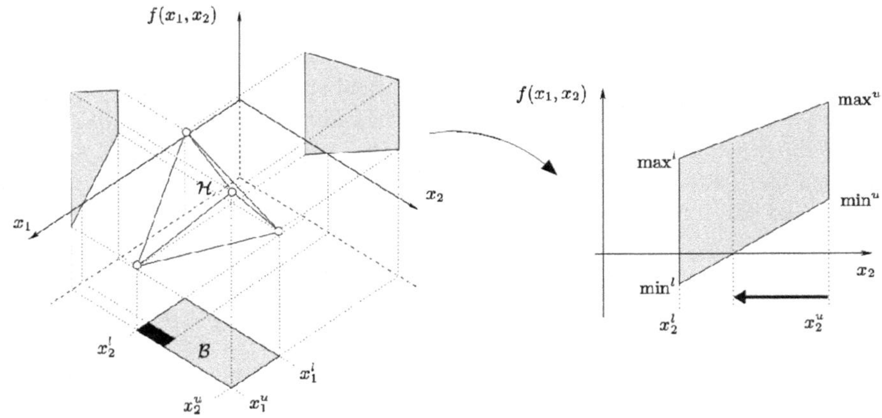

Fig. 8 Segment-trapezoid clipping. Left: in this two-variable case, the graph of the multilinear function $f(x_1, x_2)$ in the domain box $\mathcal{B}$ necessarily lies inside the shown tetrahedron $\mathcal{H}$. The vertices of $\mathcal{H}$ are obtained by evaluating f in the corners of $\mathcal{B}$. The projection of $\mathcal{H}$ to a given coordinate planes defines a trapezoid. Right: from the initial range for a variable, we can prune all points for which its trapezoid does not intersect the $f(x_1, x_2) = 0$ line

(left), and intersects each of the resulting trapezoids with the line, as shown in Fig. 8 (right). Typically, these segment-trapezoid clippings reduce the ranges of some variables, giving a smaller box (the black rectangle in Fig. 8) that still bounds the root locations. Although this strategy produces less pruning than the convex hull-plane clipping, it is advantageous in practice due to its lower cost of operation.

A Gough-Stewart platform is a six-degrees-of-freedom structure composed of a moving platform connected to a base by six legs. The pose of the platform is controlled by the leg lengths. Parallel structures are used in many applications, including positioning tools [38], flight simulators [39], or radiotelescopes [40]. After some leg rearrangements (see Sect. Singularity Analysis), the forward kinematic problem of the Gough-Stewart in Fig. 9 can be formulated using two Cayley-Menger determinants [41]

$$D(1,3,4,5,6) = 0,$$
$$D(1,2,4,5,6) = 0,$$

where P_1, P_2, and P_3 are the points defining the triangle at the base of the structure and P_4, P_5, and P_6 the vertices in the triangle at the moving platform. The branch-and-prune algorithm proposed by Porta et al. [35] determines the two solutions of this problem typically in less than 0.01 s on a standard desktop computer.

A variation of the above algorithm, where the box reduction is based on the properties of the Bernstein polynomials, has been used to elucidate the valid conformations of several molecular structures [42]. This method was parallelized and run on

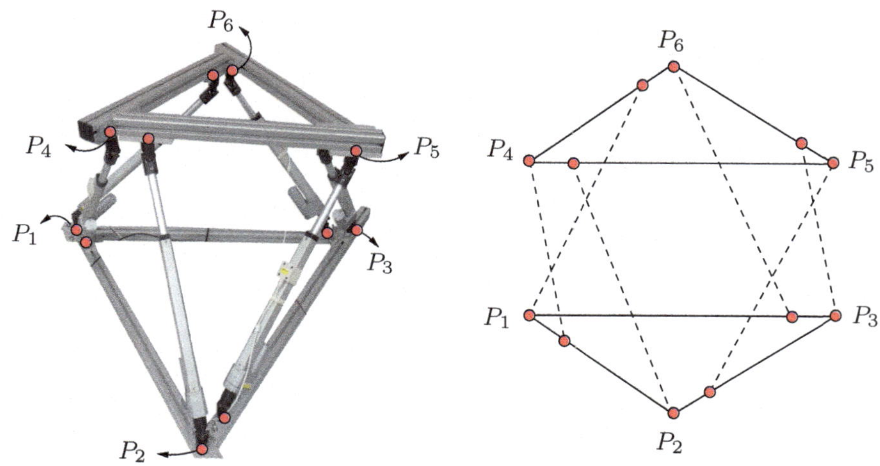

Fig. 9 Left: a hexapod structure. Right: associated distance graph where nodes are points and edges are distance constraints. Solid lines denote distances that are constant independently of the configuration

the MareNostrum supercomputer[1] to obtain the first complete description of the conformational space of the cyclooctane, which is a two-dimensional variety. To the best of our knowledge, this was the first distance geometry method able to characterize such complex solution spaces.

Distance geometry provides yet another alternative to crop the distance ranges based on the reduction and expansion of the dimension of the problem [43]. This approach is purely geometric, avoiding the algebraization of the problem. Taking the vector from P_1 to P_n as a reference, a vector $\mathbf{q} = (\bar{d}_{1,n}, \ldots, \bar{d}_{n,n})$ can be defined where

$$\bar{d}_{i,n} = \frac{1}{2\,d_{1,n}}(d_{i,n}^2 + d_{1,n}^2 - d_{i,1}^2) \tag{10}$$

is the orthogonal projection of the vector from P_n to P_i onto the reference vector. We can also define the orthogonal complement of this projection, which is a matrix $\mathbf{Q}^{\perp}$ with

$$\mathbf{Q}_{i,j}^{\perp} = d_{i,j}^2 - \frac{1}{4\,d_{1,n}^2}(d_{i,n}^2 - d_{j,n}^2 + d_{j,1}^2 - d_{i,1}^2)^2. \tag{11}$$

It can be shown that matrix $\mathbf{Q}$, with $Q_{i,j} = s_{i,j}$, is a proper Euclidean distance matrix in $\mathbb{R}^d$ if and only if $\mathbf{Q}_{i,j}^{\perp}$ with $i, j = 1 \ldots n-1$ is a correct Euclidean distance matrix in $\mathbb{R}^{d-1}$. Thus, the method projects the input distance matrix with interval ranges until the problem becomes one-dimensional, and hence consistency can be enforced using the triangular equality. The eventually reduced ranges are back-projected using the

[1] http://www.bsc.es.

intermediate vectors $\mathbf{q}$ and matrices $\mathbf{Q}^{\perp}$ to find tighter ranges for the distances in the original problem. The direct evaluation of (10) and (11) using interval arithmetics would be inaccurate due to the well-known overestimations effect of this approach [44]. However, if a function $z = g(\mathbf{x})$ is monotone in an axis-aligned domain $\mathbf{x} = (x_1, \dots, x_n)$, with $x_i \in [x_i^l, x_i^u]$, and the derivatives of g are available, its upper bound is $z^u = g(\hat{\mathbf{x}})$, where $\hat{\mathbf{x}} = (\hat{x}_1, \dots, \hat{x}_n)$ is the vertex of the domain given by

$$\hat{x}_i = \begin{cases} x_i^l & \text{if } \partial g/\partial x_i < 0, \\ x_i^u & \text{otherwise.} \end{cases}$$

The lower bound, z^l, is in a vertex defined with the opposite criterion. Thus, tight bounds for $\overline{d}_{i,n}$ and $\mathbf{Q}_{i,j}^{\perp}$ can be obtained by analyzing their respective derivatives. This method has been applied to solve the position analysis of planar and spatial mechanisms with mobility 0 and 1 [43, 45].

Bound Smoothing with Orientation Constraints

One of the major shortcomings of the approaches described in the previous section is their limited capability for encoding orientation constraints. Let us consider the regional part of the wrist-partitioned 6R robot shown in Fig. 10 (left), that is, the first three links and joints that permit locating the wrist center anywhere in the robot's workspace. Figure 10 (right) shows the formalization of this problem as a graph of distance constraints. In this example, any standard distance constraint solver would generally obtain eight different sets of compatible distances. Nevertheless, it is well known that the inverse kinematics of a 3R robot can only have up to four solutions [46]. This apparent contradiction has a simple explanation: a distance-based technique would not take into account the relative orientations of the tetrahedra defined by the sets of points $\{P_1, P_2, P_3, P_4\}$ and $\{P_3, P_4, P_5, P_6\}$. The same situation occurs in many other structures.

To address this issue, Rull et al. present a distance-bound smoothing approach that permits the incorporation of orientation constraints in the process of reducing the valid ranges of distances [47]. This approach focuses on planar problems that are formalized with the following constraints:

- For all sets of three points:

$$D(i_1, i_2, i_3) \leq 0. \tag{12}$$

- For all sets of four points:

$$D(i_1, i_2, i_3, i_4) = 0, \tag{13}$$

and

$$D(i_1, i_2, i_3; i_1, i_2, i_4) = \begin{cases} < 0 \text{ if } \sigma_{i_1,i_2,i_3}\sigma_{i_1,i_2,i_4} > 0 \\ \geq 0, \text{ otherwise} \end{cases} \tag{14}$$

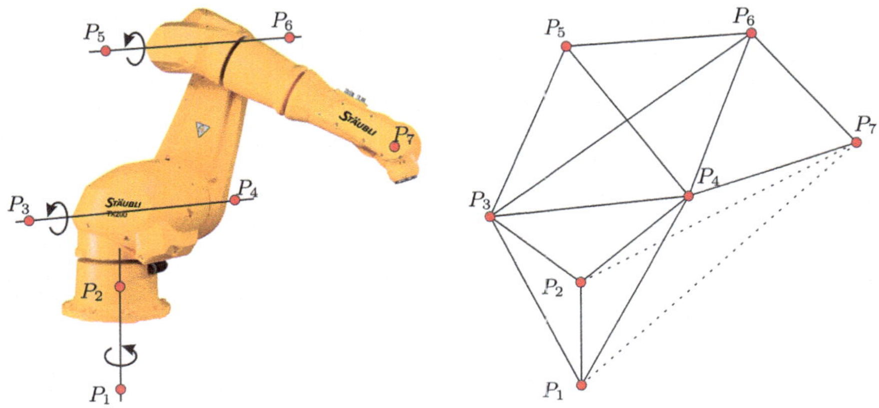

Fig. 10 Left: the configuration of the regional part of a wrist-partitioned 6R robot is determined by the location of seven points. Right: associated distance graph where nodes stand for points and edges for known distances between the corresponding points. Solid lines represent constant distances, regardless of the location of the end-effector

Fig. 11 Taking triangle P_1, P_2, and P_3 as a reference, the plane is divided into seven regions. If P_4 is bound to be in one of these regions, $s_{3,4}$ is monotone with respect to the rest of the squared distances. The boundaries separating the monotonic regions correspond to configurations where there is an alignment of three points

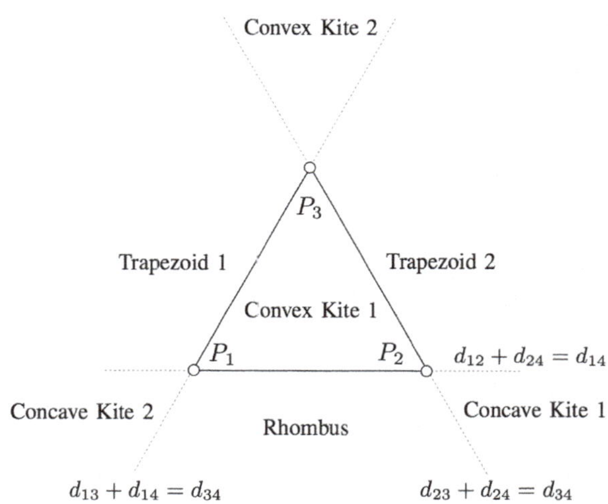

where $\sigma_{i,j,k}$ is defined as negative if points P_i, P_j, and P_k have to be arranged clockwise, and as positive otherwise.

While the expansion of (12) leads to the triangular inequality involving the distances between P_{i_1}, P_{i_2}, and P_{i_3}, (13) is nothing more than the tetrangular equality involving the six pairwise distances between P_{i_1}, P_{i_2}, P_{i_3}, and P_{i_4}.

Note that the whole set of orientation constraints in (14) cannot be fixed arbitrarily. Actually, it is possible to define a basis that determines all other orientations [48].

Since an efficient algorithm exists for tightening bounds using triangular constraints [49], it can be safely assumed that they are already satisfied by the initial ranges. Thus, the approach focuses on the analysis of (13) under orientation constraints. The expansion of this equality yields a quadratic expression in any of the involved squared distances. For instance, for the set of points $\{P_1, P_2, P_3, P_4\}$, we have

$$s_{3,4} = \frac{D(1,2,4;1,2,3)|_{s_{3,4}=0} + \sigma_{1,2,4}\,\sigma_{1,2,3}\sqrt{D(1,2,4)\,D(1,2,3)}}{D(1,2)} \tag{15}$$

or alternatively

$$s_{3,4} = \frac{D(1,2,4;1,2,3)|_{s_{3,4}=0} + 16\,A_3\,A_4}{D(1,2)}, \tag{16}$$

where A_i denotes the oriented area of the triangle defined by the ordered set $\{P_1, P_2, P_3, P_4\}\backslash P_i$, since $4A_i = \sigma_{j,k,l}\,\sqrt{-D(j,k,l)}$.

The range of $s_{3,4}$ can be tightly bounded using the monotonicity analysis presented at the end of Sect. Branch-and-Prune Methods. To apply this method to the function in (16), we need to compute the derivatives of $s_{3,4}$ with respect to $s_{i,j}$. Instead of computing these derivatives from (16), it is more convenient to obtain them from the linearization of (13), which reads as:

$$\begin{aligned} &A_1A_2\,\delta s_{1,2} - A_1A_3\,\delta s_{1,3} + A_1A_4\,\delta s_{1,4} \\ &+ A_2A_3\,\delta s_{2,3} - A_2A_4\,\delta s_{2,4} + A_3A_4\,\delta s_{3,4} = 0. \end{aligned} \tag{17}$$

Then, we have that

$$\frac{\partial s_{3,4}}{\partial s_{i,j}} = -1^{i+j}\,\frac{A_iA_j}{A_3A_4}. \tag{18}$$

As long as the sign of the oriented areas of the triangles defined by P_1, P_2, P_3, and P_4 do not change, s_{34} is monotone. Therefore, in this case, we can readily identify the vertices providing tight bounds for $s_{3,4}$ by controlling the regions where there are orientation sign changes.

Figure 11 shows a partition of the plane in regions where the orientations of the triangles defined by P_1, P_2, and P_i with $i > 3$ are constant, taking the triangle defined by P_1, P_2, and P_3 as a reference. If P_4 remains in one of these regions, the bounds for $s_{3,4}$ can be readily determined. For instance, if P_4 is in the *Rhombus* region, the patterns in Fig. 12 identify the vertices of the domain giving a tight range for $s_{3,4}$. We can identify 14 patterns, two for each region, which subsume the seven patterns used by Crippen and Havel [36].

When three points can be aligned within the allowed distance ranges, the boundaries separating the monotonic areas must be recursively analyzed. At the end of the process, tight bounds for the variable of interest are obtained.

As an alternative, a geometric approach based on projections and back-projections introduced in [45] can also be extended to take into account orientation constraints. Moreover, this method can operate in 3D problems, whereas the extension to 3D of the method introduced in [47] is not trivial.

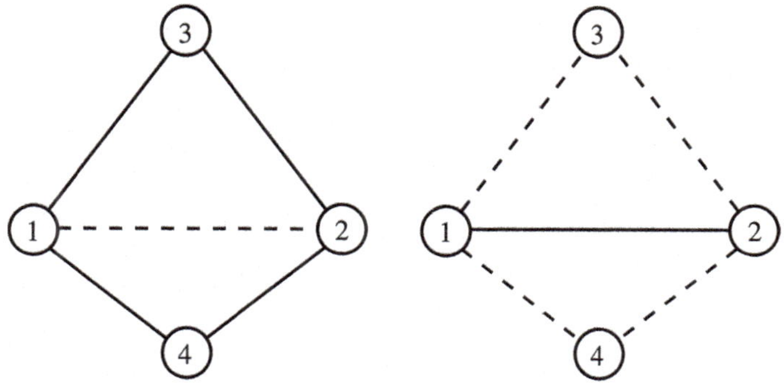

Fig. 12 The two configurations giving the lower (left) and upper (right) bounds for $s_{3,4}$ in the *Rhombus* region. Solid and dashed lines indicate distances at their lower and upper limits, respectively

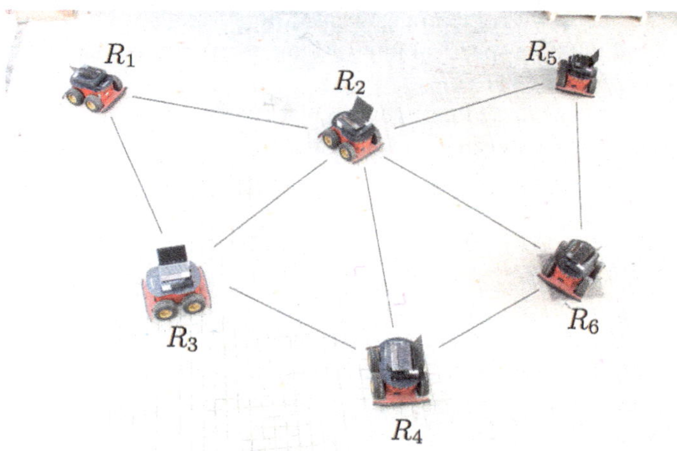

Fig. 13 A robot formation. Each robot is equipped with an ultrasound sensor to measure the distances to nearby teammates. The lines in the figure represent the distances actually measured. The orientation of the triangles is given by cameras mounted on the robots. The integration of the distance and orientation constraints permits the determination of tight bounds for the possible location of each robot

The integration of the orientation constraints opens a new range of applications for distance geometry methods, such as the coordination of robot teams, sensor data fusion, and constraint-based robot programming, to name just a few. For instance, Fig. 13 illustrates the application of this method to the mutual localization of a robot team.

Singularity Analysis

The singularity locus of any given mechanism is the set of configurations where mobility or control issues might arise. To prevent malfunctions or even structural damage, singularities must be avoided. Thus, given its relevance, singularity analysis is a major topic of research in mechanics. However, the characterization of singularities has only been achieved for particular mechanism instances or requires the use of complex computational methods [50]. In general, modifications of the mechanism parameters change the singularity locus in unpredictable ways, which hinders the analysis of new structures. Borras and Thomas, however, proposed distance geometry tools to identify singularity-invariant leg rearrangements for parallel structures, i.e., changes in the attachments of the legs to the base or the platform that do not change the singularity locus [51]. This generalizes the singularity analysis of a particular structure to all the other structures that can be defined from it with singularity-invariant leg rearrangements. Moreover, these rearrangements can be used to avoid multiple spherical joints, significantly simplifying the actual construction of the structure [41].

In a parallel structure, the linear and angular velocity of the moving platform, $\mathbf{v}$ and $\boldsymbol{\Omega}$, respectively, are related to the leg lengths by

$$\mathbf{R}_l \begin{pmatrix} \dot{l}_1 \\ \vdots \\ \dot{l}_6 \end{pmatrix} = \mathbf{J} \begin{pmatrix} \mathbf{v} \\ \boldsymbol{\Omega} \end{pmatrix} \tag{19}$$

where $\mathbf{R}_l$ is a diagonal matrix with leg lengths $l_1, \ldots, l_6$, and $\mathbf{J}$ is the matrix of non-normalized Plücker coordinates of the six leg lines. The relevant singularities for parallel structures occur when $\det(\mathbf{J}) = 0$. Now assume that we rearrange the leg attachments and that the squares of the new leg lengths, $d_1, \ldots, d_6$, are related to the previous ones by an affine relation of the form

$$\begin{pmatrix} d_1^2 \\ \vdots \\ d_6^2 \end{pmatrix} = \mathbf{A} \begin{pmatrix} l_1^2 \\ \vdots \\ l_6^2 \end{pmatrix} + \mathbf{b}. \tag{20}$$

Then, as shown by [52], the relation between the change in the leg lengths and the velocity of the platform becomes

$$\mathbf{R}_d \begin{pmatrix} \dot{d}_1 \\ \vdots \\ \dot{d}_6 \end{pmatrix} = \mathbf{A}\,\mathbf{J} \begin{pmatrix} v \\ \Omega \end{pmatrix} \tag{21}$$

and the new singularity condition is $\det(\mathbf{A}\,\mathbf{J}) = \det(\mathbf{A})\det(\mathbf{J})$. If $\det(\mathbf{A})$ is a constant non-null factor, the leg rearrangement has no effect on the singularity locus.

If $\det(\mathbf{A})$ is null, then the rearrangement introduces an architectural singularity, i.e., the resulting platform is in a singularity irrespective of its leg lengths.

A Cayley-Menger determinant can be used to derive the affine relation between the leg lengths before and after the rearrangement. For instance, consider the situation in Fig. 14 (left), where the anchor point P_1 is displaced to a new position P_4 along the line supported by P_1 and P_2. Since in this rearrangement the four points remain coplanar, then

$$D(1,2,3,4) = \begin{vmatrix} 0 & 1 & 1 & 1 & 1 \\ 1 & 0 & (d_{1,4}+d_{2,4})^2 & s_{1,3} & s_{1,4} \\ 1 & (d_{1,4}+d_{2,4})^2 & 0 & s_{2,3} & s_{2,4} \\ 1 & s_{1,3} & s_{2,3} & 0 & s_{3,4} \\ 1 & s_{1,4} & s_{2,4} & s_{3,4} & 0 \end{vmatrix} = 0. \qquad (22)$$

Expanding this determinant, we obtain

$$d_{2,4}\, s_{1,3} + d_{1,4}\, s_{2,3} - (d_{1,4}+d_{2,4})\, s_{3,4} - d_{1,4}\, d_{2,4}\, (d_{1,4}+d_{2,4}) = 0 \qquad (23)$$

which defines an affine relationship between the leg lengths before and after the rearrangement with

$$\mathbf{A} = \begin{pmatrix} \frac{d_{2,4}}{d_{1,4}+d_{2,4}} & \frac{d_{1,4}}{d_{1,4}+d_{2,4}} & \dots & 0 \\ 0 & 1 & \dots & 0 \\ \vdots & \vdots & \ddots & \vdots \\ 0 & 0 & \dots & 1 \end{pmatrix}. \qquad (24)$$

Since $\det(\mathbf{A}) = d_{2,4}/(d_{1,4}+d_{2,4})$, the proposed change in the anchor point location is a singularity-invariant leg rearrangement.

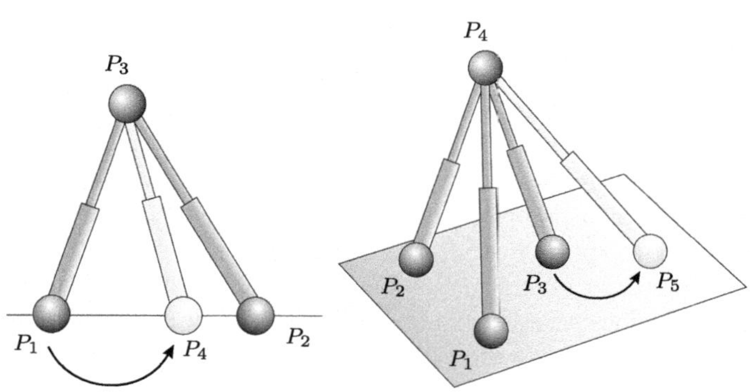

Fig. 14 Singularity-invariant leg rearrangements. Left: rearrangement along a line connecting two anchor points. Right: rearrangement in a plane defined by three anchor points

Figure 14 (right) shows another possible leg rearrangement, where a leg attachment is moved on the plane defined by the anchor points of three legs. In this case, the conditions to be held is

$$D(1, 2, 3, 4, 5) = 0 \tag{25}$$

which can be rewritten to

$$\begin{aligned} &D(1, 2, 3; 2, 3, 5)\, s_{1,4} - D(1, 2, 3; 1, 3, 5)\, s_{2,4} \\ &+ D(1, 2, 3; 1, 2, 5)\, s_{3,4} - D(1, 2, 3)\, s_{4,5} + C = 0 \end{aligned} \tag{26}$$

where C is a constant that does not depend on the distances involving P_4. This leads to a singularity factor

$$\det(\mathbf{A}) = \frac{D(1, 2, 3; 2, 3, 5)}{D(1, 2, 3)}, \tag{27}$$

which is independent of the structure configuration and, thus, defines a singularity-invariant leg rearrangement.

Note that other leg rearrangements are possible. The advantage in using distance geometry in their derivation is that, in general, the singularity factors have a direct geometric interpretation.

Path Planning

For structures with high mobility, the comprehensive description of both their configuration spaces and their singularity loci is unfeasible. Fortunately, in these cases, research efforts usually focus on path planning problems, i.e., problems consisting in determining how to move the structure from an initial to a goal configuration, while avoiding collisions or singularities [53, 54], although this second aspect is less commonly treated in the literature [55].

For tree-like structures, the configuration space is parametric, which greatly simplifies the problem. However, when kinematic loops appear in the problem (see Fig. 15), the configuration space becomes a manifold embedded in the ambient space formed by the joint variables [56]. Actually, kinematic loops appears in many relevant problems, such as complex manipulation problems [57], parallel robot path generation [15], grasp planning [58], and surgery planning [59].

Under the presence of kinematic constraints, standard formulations produce involved configuration spaces. For some families of structures, though, distance constraints produce much simpler representations of the configuration spaces, in which the path planning problem can be easily solved.

Trinkle and Milgram analyze the configuration space of planar chains with revolute joints formalized with joint angle parameters, but also use concepts from distance geometry [60]. Using the notion of long link, they prove that the configura-

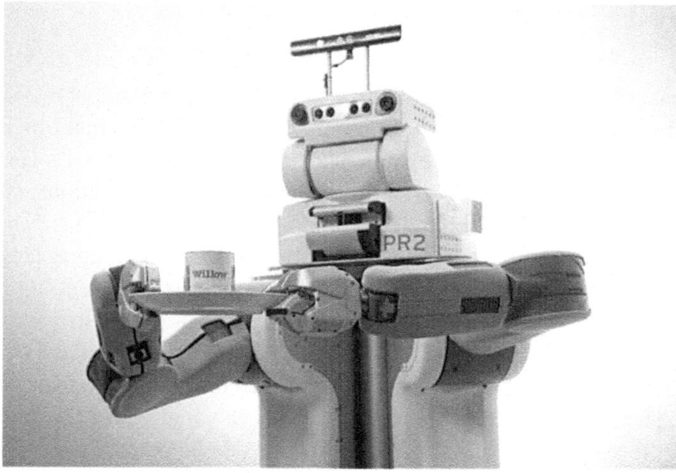

Fig. 15 A two-armed service robot holding a plate. The arms must be in contact with the plate, and the plate must remain horizontal to avoid tilting the coffee mug on it. A loop of kinematic constraints is thus generated

tion spaces of such mechanisms are connected if and only if they do not have three long links. Otherwise, their configuration space has two components, and each is toroidal. They then propose a path planning algorithm differentiating between these two cases. When the configuration space has two components, the first task is to discern whether the two configurations to be connected are in the same component. If not, the path planning problem cannot be solved. For configurations in the same component, one link is used to drive the mechanism, while the remaining links comply in a series of *accordion* moves that can be proven to connect the configurations of interest. The final path is not optimal, but the algorithm is shown to be very efficient for significantly long loops.

Han et al. [61] propose a distance-based formulation for planar closed chains with revolute joints, where a configuration is represented by the set of distances from a point on the base link to the end-points of the rest of links, complemented with a set of triangle orientations. In this parameter space, the loop closure constraints become linear inequalities, and the configuration space becomes practically piecewise convex. The boundaries between the different convex regions of this space are given by particular alignments of points, i.e., changes in the orientation of the triangles. Thus, to connect any two given configurations, one need only identify the boundaries to be crossed and define a piecewise linear path between them.

The approaches by Trinkle and Milgram [60] and by Han et al. [61] can both be generalized to spatial mechanisms with spherical joints. However, no general joints or joint limits can be encoded in these approaches, which hinders their general applicability.

Summary

Distance-based formulations provide new insights into fundamental problems. In the context of algebraic methods, the number of variable eliminations is reduced to the point that, in many cases of practical interest, they are no longer required. Numerical methods also benefit from the use of distance formulations, since trilateration approaches naturally follow from them, and they provide a rich set of tools to reduce the search space in the context of branch-and-prune approaches. Moreover, distance formulations allow one to define singularity-invariant leg rearrangements, which significantly expand the utility of previous singularity characterizations in Gough-Stewart platforms. This enables the construction of active structures with well-characterized singularity loci and without complex mechanical pieces such as double spherical joints. Finally, in many cases, the analysis of the configuration spaces and the associated path planning problems can be certainly simplified when adopting a distance-based formulation.

Despite the extensive history of distance geometry, it has only recently been introduced for the position and motion analysis of structures. Given the success of the cases presented in this survey, we expect that distance-based formulations and their associated tools will soon be recognized as a powerful alternative for addressing the complex geometric problems arising in this discipline. Finally, the distance geometry methods described in this survey are able to solve a wide variety of problems, including trilaterable and non-trilaterable cases, linkages with a coupling number higher than 1, and even flexible graphs. We expect that these methods will find application in domains other than those presented here.

Acknowledgements This work has been partially supported by the Spanish Ministry of Economy and Competitiveness under project DPI2014-57220-C2-2-P.

References

1. Miura K (1984) Variable geometry truss concept. Technical report 614, The Institute of Space and Astronautical Science
2. Hughes PC, Sincarsin WC, Carroll KA (1991) Trussarm–a variable-geometry-truss manipulator. J Intell Mat Syst Struct 2(2):148–160
3. Chirikjian GS, Burdick JW (1994) A hyper-redundant manipulator. IEEE Robot Autom Mag 1(4):22–29
4. Sultan C, Corless M, Skelton RE (2000) Tensegrity flight simulator. J Guid Control Dyn 26(6):1055–1064
5. Dadone P, Lacarbonara W, Nayfeh AH, Vanlandingham HF (2003) Payload pendulation reduction using a variable-geometry-truss architecture with LQR and fuzzy controls. J Vib Control 9(7):805–837
6. Stoughton RS, Tucker JC (1995) A variable geometry truss manipulator for positioning large payloads. In: American Nuclear Society meeting on robotics and remote systems
7. Finistauri AD, Fengfeng X (2009) Type synthesis and kinematics of a modular variable geometry truss mechanism for aircraft wing morphing. In: International conference on reconfigurable mechanisms and robots, pp 478–485

8. Miura K, Furuya H, Suzuki K (1985) Variable geometry truss and its application to deployable truss and space crane arm. Acta Astronaut 12(7):599–607
9. Kurita K, Inoue F, Furuya N, Shiokawa T, Natori M (2001) Development of adaptive roof structure by variable geometry truss. In: International symposium on automation and robotics in construction, pp. 1–6
10. Denavit J, Hartenberg R (1955) A kinematic notation for lower-pair mechanisms based on matrices. Trans ASME J Appl Mech 23:215–221
11. Porta JM, Ros L, Thomas F (2005) On the trilaterable six-degree-of-freedom parallel and serial manipulators. In: IEEE international conference on robotics and automation, pp 960–967
12. Rojas N, Thomas F (2013) The univariate closure conditions of all fully-parallel planar robots derived from a single polynomial. IEEE Trans Robot 29(3):758–765
13. Rojas N, Thomas F (2013) The closure condition of the double banana and its application to robot position analysis. In: IEEE international conference on robotics and automation, pp 4641–4646
14. Manocha D, Canny J (1994) Efficient inverse kinematics for general 6R manipulators. IEEE Trans Robot Autom 10:648–657
15. Merlet JP (2000) Parallel robots. Springer
16. Guest S (1994) Deployable structures: concepts and analysis. PhD thesis, Cambridge University
17. Rosales C, Porta JM, Suárez R, Ros L (2008) Finding all valid hand configurations for a given precision grasp. In: IEEE International conference on robotics and automation, pp 1634–1640
18. Rodríguez A, Basañez L, Celaya E, (2008) A relational positioning methodology for robot task specification and execution. IEEE Trans Robot 24(3):600–611
19. Porta JM (2005) CuikSLAM: a kinematics-based approach to SLAM. In: IEEE international conference on robotics and automation, pp 2436–2442
20. García de Jalón J, Bayo E (1993) Kinematic and dynamic simulation of multibody systems. Springer
21. Bettig B, Hoffmann CM (2011) Geometric constraint solving in parametric computer-aided design. ASME J Comput Info Sci Eng 11:021001
22. Wedemeyer WJ, Scheraga H (1999) Exact analytical loop closure in proteins using polynomial equations. J Comput Chem 20(8):819–844
23. Cox D, Little J, O'Shea D (1997) An introduction to computational algebraic geometry and commutative algebra, 2nd edn. Springer
24. Raghavan M (1993) The Stewart platform of general geometry has 40 configurations. ASME J Mech Des 115:277–282
25. Rojas N (2012) Distance-based formulations for the position analysis of kinematic chains. PhD thesis, Institut de Robòtica i Informàtica Industrial
26. Wohlhart K (2009) Position analyses of open normal Assur groups A(3.6). In: ASME/IFToMM Int Conf Reconfig Mech Robot, pp 88–94
27. Rojas N, Thomas F (2013) Application of distance geometry to tracing coupler curves of pin-jointed linkages. ASME J Mech Robot 5(2):021001
28. Porta JM, Thomas F (2017) Closed-form position analysis of variable geometry trusses. Mech Mach Theory 109: 14–21
29. Blumenthal LM (1953) Theory and applications of distance geometry. Oxford University Press
30. Porta JM, Ros L, Thomas F (2005) Inverse kinematics by distance matrix completion. In: International workshop on computational kinematics
31. Lavor C, Liberti L, Maculan N (2006) The discretizable molecular distance geometry problem. Technical report
32. Liberti L, Lavor C (2013) On a relationship between graph realizability and distance matrix completion. In: Migdalas A (ed.) Optimization theory, decision making, and operations research applications, vol 31. Springer, pp 39–48
33. Porta JM, Ros L, Thomas F, Torras C (2002) Solving multi-loop linkages by iterating 2D clippings. In: Thomas F, Lenarcic J (eds.) Advances in robot kinematics. Kluwer Academic Publishers, pp 255–264

34. Porta JM, Ros L, Thomas F, Torras C (2003) A branch-and-prune algorithm for solving systems of distance constraints. In: IEEE international conference on robotics and automation, pp 342–348
35. Porta JM, Ros L, Thomas F, Torras C (2005) A branch-and-prune solver for distance constraints. IEEE Trans Robot 21(2):176–187
36. Crippen G, Havel TF (1998) Distance geometry and molecular conformation. Research Studies Press
37. Rikun AD (1997) A convex envelope formula for multilinear functions. J Glob Optim 10:425–437
38. Ting Y, Yu-Shin YC, Jar HC (2004) Modeling and control for a Gough-Stewart platform CNC machine. Int J Robot Syst 21(11):609–623
39. Cappel KL, Marlton N (1967) Motion simulator. U.S. patent 32 95 224
40. Su Y, Duan B, Nan R, Peng B (2003) Mechatronics design of stiffness enhancement of the feed supporting system for the square-kilometer array. IEEE/ASME Tranactions on Mechatronics 8(4):425–430
41. Rojas N, Borràs J, Thomas F (2012) The octahedral manipulator revisited. In: IEEE international conference on robotics and automation, pp 2293–2298
42. Porta JM, Ros L, Thomas F, Corcho F, Cantó J, Pérez JJ (2007) Complete maps of molecular-loop conformational spaces. J Comput Chem 28(13):2170–2189
43. Thomas F (2004) Solving geometric constraints by iterative projections and back projections. In: International conference on robotics and automation, pp 1789–1795
44. Alefeld G, Herzberger J (1983) Introduction to interval computations. Academic Press, Orlando, Florida
45. Porta JM, Thomas F Sensor localization from distance and orientation constraints
46. Thomas F (2014) Computing cusps of 3R robots using distance geometry. In: International symposium on advances in robot kinematics
47. Rull A, Porta JM, Thomas F (2014) Distance bound smoothing under orientation constraints. In: IEEE international conference on robotics and automation, pp 1431–1436
48. Thomas F (1995) An approach to the movers' problem that combines oriented matroid theory and algebraic geometry. IEEE Int Conf Robot Autom 3:2285–2293
49. Havel T (1995) Distance geometry, pp 1701–1710. Wiley, New York
50. Bohigas O, Zlatanov D, Ros L, Manubens M, Porta JM (2015) A general method for the numerical computation of manipulator singularity sets. IEEE Trans Robot 30(2):340–351
51. Borràs J (2011) Singularity-invariant leg rearrangements on Stewart-Gough platforms. PhD thesis, Institut de Robòtica i Informàtica Industrial
52. Borràs J, Thomas F, Torras C (2010) Singularity-invariant leg rearrangements in doubly-planar Stewart-Gough platforms. In: Robotics science and systems
53. Choset H, Lynch K, Hutchinson S, Kantor G, Burgard W, Kavraki L, Thrun S (2005) Principles of robot motion: theory, algorithms, and implementations. MIT Press
54. LaValle SM (2006) Planning algorithms. Cambridge University Press, New York
55. Bohigas O, Henderson ME, Ros L, Manubens M, Porta JM (2013) Planning singularity-free paths on closed-chain manipulators. IEEE Trans Robot 29(4):888–898
56. Lavalle SM (2011) Motion planning. Part I: the essentials. IEEE Robot Autom Mag 18(1):79–89
57. Siméon T, Laumond JP, Cortés J, Sahbani A (2004) Manipulation planning with probabilistic roadmaps. Int J Robot Res 23(7–8):729–746
58. Rosales C, Porta JM, Ros L (2013) Grasp optimization under specific contact constraints. IEEE Trans Robot 29(3):746–757
59. Ballantyne G, Moll F (2003) The da Vinci telerobotic surgical system: virtual operative field and telepresence surgery. Surg Clin North Am 83(6):1293–1304
60. Trinkle JC, Milgram RJ (2001) Motion planning for planar n-bar mechanisms with revolute joints. IEEE/RSJ Int Conf Intell Robot Syst 3:1602–1608
61. Han L, Rudolph L, Blumenthal J, Valodzin I (2008) Stratified deformation space and path planning for a planar closed chain with revolute joints. In: Akella S, Amato NM, Huang WH, Mishra B (eds.) Algorithmic foundation of robotics VII, Springer tracts in advanced robotics, vol 47. Springer, pp 235–250

New Trends in Residential Automation

José Reinaldo Silva, Javier Martinez Silva, Celina Pereira, Camelia Avram and Sergiu Dan-Stan

Abstract Home automation is a promising market application and a very interesting area of research in the field of automation, where human–machine integration is a key issue. The history of home automation, however, can be traced through various phases since its inception: first, with concepts and equipment inherited from industry automation, then to a phase in which new tools were developed but problems were encountered with design, and finally to a new era where the challenge is to further explore a multidisciplinary approach that conceives of home automation as a service, with a different model for the interaction between the home and its users. In this new phase, the design challenges are quite different, as is the motivation for implementing new architectures for building and residential automation (B&RA). In this article we present a direct comparison between the last two phases of home automation, as described above, and propose a new methodology for designing B&RA systems based on an anthropocentric automation approach. A design discipline for improving integration is applied to models developed at the beginning of this century, achieving good results. This same model is then compared with a new approach, revealing new challenges that could arise in both integration and service. To better understand all of the differences, the older approach is presented along with a case study implemented in Portugal. It

J. R. Silva (✉) · J. M. Silva
Polytechnic School, Mechatronics Department, University of São Paulo,
São Paulo, Brazil
e-mail: reinaldo@usp.br

C. Pereira
Mechanical Engineering Department, University of Minho, Braga, Portugal

C. Avram
Faculty of Automation and Computer Science, Technical University of Cluj-Napoca,
Cluj-Napoca, Romania

S. Dan-Stan
Faculty of Mechanical Engineering, Technical University of Cluj-Napoca, Cluj-Napoca,
Romania

E. Ottaviano et al. (eds.), *Mechatronics for Cultural Heritage and Civil Engineering*,
Intelligent Systems, Control and Automation: Science and Engineering 92,
https://doi.org/10.1007/978-3-319-68646-2_6

introduces new concepts and challenges for B&RAs in implementing sustainable houses and buildings that are also capable of generating energy—as in the urban smart grid. Another important challenge is improving integration with users in terms of service systems.

Introduction

Home automation first became possible in the late 1970s, when personal computers (PCs) started entering home spaces. Yet the first significant concepts and applications related to HA did not appear until the period between 1998 and the early 2000s, in association with the notion of the smart home. Domestic technologies, home networking, and other gadgets then began to appear on store shelves [1].

Various ideas can be linked to the home automation domain. In fact, a smart home system can integrate electrical devices and techniques for building smart systems and controlling domestic applications such as electrical equipment and home entertainment systems. Such systems enable the automation of a building by managing lighting control, power control, home tele-monitoring, home networks, and other applications. Smart home devices can be connected via a computer network so that they can be controlled using a system such as a PC. These devices can also facilitate remote access using the Internet. Integration of information technology with home systems provides safety benefits and energy efficiency, as well as other conveniences.

Taking into account the high potential of residential automation, its diffusion due to its "youth" or its high-tech nature is reflected in its life cycle from technology adoption to large-scale acquisition and use [2].

If the meaning of building and residential automation (B&RA) could be captured in a single word, that word would be "integration". Integration denotes the flow of information among different subsystems involved in the sharing of control and the respective actions that characterize the system [3]. A simple way to understand the importance of integration is by examining the information system adopted and the possible failures and flaws associated with this system. As mentioned in the Standish CHAOS Report [4–7], there is a tendency toward an increased number of projects that present a series of gaps, largely because of failure to consider and to involve the final user in the tasks of designing the B&RA system. As a result, specifications and requirements are incomplete or misunderstood, and parts of them change during the process of design and implementation.

Considerable work must be done in this domain, and many projects involving B&RA systems have been developed. In efforts to attract the attention of potential users, some of these have demonstrated the advantages of using B&RA systems, primarily related to enhancing comfort and quality of life, reducing energy consumption, and increasing environmental sustainability.

One of these projects, for example, the *TRON Intelligent House* in Japan [8], was a pioneering work in this domain. This project was mainly devoted to providing increased comfort, and in fact obtained amazing results from this point of view.

The definition of "comfort" (initially proposed as an aspect of B&RA) has been extended to other concepts in this domain including security, energy efficiency, and sensitive aspects related to geographical and environmental factors. This can be observed in the *House of the Future* in Ireland [9], *e2Home* in Denmark and Sweden [10], the *MIT House* in the USA (MIT 2000), or *the interactive house of the future* in Portugal [11].

A more detailed study in the context of this work was conducted in Brazil and Portugal, which examined the respective issues related to the use and design of B&RA systems in each country. Indeed, there are huge differences between Portuguese and Brazilian realities and contexts: Portugal is a small country with 10 million inhabitants, with a small market, where families find it difficult to acquire B&RA systems, and with weather that is neither too cold nor too hot the entire year. Here, families typically acquire B&RA systems for comfort reasons (temperature and light control, music inside the house). Brazil, on the other hand, is a huge country of more than 200 million. The factors motivating families to acquire B&RA systems are largely related to security, and also reflect the fast-paced development of private neighbourhoods (e.g., 800–1000 houses), with a few crucial facilities consisting of water distribution networks, for instance, or other relevant features of integrated systems for common purposes. Thus, this type of device must be designed for and adapted to a very different reality from that of Portugal.

Even in such different contexts, B&RA systems are crucial for the well-being of families and individuals, and any B&RA system can be designed (using universal and systematic methodologies) for specific purposes (considering different needs) of specific clients. This is chiefly why systematic and flexible design methodologies are so important, and must consider all aspects that are relevant for users, including security, comfort, and energy efficiency [12].

Indeed, energy efficiency has become a very important consideration in designing and developing B&RA systems. In fact, a medium-sized family in the European Union (EU) can save between €200 and €1000 per year, depending on factors such as weather or geographic location), using an adequate B&RA system. Based on these numbers, it is estimated that the EU could save about 20% in energy consumption each year—about €60 billion—with appropriate policies governing the use of B&RA systems.

In 2001, the European Commission (EC) initiated and adopted the Directive on the Energy Performance of Buildings (EPBD), Directive 2002/91/CE, whose main goal was (from Directive 2002/91/EC published by Official Journal of European Community):

Article 1/4.1.2003: "*The objective of the Directive is to promote an improvement in the energy performance of buildings in the Community, taking into account outdoor climatic and local conditions, as well as indoor climate requirements cost-efectiveness.*"[1]

According to data provided by Hager [13], implementation of a series of B&RA systems resulted in savings of up to 50% of energy used for heating and up to 40% used for lighting.

In order to improve the efficiency of a B&RA system, in this chapter we propose a design methodology that considers an anthropocentric automation approach. Next, Sect. Design Process and Requirements for Buildings and Residential Automation Systems highlights some aspects to be considered in designing B&RA systems, and from a practical point of view discusses a project in Portugal where these concepts have been used. Section New Challenges and the Detachment from Industry presents the key concepts related to new challenges and the break from industry in the design of B&RA systems. Section Towards a Design Approach to Home Automation: The Requirements Problem discusses key points to consider in a design approach for home automation systems, mainly related to the requirements of the problem. Section A New Proposal for Home Automation Based on Goal-Oriented Requirements then discusses possibilities and solutions for dealing with the requirements of this new modelling approach and presents a case study. Lastly, this chapter summarizes the key conclusions from this work.

Design Process and Requirements for Buildings and Residential Automation Systems

In designing a B&RA system, there are important factors to consider: geographical location is a key consideration, and another important issue is handling of information and its integration within the B&RA system [3]. Integration involves the creation of an information system in which information that flows among B&RA subsystems and information on specific actions of the system are integrated and shared in a simple and efficient manner.

The Standish Group [7] has long been involved in studying the impact of design on both successful and unsuccessful projects related to information handling and management in the USA. The results of their work and its evolution through the years are reproduced in the *CHAOS Report* [4–7].

In the first *CHAOS Report* in 1994, it was noted that only 16% of analysed projects globally were able to be finished within their respective previously defined deadlines and under the previously imposed budget limits, even if the requirements implemented were according to the requirements that had been previously defined.

[1]Directive 2001/91/EC from the European Parliament Council of 16 of December of 2002 on energy performance of buildings. Published in the Journal of European Communities, vol. 46, January 4, 2003.

An analysis of other results (from 2000, 2006, and 2009) shows that these percentages increased to 28% in 2000 (*Extreme CHAOS,* in 2000), 34% in 2002, 35% in 2006, and about 32% in 2009 (*CHAOS Summary, 2009*).

According to the same organization, the main obstacles to meeting previously defined deadlines and budgets were a failure to include the perspective of the user in the design, incomplete requirements and specifications, and changes to requirements and specifications during the design phase—and sometimes even during the implementation of the final technological solution [5].

All these aspects are worrisome for users, especially those concerning changing and/or incomplete requirements. Friedrich [14] also indicated this phenomenon in his work, and attributed it to lack of knowledge elicited from the user. He described a methodology for eliciting the user's tacit knowledge and concluded that the problem of absence of tacit knowledge from users was a major reason for unmet deadlines and/or budgets for this kind of project. This issue may also account for failed or incomplete implementation of certain functions defined for the final system.

The Standish Group statistics illustrate results obtained in the domain of software development. However, these results are valid for B&RA systems as well, as they require software integration and development. Statistics show that design failures are mainly attributable to errors that occur in the early stages of the design process.

Integration plays a major role in B&RA systems. The correct flow of (integrated) information defines B&RA as an intelligent system. For this reason, the intelligence level of a B&RA system is directly related to the level of integration of the different subsystems, as well as the capacity for providing a high return on the initial investment [15]. In addition, the concept of "intelligent" building is not related as much to "last-generation technology", but more to the performance achieved by B&RA in use scenarios [16].

In order to reduce costs when designing B&RA systems, both uniform control systems and sharing of resources are recommended. The design process of a B&RA system starts with identifying client needs and ends when all documentation concerning the development, realization, and installation of the system has been prepared.

Requirements for B&RA Systems

Requirements are the basis for defining all functions, services, and operational restrictions of the B&RA system's behaviour. Well-defined requirements are the main determinant of the success of the design tasks to be developed: ambiguities are reduced or eliminated, and the definition of specifications can be more comprehensive and more realistic. Generally, requirements are written in natural language and complemented with models or tables, and their role is to ensure the correct characterization of the system to be developed.

User requirements and system requirements have different levels of detail: requirements for users are more abstract, and system requirements are more specific. Both sets of requirements are useful in defining the technical specifications for the system.

Defining requirements at different levels of abstraction is a very interesting approach, because users and B&RA developers will see them from a different perspective: users are interested in the behaviour of the system as they perceive it, whereas systems developers are more interested in the true behaviour of the system in different application scenarios.

Three types of requirements—functional, non-functional, and domain requirements—are presented below.

Functional Requirements

These describe the services provided by the B&RA system, and the reaction and behaviour of the system in response to inputs in different behaviour scenarios. Functional requirements must identify both the desired and the undesired behaviour of the system, as well as characteristics of the system's behaviour properties.

Non-Functional Requirements

These requirements represent restrictions on the services or functions provided by the B&RA system, and include design, organizational, and external requirements. This set of requirements typically imposes restrictions on the development of the B&RA system, and sometimes on the design methodology that must be used.

Application Domain Requirements

These requirements are related to real applications and real system performance rather than specific user needs, and they primarily involve attributes of and restrictions on the system behaviour. This set may contain both functional and non-functional requirements, and they are commonly characterized by specific terminology or references for the concept of application domain.

Requirements Concepts

In reality, the distinction among types of requirements is not as well-defined as these definitions may suggest. For instance, a requirement related to safety as defined by the user seems at first glance to be a non-functional requirement, but in fact, when detailed, it can be decomposed into other requirements that can be functional. An illustration for improving safety, for instance, is the need for authenticating the user of the B&RA system [17].

Based on requirements in the field of engineering, it is assumed that the process of definition consists in elicitation, analysis, and documentation of the system. This complex process is based on methods, tools, different viewpoints, human factors, and information sources, some of which relate to specific scientific and technological domains, such as designers, specialists, stakeholders, and normal users.

From a high-level abstraction point of view, the global requirements engineering process can be divided into three sub-processes: elicitation (extraction of knowledge from several sources), specification (conversion of requirements to technological needs), and validation (verification that the requirements meet the needs and desires of the users).

A series of methodologies have been developed to correctly and precisely define these requirements. Among them, the method defined in *Volere Requirements Resources* (Robertson and Robertson, 1999) is well suited for these purposes. This methodology underlines the definition of the requirements and is the critical point for the success of the process design of a B&RA system.

Another methodology was developed in 2004, in which some of the authors of this chapter proposed a similar architecture based on a generic element called "integron" [18].

In 2009, some of the authors of this chapter revisited the same point with a new version (at that time) of the same element, focusing on the arrangement of the system as a whole and on the flow of information instead of simply distribution of control. Such an approach was proposed to be more efficient than others, and would better fit the requirements in a heterogeneous environment such as residential automation. This methodology developed in 2009 was applied in a real case study in Portugal by the same team. A detailed description of the application of the methodology, as well as a scientific explanation of the reason for choosing that method for the treatment of the case study, can be found in [3]. This case study is considered for the presentation and justification of the concepts discussed below, towards a new paradigm and new challenges in designing B&RA systems.

New Challenges and the Detachment from Industry

Although we can recount several developments in residential automation, we entered the twenty-first century with domotic models that were not integrated with humans, most of them based on the old concept of "external machines" to which we should give orders. A domotic intelligent system that could evaluate the current situation and react to it in order to provide more comfort and safety to the user was a concept that had not yet been considered.

Since the early 2000s, the real question has been whether we have experienced a paradigm shift marking the complete separation of domotics from industrial automation, with the development of new methods and equipment tailored to B&RA. Such a change would be based on the concept of human-centred automation [19]—which was not really new at the beginning of this century. In fact, in 1996, Gomes and Steiger-Garção published a paper [20] directly questioning the proper approach to domotics: what they called "automatics" was directly related to the known (industrial) automation, and contrasted with an anthropocentric approach, where the human (user) was the focus of the (design) process. This dilemma became untenable in the new century, and the challenge since then has been to effect a departure from the industrial process and approach B&RA from a new perspective within the domain of anthropocentric automation.

An Anthropocentric Approach to Domotics

The unique challenge that B&RA brings to the scene is the role of modelling and design in this area. Previously, the role of design was merely to support an instrumentation process—more than digitalization or virtualization. Therefore, further changes to the project would imply a rework or a sharp division into small environments, at times resulting in redundancy of functionality, equipment, or both. A house—more than a building—requires the assembling of different small environments automated separately or simply instrumented to attend to prescriptive functionalities.

A change in this approach would start with the introduction of the concept of "instrument-mediated activity" [21], which focuses more on the action performed than the use of an (automated) instrument itself. Rabardel and Beguin [21], who are critical of classical design, define an instrument as a combination of an artefact and a subject, where it is "…the subject who grants it the status of a means for his/her action".

Using a more systemic approach, we would say that B&RA is based on integrated actions involving humans (users) and machines (artefacts), in which the latter are simply resources that require human interaction to choose a subject and contribute to the action. A new B&RA design may involve the modelling of processes (partially ordered sets of actions) to achieve a goal that would please the user.

Goal orientation is a novel feature in the design of B&RAs, and thus it must be very clearly defined. Aside from its obvious meaning as an objective to be attained, we should expand the meaning of "what would please the user" beyond trivial comfort. In fact, one of the challenges of B&RA in the new century is its utility and sustainability.

Another significant challenge is understanding the nature of B&RA as a system of systems, which we develop in the next section.

Towards a Systemic Approach to Building and Residential Automation

Another important point in B&RA design is that building and home automation is now a system of systems [22]. According to the International Council on Systems Engineering (INCOSE),[2] a "System of Systems (SoS) is a general system (or System of Interest) whose elements are managerially and/or operationally

---

[2]INCOSE is an international institution that promotes reference models and studies in systems engineering for academia and the practitioners. The institution has established as its main objective the transformation of systems engineering to a model-based discipline. More information can be found at: www.incose.org.

independent systems. These interoperating and integrated collections of constituent systems usually generate results unachievable by individual systems alone".

Thus, a system of systems is in fact a collection of independent systems, where each component encapsulates its goal, appropriate resources, and methods needed to accomplish it. This general architecture appears in modern buildings and residences, where each unit is now a small system belonging, for example, to a general urban electric utility distributed system. Also, all modern buildings and homes should be integrated within a neighbourhood system, which belongs to the city system. A similar configuration can be designed for rural areas.

Therefore, a new challenge in B&RA design is developing buildings and homes seen as units of larger systems, and the integration of such units within these larger systems is as important as attending to direct user requirements, as proposed in previous sections.

With regard to the internal components, it is also advisable to treat the components of a B&RA system as subsystems. This was proposed in a Portuguese house designed by some of the authors of this chapter, where a B&RA was defined as comprising a security system, a lighting system, a water system, and so on. In fact, if we further consider the complete integration with outside systems, dividing it into even smaller systems is warranted. An important point is that a building or a house is no longer merely a first-level system—one that simply receives "resources" from outside systems; nowadays it is hypothetically involved in generating energy, or contributing with small reserves of water to emergency situations. Thus, it is not a one-way relationship, where resources and services converge at the home or building as a final destination. Modern domotic systems can provide energy, as in an urban smart grid, contribute to a joint security program, or integrate a health-care network [23–26].

Home automation can be implemented in a set of connected houses and buildings [27], and should comprise an intelligent system of systems where resources and services are received and provided at the same time. This aligns with the concept of service proposed by service science [28].

The notion of service is proposed here as a sequence of partially ordered actions, where the last action is coupled with the end user with whom it is exchanging information and resources. The sequence of actions and resources can be delivered by an automated system (or a system of systems) [29].

Therefore, we can think about B&RA as a service system of systems, where services are exchanged between the environment and the building and residential units. The challenge is in how to design this new arrangement.

Towards a Design Approach to Home Automation: The Requirements Problem

Both challenges outlined in the last section point to a new architecture for B&RA that is a metaphor of a proposal made by some of the authors for manufacturing and industrial systems [30–32], and could also be applied to Industry 4.0 [33].

In order to explore this trend as inspiration for a new design discipline for B&RA, we will first analyse the metaphorical similarity between B&RA and manufacturing design, starting with the early phase. Differences from the classical approach based on functionality will be highlighted to open a path for proposing a new goal-oriented method.

The Architecture Metaphor

Concepts such as the Internet of Things (IoT), e-work, and collaboration have changed the way we look at distributed design, especially (automated) systems design [31, 34, 35]. Today, components of large systems as well as small parts can be encapsulated in local systems or distributed globally. The impact of this in B&RA was already discussed in the previous section, where the system-of-systems concept was introduced.

Thus, B&RA should be viewed as a configuration of cooperative cyber-physical systems with common goals, which can be coupled in different arrangements, depending on availability and priority for achieving the social goal.[3] This is substantially different from the Unified Modelling Language (UML) architecture commonly used in residential automation, which can be viewed as a set of distributed instruments or clusters of instruments (software included) used to perform a task such as intrusion detection, or locating areas where lights are on and there is no human presence. The enhancement introduced by some of the authors of this chapter in previous B&RA projects was just a means of aggregating these components, and eventually integrating different instrumental clusters using the data that they generated—the detection of intrusion, for instance. In this case, both the security and lighting systems could also use this data. By the end of the last century, the novelty was accomplishing this integration via software.

The current novelty is the conceiving of an architecture where components of a B&RA are independent cyber-physical systems [27] which act as distributed intelligent agents [36]. But this is just a metaphor of the architecture introduced in manufacturing with Industry 4.0 [33]. Instead of a prescriptive configuration of instrumental clusters, a central control would select/take control using the criteria of capacity and availability of independent (cyber-physical) subsystems. System societies can provide data and/or services to one an other through a main service that is delivered to the final user.

Product-Service Architecture (PSA) was proposed by Silva and Nof [31] and developed as a flexible and harmonic architecture to support a design discipline [33] that will be adapted to B&RA.

---

[3]Social goal has the same meaning as in systems engineering, where it denotes a common goal to which all system components contribute.

PSA and the New Design Discipline for B&RA

Silva and Nof [31] proposed a distributed architecture for manufacturing in which independent elements together would constitute a manufacturing service environment (MaSE). A MaSE is defined as an open set of value production units—independent components—that contribute to the general processes involved in the delivery of products, services, or hybrid products/services. Production units would now be matched to independent domotic units (DomUs).

DomUS are designed to contribute to various processes controlling systems such as lighting, intrusion, electronic security, external area surveillance, internal area surveillance, energy use, energy production (for systems inserted in urban smart grids), and the rational use of water.

Both the MaSE and the domotic service environment (DoSE) generate architectures that emerge from the collaborative e-work between product units [37], where other products (besides the final solution), services, product/services, and resources are shared.

In Fig. 1, each domotic unit is denoted by du_i, $i = 1 \ldots n$, with each contributing to goals s_i and sharing resources. A sequence of achieved goals constitutes a plan that leads the domotic system to deliver resources and the final goal s_n to the user.

Thus, DomUs have the same architecture as that proposed by Dutra and Silva [33] for Industry 4.0, and retain features already described in previous sections for B&RA.

It should be stressed that we are now talking about service environments, which represent the second challenge. B&RA is concerned with the development of automated service, just like modern manufacturing. The point here, however, is that we cannot simply transfer solutions from manufacturing to domotics; rather, we

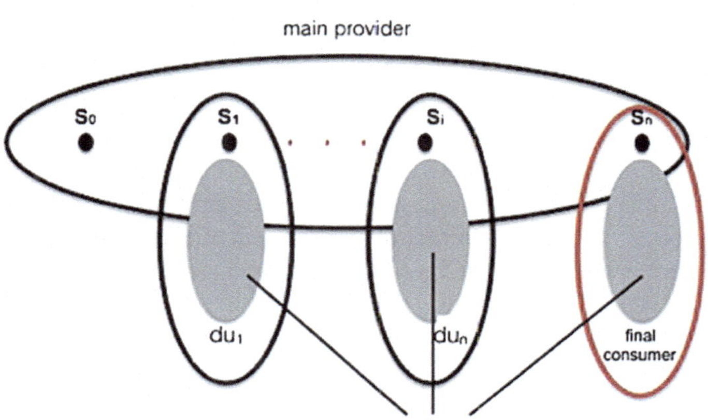

Fig. 1 DomUS system configuration for providing the a service provided by an arrangement of service systems

must recognize that both are systems of the same nature in terms of the architecture for automation. Therefore, both face similar challenges and require similar solutions.

DomUs should be instantiated with cyber-physical systems with good capacity for communication, directly or using Ethernet. The old dilemma of the uncertainty in using Ethernet is no longer a problem, and current domotic environments are fully connected by wireless networks [27, 38, 39].

On the other hand, this same characteristic of domotic systems would give rise to another problem in the design of efficient solutions that can be coupled with the end user.

The most sensitive part of the design process is simply the problem of modelling and analysing requirements. In the next section, we will address this problem and introduce the use of goal-oriented requirements engineering instead of the classical functional approach based in UML.

A New Proposal for Home Automation Based on Goal-Oriented Requirements

The beginning of the new century was marked by the emergence of new ideas in software engineering and engineering design: a new approach was proposed for eliciting, representing, modelling, and analysing requirements that was focused on goals rather than functional aspects. This also represented a good alternative for the design of B&RA, which could avoid several problems. Various techniques for formalizing requirements have been used in the context of developing critical software for industrial systems as presented in the works of Machado et al. [40], Campos et al. [41], Kunz et al. [42], and Machado et al. [43].

First, we would like to stress once more that the functional approach described at the beginning of this chapter and that was used previously has a touch of "instrumentation" and therefore contributes to preserving the product-oriented approach in B&RA. However, from what we have seen in previous sections, the challenge is in designing B&RA as a service, and even more, as a distributed architecture of independent DomUs, integrated into the main process.

It is also difficult to keep track of resource sharing using a conventional functional approach. In what follows, we will present the concept of goal-oriented requirements engineering (GORE), where requirements are elicited already attached to goals [44, 45].

In the next section we will briefly show how we could revisit the example mentioned in the first section using KAOS[4] and an academic tool developed in

[4]KAOS is an acronym that stands for Knowledge Acquisition in Automated Specification and refers to a diagrammatic schema for specification of goal-oriented requirements. A commercial

D-Lab (Brazil), where all diagrams (goal, object, operations, and responsibility diagrams) can be translated to Petri Nets. Once this is done, the task of requirement modelling and verification/validation becomes straightforward.

Case Study with Goal-Oriented Approach

The KAOS approach is a goal-oriented implementation of the GORE method, which involves a rich set of formal analysis techniques based on linear temporal logic (LTL). GORE is a sub-area of requirements engineering (RE), which addresses the use of goals for elicitation, elaboration, structuring, specification, analysis, negotiation, documentation, and requirements modification [46]. The goal of GORE methods is to justify the "why", "what", and "who" dimensions, and to ensure that requirements meet the system objectives—including those of safety and security [44].

In order to more clearly define the KAOS method, we would like to highlight first the notion of "goal-driven" [45, 47]. A goal is a prescriptive state declaring the purpose of an existing or future system whose satisfaction generally arises from the collaboration between agents with a certain responsibility over the system.[5] Therefore, goals should drive the requirements elicitation process, resulting in domain-specific requirements, avoiding the difficult balance between functional and non-functional requirements.

The KAOS method provides specifications based on formal logic—LTL—and graphical modelling representations to model requirements in terms of goals. These goals express the needs and viewpoints of stakeholders, which can eventually be managed to achieve consistent models. Thus, KAOS can be described as a multi-paradigm framework combining different levels of reasoning: semi-formal reasoning for modelling and structuring goals, and formal reasoning based on linear time logic formalism [44].

Graphically, goals are represented in the KAOS diagram by parallelograms, while requirements correspond to borders that are outlined in bold. Agents are represented by hexagons, as shown in Fig. 2. Here, we revisit the same domotic problem as at the beginning of the chapter, in the treatment of requirement goals for an automation project for a residence in Famalicão, Portugal.

A goal can be classified according to its behaviour, covering a maximal set of intended behaviours, in a declarative and implicit policy. Behaviour is defined as a sequence of events in a transition system model. Thus, a general behaviour violates

tool called Objectiv*er* was used to construct the diagrams. Here we will use a system that can transform requirements into Petri Nets also developed for some of the authors.

[5]It is important to keep in mind the match between this definition and the definition of service and PSA architecture, where partial actions are related to (caused by or the responsibility of) certain agents but contribute to the process as a whole, and will deliver a final action to the end user which will generate value.

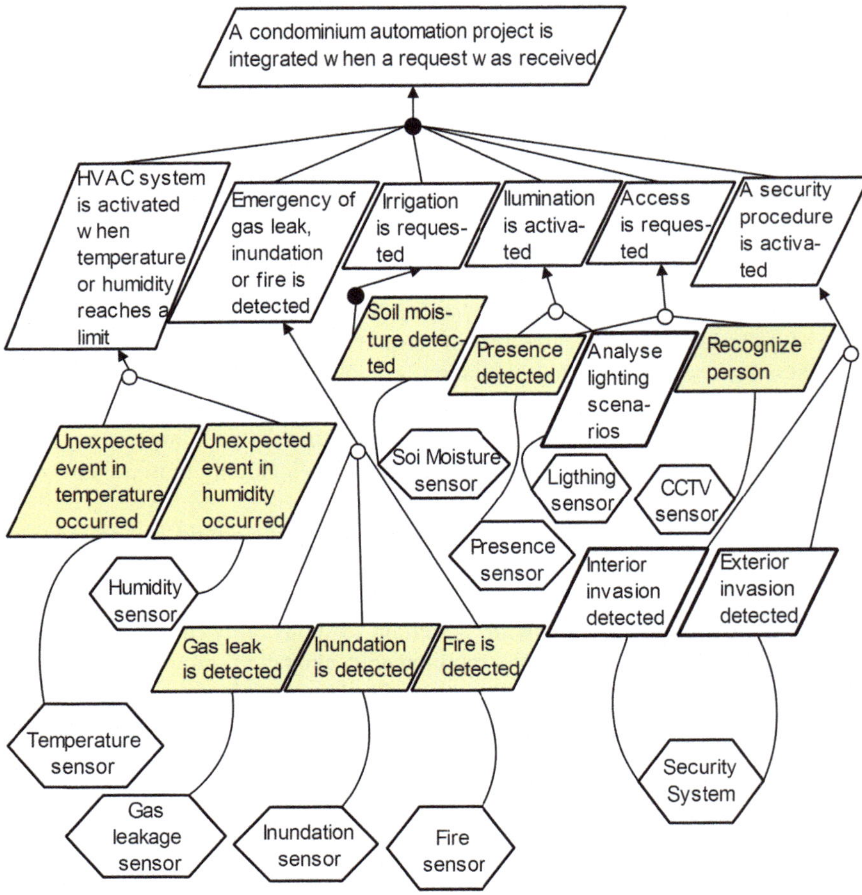

Fig. 2 Goal diagram derived for the main goal: *a condominium automation project is integrated when a request is received*

the goal specification if it does not belong to any process prescribed in the requirements.

Figure 2 shows a goal that requires integration of the sensor signal to avoid equipment redundancy. Therefore, presence sensors are used by the lighting, security, and fire systems. In addition to the heating, ventilation, and air conditioning (HVAC) system (not integrated with the sensors), the house also needed a security system for gas leaks, inundation, and soil moisture, and a closed-circuit television (CCTV) system. Note that the control system responds to sensor detection in an integrated way, coupling action against fire, for instance, with detection of human presence and inundation.

A refinement terminates when each goal becomes realizable by an individual agent (people, other software devices, robots, etc.), which means that this goal must

Fig. 3 KAOS basic elements

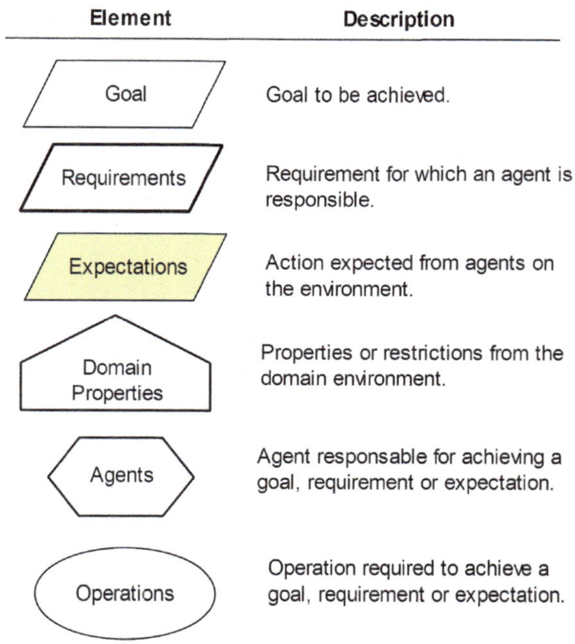

be expressible in terms of conditions that are monitored and controlled by the agent. "Parent goals" are obtained by abstraction, whereas "child goals" are obtained by refinement. Essentially, refinement is the process that results in a tree where leaf goals are assignable to single system agents.

A formal representation of a behavioural goal is defined using a temporal logic instruction:

$$C = \theta T$$

Here, C is the current condition, T is the target condition, and θ is any of the LTL operators: (Fig. 3)

- $\bigcirc$: In the next state in which T is satisfied.
- $\Diamond$: T is eventually satisfied at a future time.
- $\Box$: T is always satisfied during a future time.

Time stamp d is used to quantify these operators. Therefore, $\Diamond_d T$ means that, from the current state, T is eventually satisfied in the future before deadline d; and $\Box_d T$ means that T is always satisfied in the future up to deadline d. Now it is possible from the early requirement representation to obtain a model in a formal representation with temporally quantified constraints.

A formal LTL representation for the goal requirement *HVAC system is activated when temperature or humidity reaches a certain limit* would be:

Goal: HVAC system is called when temperature or humidity reaches a limit;

Def.: $\exists$(*ts*: *temp_sensor, hs:humidity_sensor, hvac: HVAC-System*);

(*ts.value* > *hvac.Tmax*) $\vee$ (*hs.value* > *hvac.Hmax*)

$\Rightarrow$ *enabled(hvac)*.

The KAOS approach provides a mechanism that ensures the completeness of the object model by deriving a minimum set of objects for each goal in the knowledge diagram. For instance, Fig. 4 shows the object diagram originating from the goal specification *HVAC system is activated when temperature or humidity reaches a limit*. Repeating this procedure for each goal, a final object model is obtained, as shown in Fig. 5.

Following the normal procedure proposed in the KAOS approach, after all refinements terminate, we should transfer these requirements—now called specifications—to a formal representation: LTL. Therefore, only those requirement

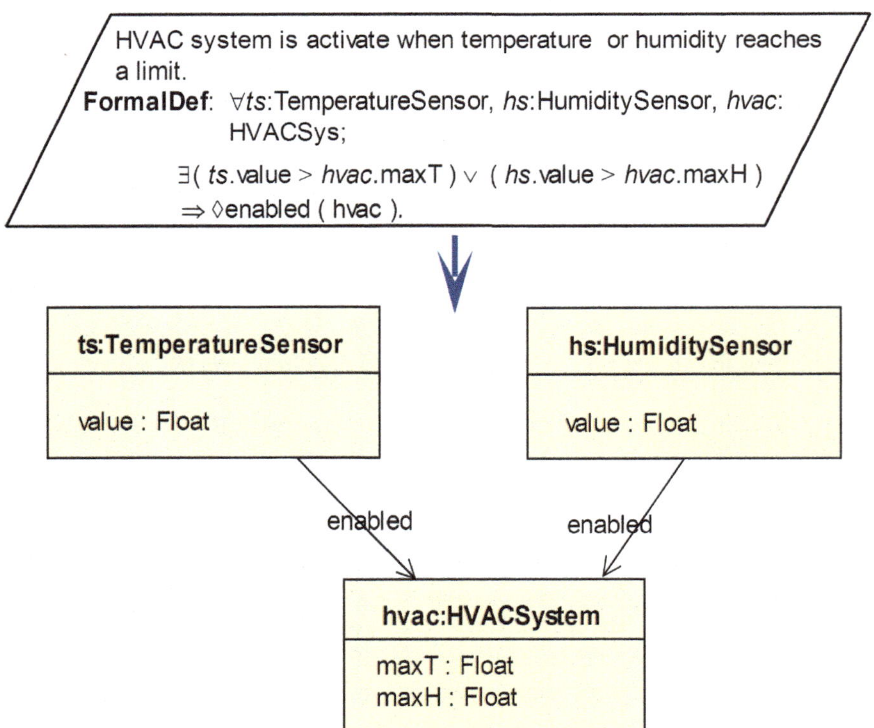

Fig. 4 Objects derived from the goal requirement *HVAC system is activated when the temperature reaches a limit*

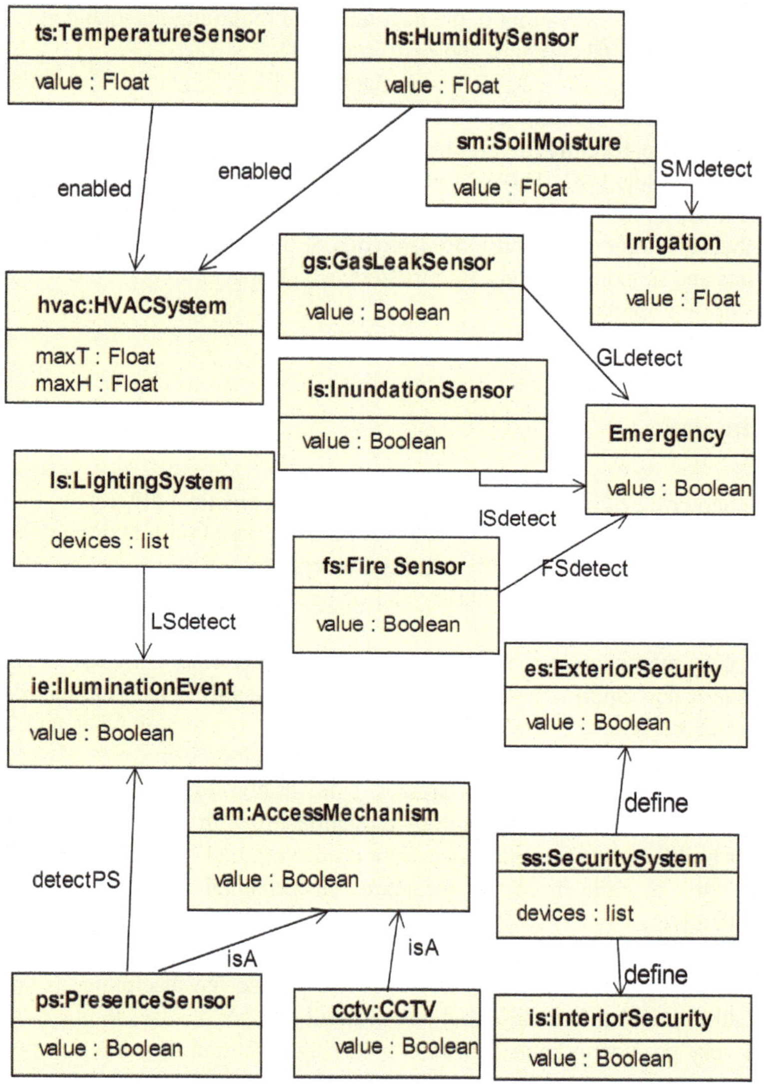

Fig. 5 Object model derived from all goal requirements

specifications that have already been analysed to remove ambiguities, contradictions, or lack of information are transferred to LTL to create formal specifications and project documentation.

This short example is enough to highlight the differences between classical approaches, shown earlier, and the new approach where requirements are based on goals. An important advantage of this technique is the agent-driven orientation and the clear representation of agent responsibility, which contributes to requirement traceability. This is a key point in the negotiation of requirements intended to ensure the satisfaction of stakeholders and end users.

One drawback to using KAOS is the lack of a clear dynamic representation of the process, which is also a key point in automation. To address this limitation, researchers have recently reported on the translation of the KAOS approach to classical Petri Nets [48]. This is a good complement, especially if compared with the design discipline presented at the beginning of this chapter.

Further developments point to transference of the KAOS diagrams to high-level Petri Nets and timed extensions of Petri Nets, which would broaden the scope of the B&RA design approach.

Conclusions

In this chapter we have presented the conceptual and practical differences between a classical design method applied to home automation based on functionality—where the main achievement is comfort—and a new approach better suited to addressing new challenges such as sustainability and complete integration, based on a service system of systems.

The new method is not an adaptation of manufacturing automation; rather, it represents a new approach that is present in manufacturing as well, especially in trends such as Industry 4.0. Thus, it is not a new version of what had already occurred in the field of home automation in the previous century.

A key issue that this method addresses is integration. First, we are talking about the integration underlying a collaborative relationship with user(s) of the automated house or building, and second, the collaboration required from all subsystems—in contrast to a collection of separate instrumental components working independently.

This challenge requires a service system approach for which design methods are still an open problem in academia. A goal-oriented design discipline, as discussed here—although it has been tested in only a small number of cases thus far—appears to be a very promising contribution, considering its sound theoretical support and positive preliminary results.

References

1. Domingues P, Carreira P, Vieira R, Kastner W (2016) Building automation systems: concepts and technology review. Comput Stand Interfaces 45:1–12
2. Pereira CS, Silva JR, Poli MA, Machado JM (2009) Requirements engineering on residential automation. Engenharia'2009, Covilhã, Portugal, Nov 25th–27th 2009
3. Silva JR, Poli MA Jr, Pereira CS, Machado JM (2009) A holonic approach to the integration of automated systems. ABCM Symp Ser Mechatron 4:442–450

4. Rubinstein D (2007) Standish group report: there is less development chaos today. SD times, software development on the web, vol 169
5. Standish Group (1995) Chaos report, technical report
6. Standish Group (2001) Extreme chaos, technical report
7. Standish Group (2009) Extreme chaos, technical report
8. Sakamura K (1987) The objectives of the TRON project. TRON Project 1987:3–16
9. UCD School of Architecture (1997) House of the future, project
10. Yeoh C-M, Tan H-Y, Kok C-K, Lee H-J, Lim H (2008) In: Proceedings—3rd international conference on convergence and hybrid information technology, ICCIT, vol 1, pp 82–87. https://doi.org/10.1109/ICCIT.2008.191
11. Alves J, Marques M, Saur I (2004) Role of networking in innovation promotion and cluster modernization: house of the future. In: 44th congress of the European regional science association: regions and fiscal federalism, Porto, Portugal, 25–29 Aug 2004
12. Chamusca A, Website—http://alexandre.blogspot.com/—Accessed in 2009
13. Abreu P (2008) "Energy Management"—Domotics, III Jornadas Tecnológicas, Publindustria, Viseu, Portugal, 27 May 2008
14. Friedrich WR, Poll JA (2007) Towards a methodology to elicit tacit domain knowledge from users. Interdisc J Inf Knowl Manage 2:179–193
15. Arkin H, Paciuk M, (1997) Evaluating intelligent buildings according to level of service systems integration. Autom Constr 6(5/6):471–479
16. Carlini J (April 1988) Integrating measuring a building's IQ. In: Bernaden JA, et al (eds) The intelligent building sourcebook. Prentice-Hall, London, p. 427–438
17. Sommerville I (2007) Software engineering. Addison-Wesley, 8ª Edição
18. Silva JR, Poli MA, Restrepo PA (2004) Towards a modeling discipline for building and residence automation. ABCM Symp Ser Mechatron 1:560–569
19. Oishi M, Tilbury D, Tomlin CJ (2016) Guest editorial special section on human-centered automation. IEEE Trans Autom Sci Eng 13(1)
20. Gomes L, Steiger-Garção A (1996) Domots are coming! or how to manage building automation in a balanced way? In: Balanced automation II, IFIP advances in information and communication technology. Springer, pp. 55–64
21. Rabardel P, Beguin P (2005) Instrument mediated activity: from subject development to anthropocentric design. Theoret Issues Ergonomics Sci 6(5):429–461
22. INCOSE (international council on systems engineering) systems engineering handbook, 4th edn. Wiley (2015)
23. Bonino D, Corno F (2016) DogOnt—ontology modeling for intelligent domotic environments. Lect Notes Comput Sci 5318:790–803
24. Rea P, Ottaviano E, Castelli G (2013) A procedure for the design of novel assisting devices for the sit-to-stand. J Bionic Eng 10(4):488–496
25. Rea P, Ottaviano E, Conte M, D'Aguanno A, De Carolis D (2013) The design of a novel tilt seat for inversion therapy. Int J Imaging Rob 11(3):1–10
26. Scalise L, Pertroni F, Cascaccia S, Revel, GM, Monteriù A, Prist M, Longhi S, Pescosolido L (2016) Implementation of an "at-home" e-health system using heterogeneous devices. In: IEEE international smart cities conference (ISC2)
27. Massoumy M, Vincentelli AS (2016) Smart connected building design automation: foundations and trends. Found Trends Electron Design Autom 10(1–2):1–143
28. Qiu RG (2014) Service science: the foundations of service engineering and management. Wiley
29. Oliveira VC, Silva JR (2015) A service-oriented framework to the design of information system service. J Sci Serv Res, Springer 7(2):56–96
30. Costa J, Carvalho N, Soares F, Machado J (2009) The fins protocol for complex industrial applications: a case study. In: Proceedings of 6th international conference on informatics in control, automation and robotics ICINCO-RA, vol 2, pp 348–354
31. Silva JR, Nof SY (2015) Manufacturing service: from e-work and service-oriented approach towards a product-service architecture. IFAC Proc Vol (IFAC-papersonline) 48(3):1628–1633

32. Silva M, Pereira F, Soares F, Leão CP, Machado J, Carvalho V (2015) An overview of industrial communication networks. Mech Mach Sci 24:933–940
33. Dutra DS, Silva JR (2016) Product-server architecture: towards a Service engineering perspective in industry 4.0. In: IFAC workshop on intelligent manufacturing systems (IMS), Austin
34. Barros C, Leao CP, Soares F, Minas G, Machado J (2013) Issues in remote laboratory developments for biomedical engineering education. In: 2013 international conference on interactive collaborative learning, ICL. Article no. 6644585, pp. 290–295
35. Leão CP, Soares FO, Rodrigues H, Seabra E, Machado JM, Farinha P, Costa S (2012) Web-assisted laboratory for control education: remote and virtual environments. Commun Comput Info Sci (CSIS) 282:62–72
36. Mehr HD, Huseyn P, Cetin A (2016) Residential activity recognition in smart homes by using artificial neural networks. In: 4th international Istanbul congress of smart grid (ICSG), Istanbul
37. Nof Y, Ceroni J, Jeong W, Moghaddam M (2015) Revolutionizing collaboration through e-work, w-business and e-service. Springer, New York, USA
38. Korkmaz I, Metin SK, Gurek A, Gur C, Gurakin C, Akdeniz M (2015) A cloud based and android supported scalable home automation system. Comput Electr Eng, Elsevier 43:112–128
39. Mahmood A, Javaid N, Razzaq S (2015) A review of wireless communications for smart grid. Renew Sustain Energy Rev, Elsevier 41:248–260
40. Machado J, Denis B, Lesage J-J (2006) A generic approach to build plant models for verification purposes. In: Proceedings—eighth international workshop on discrete event systems, WODES 2006, pp. 407–412
41. Campos JC, Machado J, Seabra E (2008) Property patterns for the formal verification of automated production systems. In: Proceedings of IFAC (IFAC-papersonline), vol 17, (1 Part 1)
42. Kunz G, Machado J, Perondi E (2015) Using timed automata for modeling, simulating and verifying networked systems controller's specifications. Neural Comput Appl. 1–11
43. Machado J, Seabra E, Campos JC, Soares F, Leão CP (2011) Safe controllers design for industrial automation systems. Comput Ind Eng 60(4):635–653
44. Van Lamsweerde A (2009) Requirements engineering: from system goals to UML models to software. Wiley
45. Mylopoulos J (2006) Goal oriented requirements engineering, part II. In: 14th IEEE international conference in requirements engineering
46. Van Lamsweerde A (2000) Requirements engineering in the year 2000: a research perspective. In: Proceedings of 22nd international conference on software engineering. ACM, pp 5–19
47. Van Lamsweerde A (2004) Goal oriented requirements engineering: a road trip from research to practice. In: Proceedings of 12th requirements engineering conference
48. Silva JM, Silva JR (2015) Combining KAOS and GHENeSys in the requirement and analysis of service manufacturing. IFAC Proc Vol (IFAC-papersonline) 48(3):1634–1639
49. Anvari-Mogaddam A, Monsef H, Rahimi-Kian A (2015) Optimal smart home energy management considering energy saving and a comfortable lifestyle. IEEE Trans Smart Grid 6(1)
50. Demeure A, Caffiau S, Elias E, Roux C (2015) Building and using home automation systems: a field study. Lect Notes Comput Sci, Springer, 9083:125–140
51. Jacobson A, Boldt M, Carlsson B (2016) A risk analysis of a smart home automation system. Future Gener Comput Syst, Elsevier 56:719–733
52. Kotonya G, Sommerville I (1998) Requirements engineering with viewpoints. Softw Eng J 11 (1):5–18
53. Miller TR (Dec 1991) Requirements engineering and analysis. Technical report, Software Engineering Institute, Carnegie Mellon University, Pittsburgh, Pennsylvania
54. SCE—certification system of energetics and quality of air on inside of buildings. Technical report, Mechanical Engineering Department, University of Minho (2008)

55. Silva JR (2014) New trends in manufacturing: converging to service and intelligent systems. In: 19th proceedings of IFAC world congress, Cape Town, South Africa
56. Vaquero TS, Silva JR, Beck JC (2013) Itsimple: towards an integrated design system for real planning applications. Knowl Eng Rev J 28(2):215–230
57. Vega AM, Santamaria F, Rivas E (2015) Modeling for home electric energy management: a review. Renew Sustain Energy Rev, Elsevier 52:948–959

Part II
Digital Technologies for Cultural Heritage

Digital Technology and Mechatronic Systems for the Architectural 3D Metric Survey

Marco Giorgio Bevilacqua, Gabriella Caroti, Andrea Piemonte and Alessandro Ariel Terranova

Abstract Over the last decade, we have seen the widespread use of digital survey technologies that have made the three-dimensional (3D) metric survey within reach of all. In the past, lengthy training was needed to use total stations and classical photogrammetry. Today, laser scanning and "new photogrammetry" allow operators with little training to produce 3D models with high spatial density in real time. These systems have therefore made 3D metric survey available to a wide audience of professionals, and have also allowed surveys to be performed with little economic investment in instrumentation. Although this evolution in survey methodologies has certainly brought great benefits, the use of these methods by operators with limited training poses some risk. The proliferation of imprecise processed 3D data, however, constitutes a digital archive of documentation which, by its nature, should be semi-automatically integrated. Issues related to reference systems, scale of representation, accuracy, and related metadata therefore become highly relevant. This paper aims to describe, by means of several case studies, the laser scanner and "new photogrammetry" survey methodologies in light of the aforementioned issues. In addition, the use of "new photogrammetry" in combination with UAV systems will be presented. The integration and miniaturization of positioning systems, attitude measuring systems, and survey instruments (cameras, laser scanners, thermal and multispectral cameras, etc.) allow, by drone flight, the creation of 3D surveys, something that was impossible several years ago without a substantial budget for the use of conventional aircraft.

Keywords Architectural survey · Laser scanning · Structure from motion UAV

M. G. Bevilacqua (✉) · A. A. Terranova
DESTEC, University of Pisa, Pisa, Italy
e-mail: mg.bevilacqua@ing.unipi.it

G. Caroti · A. Piemonte
DICI, University of Pisa, Pisa, Italy

E. Ottaviano et al. (eds.), *Mechatronics for Cultural Heritage and Civil Engineering*, Intelligent Systems, Control and Automation: Science and Engineering 92,
https://doi.org/10.1007/978-3-319-68646-2_7

Introduction

Over the last decade, survey methodologies have benefited from rapid technological progress, in which electronic and computer components have driven the development of the instrumentation. The traditional instruments used by surveyors, which comprised only optical and mechanical components (e.g., theodolites, optical levels, and staffs), have been replaced by integrated equipment in which the electronics and computer components mask the fundamental optical and mechanical ones.

The term "topography," which has described the science of surveying, has now been replaced by a new name, "geomatics," whose etymology signals the importance of the informatics component.

These new technologies offer several advantages, including greater efficiency in data collection, improved control and rigor in the geometric representation, and the potential provision of additional information in the analysis of the built environment. In fact, although the traditional schematization of 3D reality in two-dimensional (2D) representations is valid, some evaluations can be done only by a 3D visual experience, which allows the operator to navigate virtually in 3D and provides a 360° view of the object [1]. Such is the case for virtual or augmented reality experiences applied to cultural heritage [4, 8, 12]. For these reasons, the new survey technologies should be included in the field of mechatronics.

Among the numerous advantages of the current geomatics methodologies, it's worth considering the operative simplification due to process automation. However, this simplification must not lead to a non-awareness of the process. Even if electronics and informatics seem to mask the mechanical and optical characteristics of the survey instruments, these basic elements remain crucial for an effective and metrically rigorous survey. The operator must be aware that new technologies do not replace best practice in survey methods.

State of the Art

Laser scanning and photogrammetry are included among the most common 3D architectural survey methodologies. Photogrammetry, which existed as early as the nineteenth century, long before laser systems, is now experiencing resurgence thanks to the success of digital cameras and the integration of its own algorithms with those of the computer vision.

In the following section, a short description of these two methods is provided in order to facilitate an understanding of the design, survey, and data post-processing pipelines, as well as the most interesting applications in the field of architectural survey. A product common to both methods is the point cloud, which is a discrete set of points, each of which is defined by its 3D coordinates. However, the clouds produced by laser scanning and photogrammetry have very different definitions and properties. Part of the data post-processing is specific to the technique being used, but once the point cloud is created, the next phases of creating the model and texturing its surfaces are common to both methods.

Laser Scanner

The laser scanner determines each point's position using polar coordinates, referred to as a 3D Cartesian system. This requires measuring two angles—zenith and azimuth—and a distance. The laser scanner is ultimately a high-performance, motorized total station. Once the instrument is in the correct position with the angular sampling set, the instrument produces a survey of everything that is in direct view and included within the range of the laser distance measurer, with a 360° angle of view in the horizontal plane and a slightly smaller angle in the vertical plane.

The technological evolution of instrumentation has been very fast. In less than ten years, we have moved from instruments able to detect tens of thousands of points per second to those that can detect one million points per second, and from a wired connection to an external computer to wireless systems and miniaturized processors with onboard display.

There are several laser scanner technologies, allowing for varying degrees of accuracy [13]. Triangulation-based range sensors are suitable for real–or almost real-size surveys (a scale of 1:1–1:10) of small objects (statues, capitals, ornaments, etc.). Time-of-flight (TOF) range sensors are most suitable for use with architectonic scales (from 1:20 to 1:100) and ranges of several hundred meters.

The laser scanner generally associates with each point chromatic data in black and white linked to the intensity of the return signal, which is a function of the type of reflective surface. However, some instruments are equipped with a camera in which the orientation parameters related to the laser system are known. The camera acquires images in order to assign to each point an RGB [red, green, blue] color value tied to the color of the object. Although many instruments are implemented with high-dynamic-range imaging (HDRI) algorithms, integrated cameras typically are not able to provide high-quality photography comparable to images obtained with professional reflex cameras [6].

In order to plan a laser scanner survey, it is necessary to compare the required scale with the accuracy of the scanner. This accuracy depends on the distance of the object to be detected, the angle of incidence of the laser ray on the surface, the reflectance characteristics of the surface itself, and the characteristic noise of the signal emitted by the laser. If the survey object is not visible in its entirety from a single station point (for example, an external and internal survey of a building), scans must be performed from several points of view so that, in post-processing, the data will be recorded in a single reference system. For this purpose, it may be appropriate to introduce some targets in the scene that may serve either as common points visible from adjacent scans or as points of known coordinates in the same reference systems to which the individual scans can be linked.

If targets of this type are not available, it is possible to use efficient iterative closest point (ICP) algorithms to register several point clouds in a single reference system, acquiring them with wide overlapping zones [13]. One must keep in mind that the registration process involves a decrease in the overall accuracy compared with that of a single scan. The overall accuracy may depend on the accuracy of the targets or the standard deviation of the ICP process or even the accuracy of the survey network.

The point cloud of the entire scene is detected in real scale, but it has a variable density due to the selected step angle and the varying distances of the surveyed points. To isolate the point cloud of the object of interest, it is necessary to filter the data acquired by the laser scanner to eliminate all of the points that do not belong to the surface of the object. In order to efficiently manage these dense model points (sometimes there are several hundreds of millions of points) in modeling software, it is necessary to optimize the point cloud of the object of interest through a simplification and decimation (homogeneous point number reduction) controlled phase.

Structure from Motion and Multi-view Stereo

Classical photogrammetry has long been used for both aerial survey, mainly for cartographic purposes, and ground survey, using close-range photogrammetry. The algorithms and practices of photogrammetric survey are well established and can be found in many texts, including [15]. However, it has always been considered to have greater theoretical complexity, and therefore its use is limited to experienced professionals.

The advent of digital photography and the connection between the metric rigor of classical photogrammetry and efficient algorithms of computer vision have given new strength to the already theorized methods of image matching [10] and permitted the automation of the image-based 3D modeling process [18]. This new course of photogrammetry has also resulted in a greater distribution of surveys due to the modest cost of the survey instruments (cameras) and the availability of numerous low-cost software options, including some open source options, with extremely user-friendly interfaces.

It is especially in the area of photogrammetry that the widespread use of the 3D metric survey has been brought within the reach of all in recent years. Thus, even an inexperienced operator using a set of redundant images taken from different points of view and modern processing software can obtain, in a short time, a virtual realistically texturized mesh model.

Automation and the rapid production of 3D models, however, may lead to imperfections, because the model has no real metric consistency; it is only aesthetically pleasing. Accordingly, the correct planning of the photographic shots and the survey of the tie points (TPs) and the ground control points (GCPs) remains crucial.

When using a digital camera in a survey and processing 3D image-based modeling, the starting point is defining the desired scale of representation. The definition of the scale allows us to define on first approximation the ground sampling distance (GSD), which represents the size of an individual pixel relative to the object scale. In general, if the GSD is larger than the required accuracy, the representation will ensure a smaller scale than the one desired.

Next, the camera must be chosen, considering the sensor type (size in millimeters and pixels) and optics that will allow the required GSD to be obtained at the survey distance.

The camera is thus the survey instrument in this methodology, and the achievable accuracy is dependent on several factors, including the GSD, the pattern on the object to be detected, and the static nature of the scene. Another major factor, however, is the quality of the image obtained in terms of sharpness, brightness, contrast, and color saturation. Therefore, to be a good photogrammetry surveyor, one must also be a good photographer and be able to perform good radiometric post-processing of the images to optimize the subsequent 3D modeling process.

The general principle of photogrammetry holds that one can achieve a 3D survey only of the points that are visible in at least two images taken from two different points of view that have been properly oriented. The orientation of an image involves in this context both the calculation of the internal orientation parameters, including the principal point, the focal length and the coefficients of the radial and tangential distortion, and the external orientation parameters, such as the 3D position of the shooting center and the camera attitude (roll, pitch, and yaw).

One of the algorithms borrowed from computer vision is the so-called structure-from-motion (SfM) or structure-from-stereo vision (SfSV), which begins with a set of images of an object that are characterized by very large areas of overlap (70–80%), and is then able to estimate the orientation parameters of the images (self-calibrating bundle adjustment).

Compared with classical photogrammetry, it requires a large overlap between adjacent shots in both the longitudinal and transverse directions of the photogrammetric strips. Actually, the classical concept of strips is replaced by a set of overlapping images that does not necessarily have the same rigid structure as classical photogrammetry strips.

The orientation of the images is provided by identifying the so-called tie points on each image. Those are the same points appearing in more than one image taken from different points of view. The tie point coordinates are based on automatic algorithms of point detectors and descriptors of the scale-invariant feature transform (SIFT) type [16]. Sometimes, however, it is necessary to insert some points manually, such as when objects have a pattern that is not suitable for using automatic recognition algorithms.

By placing only tie points, a model still lacks the component relative to its scale and the reference system in which one wants to add it. The algorithms of SfM have solved the problem of collinearity and coplanarity in the so-called mode-free net, providing infinitely possible similar models.

Only the addition of constraints allows the model to be fixed. These constraints could be given by the survey of some GCPs, a set of known coordinates in a defined reference system.

The orientation of the images is the most important step as regards the accuracy of the metric model. For this reason, it is good practice to detect the independent points that act as control points (CPs). Verification by CPs avoids deformation of a model in its entirety or in part because some images have the wrong orientation.

After determining the image orientation, the point cloud is generated at the desired density, which is comparable to the point cloud measured by a laser scanner. The dense cloud resulting from a photogrammetric survey has associated the RGB value of the color of the images with each point.

Meshing and Texturing

When the point cloud is achieved using laser scanners or photogrammetry, it typically produces a polygon mesh model. Depending on the purpose of the survey, this model could be extremely simplified (e.g., structural analysis, dissemination, web content) or could maintain a high level of detail (e.g., building information modeling [BIM] developed at a detail level, restoration).

Either way, since the model has a density equal to that set by the operator during the 3D acquisition, it may contain in some parts redundant information (e.g., representation of a 1-m flat surface with a step point of 1 cm). Selective model simplification can lighten the model while retaining the detailed description of its geometry [9, 14].

Finally, the texture resulting from the images can be applied to the model. When a model is obtained from images, they are already properly oriented, and it is possible to create a color projection on the mesh. In the case of a laser survey with an integrated camera, the orientation is known. If the color comes from a separate photography campaign, the images must be aligned on the 3D mesh by identifying homologous points on the 2D images in order to calculate the required transformation.

Drones

Over the last few years, we have witnessed the growing use of drones in survey activity. The principal types of drones, also referred to as unmanned air vehicles (UAVs) or, as they are defined by international standards, remotely piloted aircraft systems (RPAS), are fixed-wing and rotary-wing (Figs. 1 and 2). Drones are complex systems integrating various electronic and mechanical components:

Fig. 1 Fixed-wing UAV for photogrammetric purposes

Fig. 2 Rotary-wing UAV for photogrammetric purposes

Fig. 3 Ground station and FPV system

airframe, engines, batteries, power distribution system, autopilot, satellite positioning system, inertial system, gimbal, and takeoff and landing systems [17].

Furthermore, the system comprises aircraft and control components, including a ground station, radio link, first-person view (FPV) system, and video link (Fig. 3).

However, from a geomatics point of view, what makes a UAV a survey system is the onboard presence of a camera, laser scanner, multispectral sensor, or thermal camera.

The use of drones in laser scanner surveys is an interesting but still developing field. It is characterized by high costs because it requires tactical grade positioning and attitude systems. Nevertheless, drone photogrammetry is growing and has already reached a good level of engineering. This methodology is very similar to the method of processing described above and is based on structure from motion. With a drone, one can obtain images from viewpoints otherwise unreachable without the use of manned aircraft. These images are then processed in accordance with a standard 3D image-based modeling methodology [2].

Close-range photogrammetry using aerial survey has existed for some time, employing balloons or kites as aircraft. What is novel about today's RPAS is that they have an onboard autopilot system that allows the operator to keep the airplane stable and also allows the aircraft to follow a route independently. If one adds to these features the option to trigger a camera at regular intervals or at a precise waypoint along the route, one can obtain a set of images during the survey design.

With RPAS, a flight plan is also needed, which requires setting the following parameters: flight speed, flight altitude, coordinates of the waypoint, tack mode, the direction of the bow and the gimbal on which the payload is mounted, and the image motion compensation distortion [11].

A wide variety of RPAS are currently available for civilian use in photogrammetric surveys, ranging from small aircraft weighing a few hundred grams to larger aircraft weighing several tens of kilograms. Likewise, the cameras installed on board vary, including small, lightweight action cameras, cameras for amateurs, recent mirrorless cameras, single-lens reflex (SLR) cameras, ultra-high-resolution cameras, and large format cameras [10]. The choice, of course, depends on the required degree of accuracy as well as the size and the configuration of the working area.

After a period of substantial deregulation of the sector, the majority of countries have now adopted specific regulations on the use of RPAS while waiting for the International Civil Aviation Organization (ICAO) to establish universal regulations, as it has for traditional manned aircraft. The current regulations are very restrictive on the use of RPAS in critical areas, such as areas where people are present who are not under the immediate control of the operator. In fact, many RPAS currently on the market cannot fly over roads, urban centers, or gatherings of people. An exception is the so-called mini-UAV, which weighs less than 250 g and is considered harmless.

Case History

Some representative examples of the aforementioned survey methods (laser scanner, ground-based photogrammetry, and UAV-based photogrammetry) are described below, along with some possible applications. We also present a case where the result can be obtained only by the integration of more than one survey method.

Laser Scanner

The example of a laser scanner survey shown here is part of research that aims to identify the survey methods and procedures capable of producing metric data for risk mitigation of buildings in the case of disaster [5]. In particular, it studies the potential for laser scanner data to produce 3D models useful for structural analysis and verification of buildings through structural calculation software.

Starting from the point cloud (Fig. 4) acquired during the laser scanner survey of the main building of the Lo Scoglio della Regina complex on Livorno's waterfront, different polygonal models of variable mesh density were produced, and the inputs needed to create a shell model computable from structural calculation software were generated.

Many structural verification software programs are based on the calculations of the finite-element method (FEM), which uses numerical simulation models to evaluate the response of structures to different types of actions. This method requires the structure to be suitably discretized into portions, called "finite elements". In contrast, the output of the laser scanner is an unstructured point cloud and not a structural model (mesh). This implies that the geometric data acquired by the laser must be articulated and arranged to be associated with the structural characteristics of the various elements and nodes comprising the structure so as to generate a mesh model (or shell) suitable for calculation of the finite elements.

Clearly, in structural investigations, it is important to know both the internal and external geometry of the structure and to have these two pieces of information in one 3D reference system. The entrance to this building was closed due to structural and security problems; thus, it was not possible to detect the internal geometry. Therefore, this data was retrieved from archival research, which included an overall

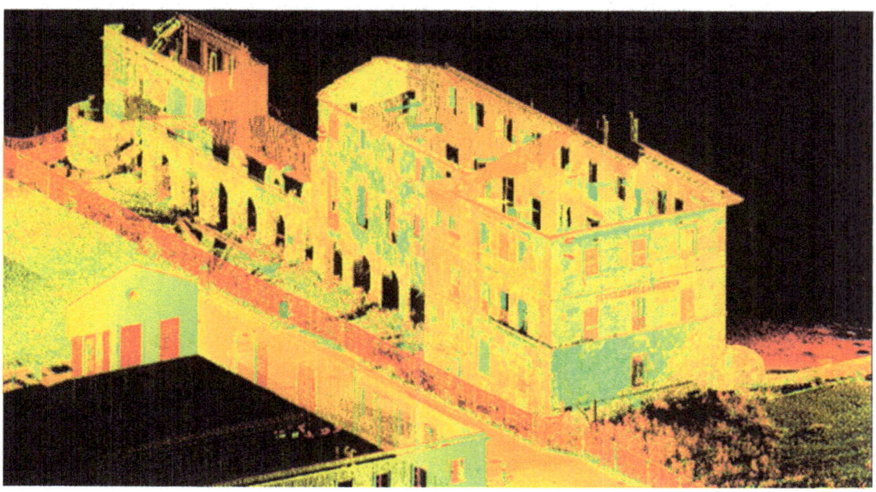

Fig. 4 Point cloud derived from laser scanner survey

Fig. 5 Integration of laser survey with internal survey retrieved from archival research

geometric survey of the building made with traditional techniques including plans, elevations, and sections in computer-aided design (CAD). These drawings, after being suitably oriented and aligned to the laser scanner survey, were used to reconstruct the internal geometry, thus achieving a "hybrid" geometric product, from which it was possible to identify floors, masonry thicknesses, stairwells, and internal divisions (Fig. 5).

We then proceeded with the meshing operation using the Delaunay algorithm from the unstructured point cloud gained from the survey. The mesh obtained from the original point clouds of high density showed a high number of polygons. By acting directly on the number of polygons, a high-density model of polygons was created for architectural use, and a low-density model of polygons characterized by a lower mesh density that maintained the formal and geometric characteristics suitable for processing with the structural analysis program was also achieved.

This model, after being properly processed, provided input for the structural calculation software (Fig. 6 left), describing more accurately the geometry of the

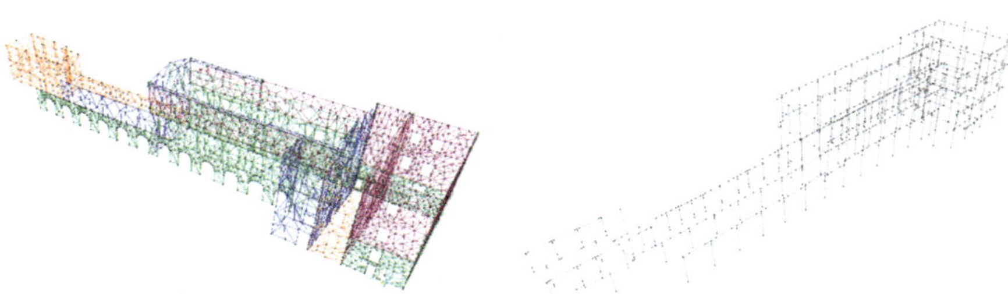

Fig. 6 Input model for structural analysis. On the left is the surveyed mesh, and on the right is the equivalent frame model

structure than the equivalent frame model (Fig. 6 right) that usually results from the interpretation of the classic topographic detector measurements in discrete points.

Image-Based 3D Modeling

Models obtained with the 3D image-based modeling methodology have a multitude of applications, both in the use of the model in its 3D component and in the production of 2D drawings. This example refers to the production of photorealistic fronts and plants.

Before the introduction of the 3D survey described above, the fronts were often created through a simple photo plan by means of individual images. Consequently, the external components of the reference plan were not to scale, were deformed and erroneously positioned, and needed to be corrected with photo editing. These readjustments were qualitative, however, and were controlled by the operator. The metric component was completely lost, or at the very least, the drawings lacked precision compared with the photogrammetric project.

If one instead utilizes a 3D model, and exports the image of the orthogonal views, one can directly obtain the orthophotos, which are, by their very nature, metrically correct. Below, we describe the survey performed to create a 3D model of the access stairway to the church in the courtyard of honor at the monumental Certosa di Calci.

Two sets of images were acquired: one with a horizontal shooting axis with the camera on a tripod to take the vertical components (Fig. 7, left), and the other with a pseudo-nadiral shooting axis with the camera placed on a telescopic bar in order to take the horizontal components (Fig. 7, right).

The distribution of some GCPs in the scene and the collimation of post-processing of an appropriate number of tie points made it possible to record the two dense clouds in a single reference system and to generate a comprehensive 3D model. The required orthophotos were then extracted by defining the model projection planes (Fig. 8).

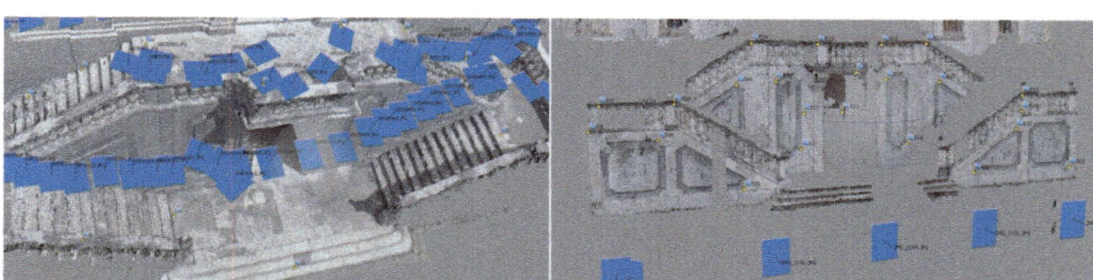

Fig. 7 Image-capture schemes

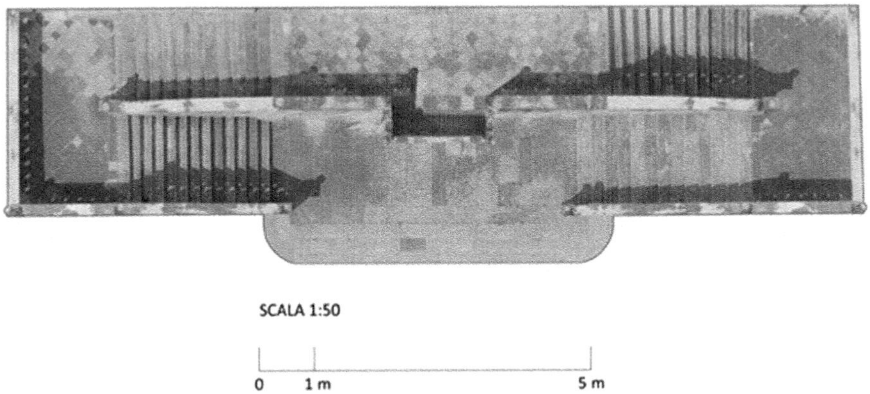

Fig. 8 Example of orthophoto extracted from 3D model

Integration of 3D Survey Methodologies

As we stated previously, both the laser scanner and the image-based methodology can provide a metrically correct 3D model. Each of these techniques, however, has its own strengths and weaknesses. The laser scanner allows for homogeneous accuracy in determining the coordinates of the point cloud, and the survey is not affected by the lighting conditions of the scene or the pattern of the detected object. However, as mentioned earlier, the integrated image-acquisition systems in the laser scanner are usually not of very high quality. In contrast, the image-based modeling techniques begin with a high-quality photographic campaign, and special care is taken regarding the image's radiometric and colorimetric aspects. However, the dense matching algorithms may not always be efficient, especially in the presence of zones with inadequate patterns.

Thus, in some applications, it's useful to integrate the two survey methodologies to ensure the homogeneous metric rigor provided by the laser scanner and the photo quality realized by a precise image-based approach.

An example of laser scanner/image-based modeling integration is described below. This example is derived from research aimed at developing a frescoed vault in real size. Specifically, the example relates to the survey of a pavilion vault of the Palazzo Roncioni in Pisa, frescoed in the eighteenth century by Giovanni Battista Tempesti [6].

By entering some GCPs in the scene, the 3D surveys obtained from the two methods were integrated into the same reference system. Even a quick analysis reveals that the laser scanner geometric model shows less noise than the other models (Fig. 9c), while that derived from images presents some problems (Fig. 9b).

Apart from the local deformations, which are due to inefficiencies of dense matching algorithms, the model resulting from the images is fairly accurate, taking the laser scanner model as a geometric reference and verifying it with cloud-to-cloud comparison techniques (Fig. 10). The orientations of the images

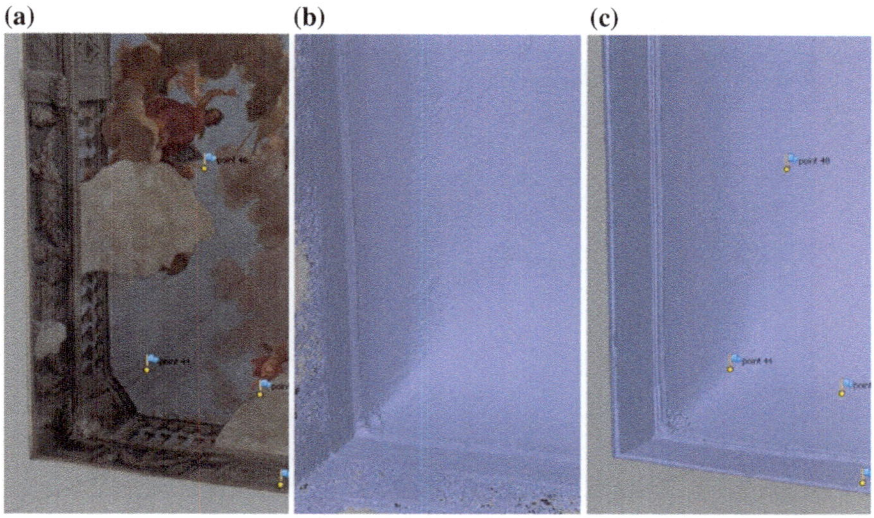

Fig. 9 **a** Textured model; **b** mesh model derived from images; **c** mesh model derived from laser scanner survey

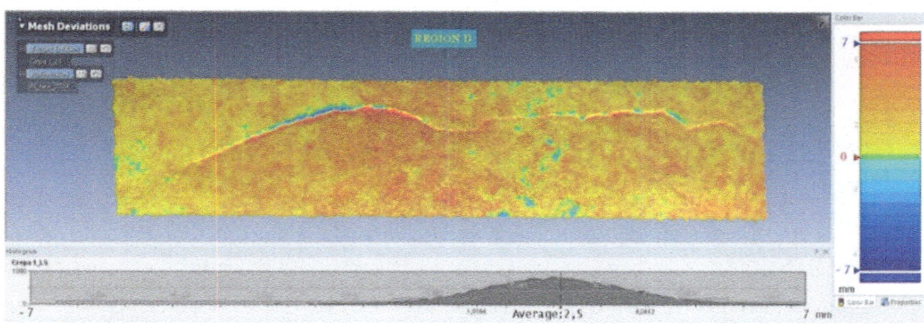

Fig. 10 Cloud-to-cloud comparison

were properly calculated, but there are problems in the extraction of certain points of the dense cloud.

However, if the images are correctly oriented, and these orientation parameters refer to the same reference system in which the laser scanner survey is inserted, it is possible to project the images captured by the image-based 3D model directly onto the laser scanner model [19] (Fig. 11).

From this model, integrating the metric rigor of the laser scanner and the high quality of the images properly oriented, it's possible to produce the planar model of the vault [3].

The real measured and modeled 3D surface deviates from the theoretical one, which was derived from the composition of different portions of cylindrical surfaces

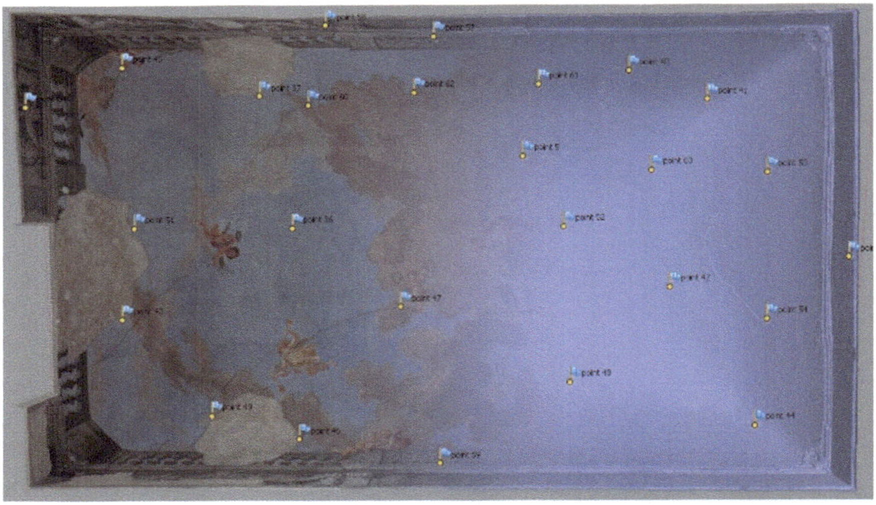

Fig. 11 Oriented by SfM image projection on the laser scanner mesh model

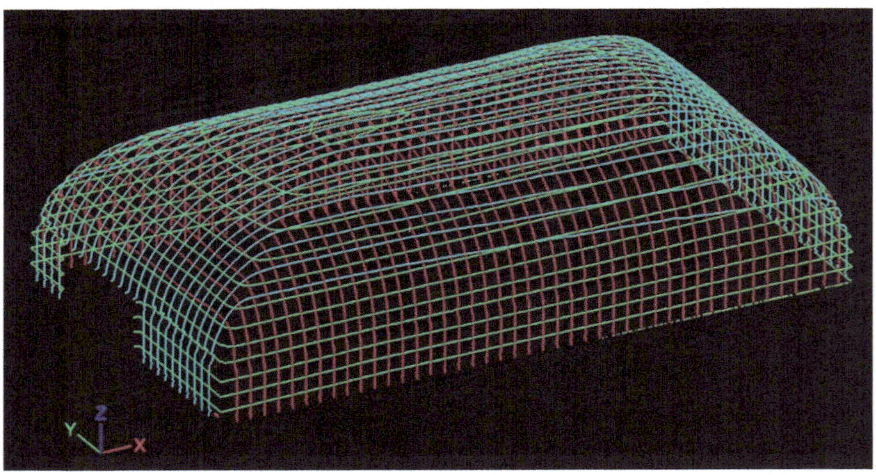

Fig. 12 Actual measured and modeled 3D surface

(Fig. 12). With the availability of a virtual copy of the actual area, one can develop this instead of a synthetic model.

The final result is a flat representation of the frescoed vault that maintains the metric rigor of the position and shape of the elements (Fig. 13), thus allowing the production of cartons in a 1:1 scale for restoration purposes (Fig. 14).

Fig. 13 Planar development of the frescoed vault

Fig. 14 1:1 Scale accuracy assessment

RPAS-Based Photogrammetry

As previously mentioned, a drone is primarily a platform that allows us to reach otherwise inaccessible points of view, without considerable logistic and economic effort. A typical use of this system in the architectural field is represented by the survey of building roofs.

The example [7] refers to the survey of the roof of the Church of San Miniato a Marcianella, near Pisa, which was performed using a multicopter with a Nikon D600 SLR camera on board (Fig. 15).

The Nikon D600 SLR can obtain a 2-mm GSD flying 20 m above the roof. It is equipped with a full-frame 24-megapixel sensor and 50-mm fixed-focal-length optics. In order to determine the scale of the model, the survey was inserted into a particular reference system, and its accuracy was verified using numerous CPs detected in the area. The resolution of the images, the accuracy of the calculation of the orientation parameters, and the quality of the dense cloud obtained resulted in the creation of the orthophotos of the roof in relation to a horizontal plane with a 1:20 scale plan, potentially usable for the analysis of the degradation and for the restoration project (Fig. 16).

Fig. 15 UAV photogrammetric survey

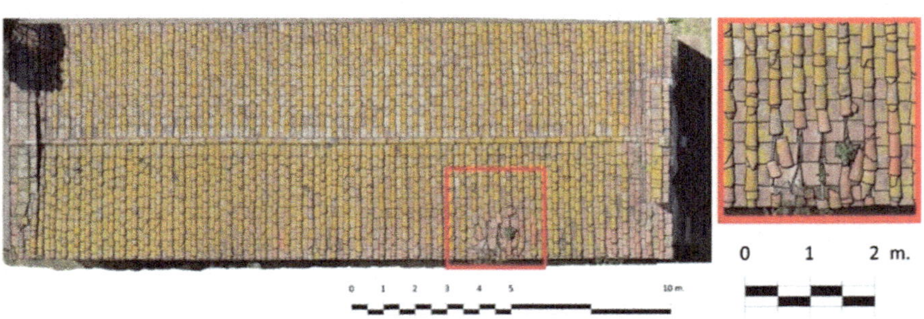

Fig. 16 Orthophoto of the roof

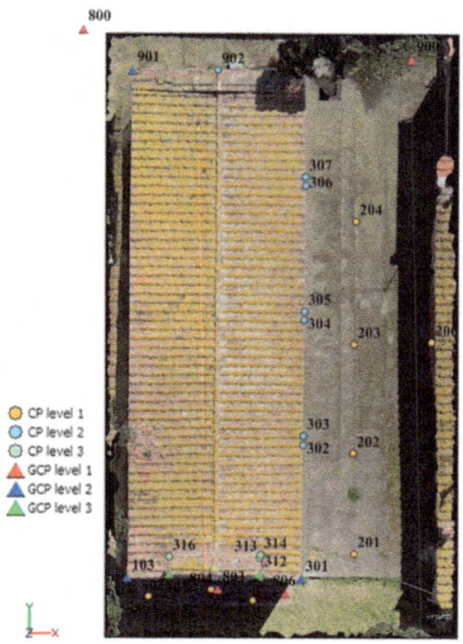

n.of GCPs	Case	n.of CPs	RMSE Z (m)	RMSE X (m)	RMSE Y (m)	RMSE XY (m)
4	A	20	0.048	0.013	0.004	0.013
4+3	B	20	0.029	0.008	0.004	0.009
4+3+2	C	20	0.030	0.007	0.005	0.009

Fig. 17 Accuracy assessment in the case of various GCP distributions

There are three important aspects of the achieved accuracy:

- Image resolution;
- Orientation parameter precision;
- Dense cloud quality.

The image resolution refers more to the photographic approach than that of geomatics. The operator must choose an appropriate shutter speed, aperture, and focus mode, taking into account the fact that shooting is performed with a moving camera. The resolution is seen and evaluated in individual frames, and the level of detail that can be analyzed on the orthophoto is then determined. The analysis of the resolution is not sufficient. One could have an orthophoto with radiometric and colorimetric quality that seems to represent the object, but in fact it reports the different portions of the object in incorrect positions and sizes for the chosen scale representation. This can occur as a result of an incorrect mesh model, which in turn can result from incorrect alignment or poor quality of dense matching. To obtain a good model, it is necessary to provide the coordinates of a number of GCPs that determine the scale and the reference system of the model itself. The provision of these GCPs can affect the overall accuracy of the model.

In the example in Fig. 17, the accuracy was evaluated by comparing the pattern coordinates with various CPs. This comparison was repeated on different models generated by inserting in the computation the GCPs arranged in three different modes: only on the ground, on the ground and on the eaves, and finally adding some points on the bell tower. From the values of the root mean square error (RMSE), it is obvious that even if the GCPs are correctly positioned, it is possible to create models of varying accuracy from the same processed images.

These RMSE values also highlight the fact that operators must be aware that the resolution of an image (in this case the individual pixels represented a portion of a 2-mm object) does not provide a direct indication of the accuracy of the final model (here it is on the order of 1 cm and in altimetry of a few centimeters). Therefore, it is necessary to distinguish between the accuracy of the image and the orthophoto resolution.

Conclusion

The 3D survey methodologies presented herein are currently in use thanks to the rapid technological development that has taken place within the hardware and software fields. The mechatronics component of the current approach to geomatics survey is evident. As has been shown, however, it offers both advantages and disadvantages. Among the advantages is certainly the partial automation of survey processes and data post-processing. However, automation can actually lead to some flaws, including a lack of awareness of the potential and realized accuracy. Automation, especially in the post-processing phase of a photogrammetric survey, has brought the widespread use of the 3D survey within reach of all. Even inexperienced operators can acquire images, which can be obtained with something as simple as a camera phone, insert them as input in SfM software without worrying about how they were acquired or how to provide the appropriate set of control points, and obtain a 3D model with a few clicks. For pure qualitative analysis on a small scale, these models may seem appropriate, but the resulting models typically include obvious anomalies when analyzed on a large scale. One problem is that an inexperienced operator can attribute these anomalies to a software malfunction rather than the inappropriate raw data that was acquired. Furthermore, automation can incorrectly suggest that the operator is not required for the design phase.

This widespread use of 3D surveys, however, is leading to the rapid generation of spatial data, such as point cloud, mesh patterns, textured models, and BIM. It is important that they constitute a substantial archive from the metric point of view, or at least that one has the elements in order to evaluate the accuracy. It becomes crucial that these 3D models are accompanied by all the metadata that describe the genesis and validations.

References

1. Bevilacqua M, Piemonte A, Caroti G, Martínez-Espejo Zaragoza I (2015) An analysis of the evolution of architectural survey by comparison of two surveys of Porta all'Arco (Volterra, Italy): from 1996 to 2014. In: Proceedings of 7th international conference on contemporary problems of architecture and construction, Fabbrica della Conoscenza, vol 58, pp 573–578. ISSN 2464-9678, ISBN 978-88-6542-431-5
2. Bolognesi M, Furini A, Russo V, Pellegrinelli A, Russo P (2014) Accuracy of cultural heritage 3D models by RPAS and terrestrial. In: The international archives of the photogrammetry, remote sensing and spatial information sciences, vol XL-5, ISPRS technical commission V symposium, pp 113–119. https://doi.org/10.5194/isprsarchives-XL-5-113-2014
3. Bevilacqua MG, Caroti G, Zaragoza IM-E, Piemonte A (2016) Frescoed vaults: accuracy controlled simplified methodology for planar development of three-dimensional textured models. Remote Sens 8(3), 239. https://doi.org/10.3390/rs8030239
4. Capriuoli F, Colangelo D, Curto L, Fileri D, Guariglia A, Sansanelli VM, Santarsiero P (2015) La ricostruzione 3D in realtà virtuale del Piano Nobile del Palazzo del Quirinale. Geomedia no 2, pp 26–28. ISSN 1128-8132
5. Caroti G, Franconi A, Piemonte A (2012) Metodologia di elaborazione dati laser scanner per la generazione di modelli utili al calcolo strutturale. In: Proceedings of 16a Conferenza Nazionale ASITA, pp 383–390. ISBN 978-88-903132-7-1
6. Caroti G, Martínez-Espejo Zaragoza I, Piemonte A (2015) Range and image based modelling: a way for frescoed vault texturing optimization. Int Arch Photogramm Remote Sens Spat Inf Sci—ISPRS Archives 40(5W4):285–290. https://doi.org/10.5194/isprsarchives-XL-5-W4-285-2015
7. Caroti G, Martínez-Espejo Zaragoza I, Piemonte A (2015) Accuracy assessment in structure from motion 3D reconstruction from UAV-born images: the influence of the data processing methods. Int Arch Photogramm Remote Sens Spat Inf Sci—ISPRS Archives 40(1W4):103–109. https://doi.org/10.5194/isprsarchives-XL-1-W4-103-2015
8. Chiabrando F, Donadio E, Rinaudo F (2015) SfM for orthophoto generation: a winning approach for cultural heritage knowledge. Int Arch Photogramm Remote Sens Spat Inf Sci—ISPRS Archives XL-5/W7:91–98. https://doi.org/10.5194/isprsarchives-XL-5-W7-91-2015
9. Cignoni P, Montani C, Scopigno R (1998) A comparison of mesh simplification algorithms. Comput Graph 22(1):37–54. ISSN 0097-8493
10. Colomina I, Molina P (2014) Unmanned aerial systems for photogrammetry and remote sensing: a review. ISPRS J Photogramm Remote Sens 92:79–97. https://doi.org/10.1016/j.isprsjprs.2014.02.013
11. Eisenbeiss H (2009) UAV photogrammetry. Dissertation ETH No.18515, Institute of Geodesy and Photogrammetry, ETH Zurich, Switzerland, Mitteilungen 105
12. Gonizzi Barsanti S, Caruso G, Micoli LL, Covarrubias Rodriguez M, Guidi G (2015) 3D visualization of cultural heritage artefacts with virtual reality devices. Int Arch Photogramm Remote Sens Spat Inf Sci—ISPRS Archives XL-5/W7: 165–172. https://doi.org/10.5194/isprsarchives-XL-5-W7-165-2015
13. Guidi G, Remondino F (2012) 3D modelling from real data. In: Modeling and simulation in engineering, pp 69–102. ISBN 978–953-51-0012-6. https//doi.org/10.5772/30323
14. Hoppe H (1996) Progressive meshes. In: Proceedings of SIGGRAPH96, pp 99–108. ISBN 0-201-94800-1
15. Kraus K (1993) Photogramm, vol.1, p 397. ISBN: 3427786846
16. Lowe D (2004) Distinctive image features from scale-invariant keypoints. Int J Comput Vision 60:91–110
17. Nex F, Remondino F (2014) UAV for 3D mapping applications: a review. Appl Geomat 6:1–15. https://doi.org/10.1007/s12518-013-0120-x

18. Remondino F, Spera MG, Nocerino E, Menna F, Nex F (2014) State of the art in high density image matching. The Photogramm Rec 29(146):144–166. https://doi.org/10.1111/phor.12063
19. Robleda PG, Caroti G, Martínez-Espejo ZI, Piemonte A (2016) Computational vision in UV-Mapping of textured meshes coming from photogrammetric recovery: unwrapping frescoed vaults. Int Arch Photogramm Remote Sens Spat Inf Sci—ISPRS Archives 41:391–398

Cultural Heritage Documentation, Analysis and Management Using Building Information Modelling: State of the Art and Perspectives

Filiberto Chiabrando, Vincenzo Donato, Massimiliano Lo Turco and Cettina Santagati

Abstract This chapter presents current issues relating to the use of a building information modelling (BIM) approach in the field of cultural heritage, which is better known as historic-BIM (H-BIM). Technological innovation in the field of automatic metric data acquisition (e.g., 3D laser scanning, digital photogrammetry techniques) requires data processing to produce a coherent parametric model that is congruent with metric survey information. Working with historical artefacts, this process becomes very complex, because the existing tools used to support BIM methodologies are mainly oriented towards new design interventions. From this perspective, the aim of our research is to explore the state of the art of current data acquisition techniques and their integration to obtain a master model, which is defined in the literature as an "inventory BIM model". We will analyse the different scan-to-BIM approaches aimed at creating building object model (BOM) components that, using different acquisition techniques to obtain geometric data, allow the user to define the level of accuracy with which they are generated; in addition, the integrated development of new technologies for existing management (in terms of data enrichment of non-geometric information) will be analysed. Future research perspectives in this field are directed toward the design of a mechatronic system that optimizes computing systems, thus improving the efficiency and effectiveness of the entire process, and making use of a wealth of interdisciplinary knowledge in order to arrange a hierarchy of knowledge that can be shared with other professionals involved in interventions for the protection of cultural heritage.

F. Chiabrando · M. Lo Turco
DAD, Politecnico di Torino, Turin, Italy

V. Donato
DISEG, Politecnico di Torino, Turin, Italy

C. Santagati (✉)
DICAR, University of Catania, Catania, Italy
e-mail: cettinasantagati2@gmail.com

E. Ottaviano et al. (eds.), *Mechatronics for Cultural Heritage and Civil Engineering*,
Intelligent Systems, Control and Automation: Science and Engineering 92,
https://doi.org/10.1007/978-3-319-68646-2_8

Introduction

The development of semantically rich digital facility models, known as building information modelling (BIM), is garnering increasing attention in the architecture, construction, engineering and facility management (AEC/FM) community due to their ability to enhance communication between different stakeholders involved in the various stages[1] of the building process [1].

For the past several years, most countries in northern Europe have been involved in the widespread implementation of BIM technology, especially in the domain of new construction. Numerous studies describe methodologies, processes, case studies and standards in the field of BIM [2]. Examples can be drawn from countries such as the United Kingdom [3, 4] and Scandinavian countries such as Finland [5] and Norway [6]. Other relevant examples can be found in the United States [7], Singapore [8], Australia [9] and Canada [10].

Like many other countries, Italy faces a challenging situation: on the one hand, it is in the midst of a severe economic crisis, and on the other, there is a compelling need for the management of large buildings placed in historical contexts, which are often subject to artistic and landscape protection by the Cultural Heritage Ministry. The recent sequence of earthquakes in Italy (Aquila 2009, Emilia 2012, Amatrice 2016, Visso 2016, Norcia 2016) demonstrates the seismic vulnerability of the entire territory and the need to preserve and protect the historical built heritage.

In this respect, it seems clear that the efforts should be directed towards the restoration and management of existing buildings. If we consider that many of the historical centres of Italian towns are UNESCO sites, it is clear that the integrated development of standards and techniques for acquisition, processing and data integration is needed.

Cultural heritage sites and our valuable historical architectural heritage require high-resolution 3D models in order to achieve a significant added value from their digitalization [11]. These models are increasingly available thanks to the rapid technological progress in the field of acquisition, methods based on laser scanners and/or digital photogrammetry [12, 13], and several international projects such as those promoted by the CyArk organization. However, very little research has been undertaken to explore the advantages and criticalities of BIM methodologies in the cultural heritage domain [14–17] in order to characterize and improve the accuracy, precision, quality and graphic/alphanumeric representation and data enrichment in compliance with the acquired data.

We are essentially living during one of the most significant periods of technological innovation in all of history. The most recent research perspectives in the field of integrated survey procedures are directed towards the design of a mechatronic system that optimizes computing systems, improving the efficiency and

[1]The latest European guidelines concerning building information modelling (BIM) (2014/24/EU) require EU member states to adopt systems that allow the exchange of information through interoperable platforms using open exchange file formats.

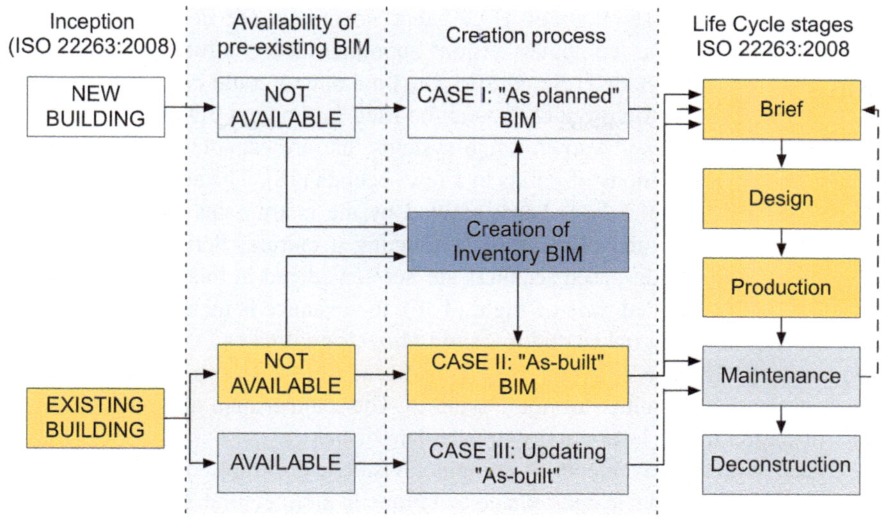

Fig. 1 The H-BIM process in relation to the life-cycle stages

effectiveness of the entire process and making use of a wealth of interdisciplinary knowledge in order to structure the hierarchy of knowledge to share with other professionals for the protection of cultural heritage (Fig. 1)[2].

This chapter is structured as follows: after presenting an overview on data acquisition, processing and integration, the chapter will deal with other relevant issues in the field of parametric modelling and semantics, level of accuracy, data enrichment and organization, graphic and numeric representation, and sharing (interoperability), after which it will conclude with a discussion of future research directions.

Data Acquisition

The digitalization of cultural heritage starts with a survey phase, the result of which is a digital model that is a virtual replica of the object of study and is as detailed and accurate as current technologies permit. Thus, the reverse engineering process provides a virtually scaled model that is coherent in metric terms and enriched with the information regarding the aesthetic appearance of the material and reflectance properties.

The latest available technologies and methods in the field of cultural heritage 3D data acquisition take advantage of two approaches: active sensors such as the well-known terrestrial laser scanning (TLS), and passive methods such as digital

[2]Elaborated on the basis of Volk et. al. [17]

photogrammetry (close-range or UAV) that nowadays are connected with the automatization of the computer vision approach using structure-from-motion (SfM) techniques. In both TLS and SfM, the final output could be a coloured point cloud, which is a metric product that can be used in the next 3D modelling phase.

TLS systems are well-known range systems that are capable of measuring the 3D coordinates of millions of points in a few seconds [18]. The acquisition strategy is based on the use of a laser beam emitted by the instruments using one of two main approaches for the purpose of conducting a cultural heritage survey [19]. (Note that triangulation-based scanners are not considered in this classification). In the first approach, called time-of-flight (ToF), the distance is measured by shooting a laser beam out to an object and measuring how long it takes for the laser beam to bounce back. The second strategy, called phase-shift, utilizes a constant beam of laser energy that is emitted from the scanner. The scanner then measures the phase shift of the returning laser energy to calculate distances.

Today, with improvements in computer vision algorithms, digital photogrammetry is a more affordable and reliable technique in architectural 3D acquisition and processing. It provides a low-cost solution (especially packages capable of fully automatic 3D reconstructions) against the initial investment costs for TLS equipment, including instruments and processing software. Packages such as Agisoft PhotoScan, 3D Zephyr, Pix4D, ReCap 360 and ContextCapture MicMac are able to reconstruct a textured 3D model (point cloud and mesh) starting from an appropriate data set of images taken according to photogrammetric criteria.

Apart from a comparable metric precision, both techniques have singularity and present advantages and disadvantages in accordance with objects of particular shape and morphology.

Starting from the assumption that both techniques are able to deliver a point cloud (i.e., the first product needed for realizing a correct 3D model that can be input into a BIM), a comparison could be focused in terms of acquisition strategy, time for data processing, accuracy and cost.

The first step for both the techniques is the acquisition project. This part is planned according to the shape of the object and needs to be realized following several criteria that can be summarized from the TLS acquisition:

- Overlaps of adjacent scans need to be greater than 30%.
- Scan density, which is connected to the type of laser used and the minimum detail acquired, needs to be represented.
- Angle of incidence is needed; this is related to the distance between the laser and the acquired object and to the orientation angle.

On the other hand, the photogrammetric acquisition rules for point cloud generation could be summarized as following:

- Overlaps between images should be more than 80%.
- Normal and convergent acquisition at different distances.

- Evaluation of the ground sample distance (GSD)[3] according to the scale of the final representation.

The next step relates to data processing: extracting the final point cloud using the TLS is not computationally expensive because the instrument needs to acquire the 3D coordinates derived from the images. Data processing using the photogrammetric approach, on the contrary, is very time-consuming, since it utilizes a dense cloud production, especially if is performed at a high level of detail, which requires a large amount of computational time.

In terms of accuracy, it is important to note that both techniques for obtaining metric products need to be joined with traditional total station survey in order to reference the acquired object. This is then used to correctly perform the bundle block adjustment in the photogrammetric process and to evaluate the final point clouds using the measured point as ground truth (i.e., checkpoints). According to several studies, point clouds achieved by the photogrammetric process are comparable to those surveyed by laser scanner [20–25]. With laser scanning, the sharpness of the edges of the point clouds is better preserved than in those automatically obtained with digital photogrammetry algorithms, which suffer from a certain smoothness. Moreover, SfM packages often tend to automatically fill the holes in the point cloud by introducing additional geometry (which theoretically corresponds to the occlusion parts in laser scanning models); these must be deleted from the model to avoid confusion during the 3D modelling phase.

Finally, we should address the cost of both systems. The TLS is not a low-cost instrument, whereas the photogrammetric approach is supported by the open source community (several open source software applications for point cloud generation are available today), and is thus cost-efficient. For these reasons, photogrammetric techniques are now more often used for survey architectural elements and details (doors, capitals, plasters, frames) that can be represented with sufficient resolution and accuracy.

Data Processing and Integration

The first result of a survey campaign using TLS is the collection of a certain number of scans, each containing millions of measured points. After the point clouds are acquired, they have to be processed for use in 3D modelling.

The processing steps are well known, and the procedure is quite straightforward.

[3]GSD = [pixel size * *H*]/*f*, where the pixel size is the dimension of the pixel on the sensor, *H* is the distance between the camera and the object, and *f* is the focal length. The common software for data processing are able to extract at a higher level a point each pixel.

The first step of the process consists in the following:

- Point-cloud colouring.
- Scan registration.
- Georeferencing/referencing (according to the reference system used).
- Filtering/export of data (interoperable file).

This approach is actually carried out using the software delivered by the scanner and, with minimal differences, is the same for all the main scanner manufacturers. Point-cloud colouring is the first step that assigns an RGB value to each measured point. The RGB values are extracted from the images acquired by the camera coupled to the TLS. Typically, the camera is embedded in the instruments, and only a few instruments (e.g., Riegl) use a camera mounted on the top of the device. Using an external camera, it is possible to obtain images with a higher resolution compared to the camera mounted in the instrument, yet only the high-dynamic-range (HDR) sensors inserted in the new TLS (such as the Faro Focus 3D or the Z + F IMAGER 5010C) can acquire images with a resolution compared to the external one. The scan registration merges multiple images to a common point cloud, and here it is possible to follow two paths: registration based on the shape of the object using an iterative closest point (ICP) approach [26, 27], or registration using the marker positioned on the object.

With the ICP approach, it is important to insert some markers as well, since these are important (but not compulsory, as it is possible to use natural points) for the next step of georeferencing the registered point clouds.

To do this, a topographic network is first created that defines the reference system. Starting from the known points, a side-shot approach is used to measure the markers for scan georeferencing.

Naturally, it is possible to use these points to evaluate the accuracy of the registration phase and to obtain a better optimization of the process. The overall model is composed of multiple overlapping point clouds, which represents the first material and a geometric description of the analysed building. Nevertheless, it has to be considered a discrete model, a numerical sampling of 3D points derived from the object; even if it is a dense point cloud, it does not necessarily mathematically describe all of the targeted surfaces.

On the other hand, the photogrammetric workflow (using the SfM approach) consists of the following steps:

- Tie-point extraction (using SIFT or similar algorithms [28, 29]).
- Image orientation—relative orientation.
- Markers (used as ground control points); the measured points are placed on the images where are visible).
- Adjustment, absolute orientation and camera calibration [30, 31].
- Dense cloud generation [32].
- 2D and 3D model realization.

It is important to underscore at this step that in both of the methodologies, the use of measured markers is preferred in order to control the accuracy of the achieved products and to allow a complete integration, since the final outputs are in a common reference system.

As a general rule, the simplest way to obtain traditional representations (orthophotos) associated with the chromatic information, consists of extracting a triangular mesh. Then, the corresponding image needs to be calibrated and projected on that mesh in order to achieve a textured model.

The creation of the mesh can be achieved using different processing software. The procedure can follow two paths: the first one consists of the creation of a highly defined mesh, which, however, does not take into account the morphological characteristics of the surfaces; the second process can be used to create an intelligent mesh whose generative algorithms refer to the calculation of angular discontinuities according to the thresholds defined by the user.

The textured mesh models, however, present a number of issues—for example, they require a large file size. This is directly related to the resolution and accuracy of the point clouds. There are several techniques available to achieve "lean" digital models, such as the decimation of the mesh [33] or its re-topology (that is, rebuilding an existing mesh with more or less the same volume and shape but with a different mesh layout) by using sub-displaced modelling [34–36]. Using this method, these models can also be used for both online or desktop interactive visualization [37].

Another relevant issue is related to the correct geometric description of the generated surfaces. Although it is possible to obtain very high-resolution mesh, it lacks any kind of information before the critical interpretation.

The primary step is to semantically subdivide the point cloud and to identify and classify each building (or building element) according to a shared ontology of historical buildings that defines the hierarchies between the whole and its parts. Point clouds can be subdivided according to several levels of detail that are coherent with rules of architectural representation at different scales and could be resampled in relation to the requested level of detail of the project.[4]

Parametric Modelling and Semantics

Accurate and detailed instrumental metric acquisition of data and their subsequent integration (occlusion areas) generate a database that visually represents the involucres of the acquired surfaces. However, such information has no intelligence until it is semantically interpreted and geometrically parameterized.

[4]For example, the wall component (whenever it is not necessary to describe their conservation status) could be resampled with a step of 2 cm, while details in doorways or windows could maintain high resolution.

Scientific reports about the restitution process in a BIM environment applied to cultural heritage can follow different procedures.

Interesting work has been conducted by Armeni et al. on semantic automatic recognition in a point cloud [38]. Through complex algorithms that interpret the statistical presence of a point in a particular alignment, the software can identify the separation of spaces and automatically recognize objects.[5]

In any case, this historical architecture knowledge-based approach is fundamental for recognizing single architectural components, and the understanding of their links and relationships with the whole architectural building is crucial for achieving a proper classification and semantic decomposition of the architectural elements and the different components/details that form them.

One very interesting approach is the use of historical treatises (e.g., Palladio's), where this semantic decomposition is well structured and explained. This procedure can be used to move from the general to the specific and vice versa using knowledge about the relations between the parts. It is thus possible to create a library of profiles and components for adaptation and reuse in other projects [39–42]. Another possibility is based on the direct interpretation of survey data, once again recognizing and identifying the architectural components through their critical interpretation (geometric analysis of typical shapes, compositional rules, positioning and orientation constraints) to formalize the semantic structure and make it explicit in an analytical parametrized language. Furthermore, keeping in mind the goal to create a library of shared and reusable architectural components, a formal and typological classification is needed in order to detect the regulatory rules, invariants and the variation in the architectural style [11, 43, 44].

However, not all architectural components (e.g., vaults or irregular components) can be correctly modelled into commercial platforms [45]. In this case, a great support for achieving this goal could be given by shape grammar and procedural modelling, which requires significant computer science competence [46]; another possibility is to use reverse modelling software to create a non-uniform rational basis spline (NURBS) surface and them import that surface into the BIM platform [47, 48].

"Reasoning" for architectural components means to carry out "local" modelling to create semantic-aware libraries; otherwise, in a "global" modelling approach, each modelled element is directly referred to the point cloud.

Therefore, "global" modelling requires the referencing of the point cloud into a unique coordinate system; thus there will be a direct correspondence between the topographic environment reference system of the geometric database (point cloud) and the reference system used for parametric modelling.

"Local" modelling does not require knowing that relationship, but it is necessary that the cropped point cloud is aligned with the reference planes used for modelling of the component.

[5]More information could be found at: http://buildingparser.stanford.edu/method.html.

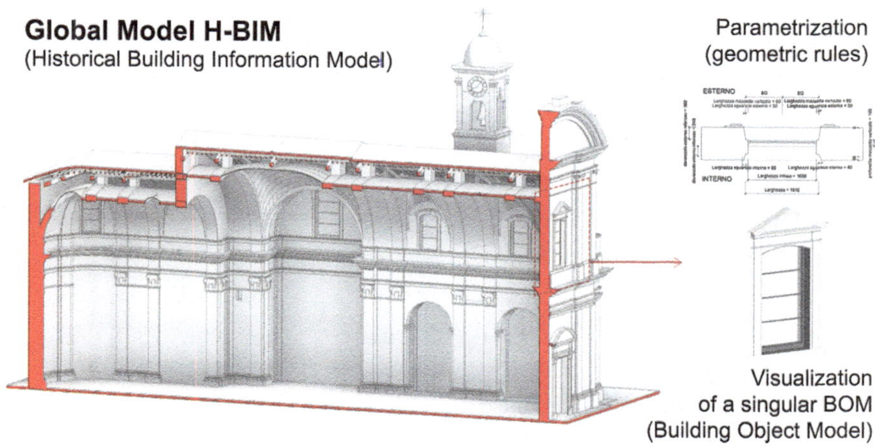

Fig. 2 H-BIM of Poggio Rusco Church (Mantua–Italy)

The global modelling (Fig. 2)[6] approach is usually preferred to overcome the difficulty in using point clouds within a commercial platform library interface (e.g., the Revit family). Over the past few years, several plug-ins (e.g., Cloudworks, PointSense, PointCab4Revit, Scan2BIM) have been released and can manage point clouds in both a project interface and a single-component creation interface. These plug-ins can easily extract x-ray orthoviews from the point cloud to crop and segment the point cloud in real time. They can also measure the deviation between the point cloud and the 3D-modelled geometry. This way, both the modelling and assessment phases are facilitated.

A second distinction allows us to reflect on a different approach arising from the experiences of several research groups who investigated H-BIM: assuming that 3D geometric parametric modelling of architectural elements is time-consuming, should an object in an H-BIM be geometrically parametric or not? What seems to be a provocation actually underscores a different approach to the issue of modelling existing artefacts (or reduced portions).

A rigorous approach [47] assumes flexibility and reusability of the different components using the geometric parameterization of dimensions, relationships typical of systemic and cross-parameterization operations. But is it always useful? There are many unique pieces in classical architecture that do not require such flexibility. This is the case with several decorations that depict anthropomorphic and zoomorphic elements: altars. Or again, we refer to existing elements that do not find their correct categorization for some building component reverse modelling, as in the case of extremely irregular-surfaces vaults.

---

[6]The H-BIM model is part of a research study conducted at University of Florence. The research group is composed by the following: Biagini C., Capone P., Donato V. and Facchini N. In particular, the H-BIM model was developed as part of a master's thesis by Nora Facchini [49].

These are the best samples to demonstrate the added value of new BIM approaches, dealing with the implementation of the alphanumeric component and going beyond the typical static nature of those particular geometric conformations. In this respect, the international guidelines and standards define a clear distinction between graphic content and a minimum level of reliability of the information associated to the modelled components. This distinction can be applied to the project related to the surveyed elements, as described in the following paragraph.

Measuring the Level of Accuracy

In the architectural survey field, the geometric accuracy of digital models refers to the assessment of the deviation between the real model and its virtual representation. For a BIM model, the process is conducted by measuring the deviation that the H-BIM models, and in particular of BOMs, has in relation to the survey data (Figs. 3 and 4) [11, 49, 50].

However, it is possible to refer to another type of level of accuracy, which is perhaps less intuitive, that takes into account the cognitive factors of an object. If we consider the structural analysis of a historical masonry, for instance, the verification is calibrated through a "confidence factor" that allows one to overestimate the forces in the case of missing information. For each case, it is possible to classify the artefact and to associate a particular mechanical characteristic. In that sense, the confidence factor can be amplified or reduced without using an invasive procedure that could irreversibly harm the historical masonry. Therefore, it refers to the quantification of the levels of accuracy of the knowledge directly linked to the virtual elements.

Fig. 3 Point cloud comparison with a historical building object model (H-BOM)

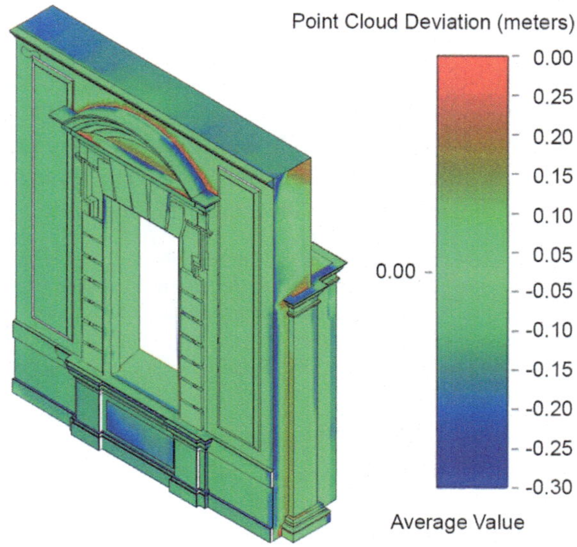

Fig. 4 Sample of point cloud deviation for a building object model (H-BOM)

Considering the level of geometric accuracy related to the architectural survey, the references provided by regulations and international standards are very poor [17]. An example of level of accuracy definition was set in 2012 from COBIM2012 —series 2 [51]. In this document, the authors establish the concept of an Inventory BIM model, and the level of accuracy is defined as follows:

> Accuracy levels of Inventory model—Requirement—The structures of old buildings are almost always somewhat slanted, sloping, curved or otherwise inexact in their geometry. Striving for "absolute" accuracy in the Inventory model is not appropriate […]. The allowed measurement deviations for the Inventory model are: 10 mm on corner points of building elements, 25 mm on surfaces, e.g. walls and floors, 50 mm for old irregular structures such as roof structures.

Therefore, the definition of the level of accuracy is mainly related to "geometric" requirements, according to the inventory BIM rules. This definition does not take into account particular instances of simplified graphic detail associated with specific H-BIM models used in facility management (FM) or energy analysis.

It has to do with the old dilemma of the "representation scale" of the 3D model, partly overtaken in the BIM environment, considering the different levels of detail. It is therefore evident that according to the different uses of a BIM model [52], we can associate a different level of accuracy.

An interesting approach that could associate a level of accuracy in accordance with possible workflow for an H-BIM was elaborated on the basis of guidelines provided by Penn State University [52, 53]. The work individuates 25 "BIM uses", organized by design development phase. BIM uses are categorized into five primary purposes: gathering, generating, analysing, communicating and realizing.

Although the document is complex, no references are available in terms of the accuracy of the metrics and the level of development (LoD), which are explained later. In our opinion, an effort can be made to identify the relationship between H-BIM's uses and accuracy because, actually, there is no reference on the existing building topic. A first attempt was proposed in the work of Biagini and Donato [11].

Another international work that deals with the level of accuracy is the GSA document, volume 3 [54]. In Sect. 2.3 ("Types of Deliverables from 3D Data") and Table 2, tolerance as a function of deliverables is reported. Moreover, in the document, the definition of "tolerance" in relation to the BIM model is:

> the allowable dimensional deviation in the deliverable from truth (…). Some examples of tolerances are: (1) Point cloud: the distance between two points in a point cloud as compared to the true distance between the same two points in the actual scene should be less than or equal to the specified tolerance, (2) Plan: the difference between the length of a wall length in a 2D plan and the actual wall length should be less than the specified tolerance [[54] p. 5–6].

An elaboration of international and local standards is missing and must be implemented. In this regard, the work group UNI/CT 033/GL 05 of the Italian Organization for Standardization (UNI) is processing documents on the LoD of historical buildings, in particular with reference to the UNI-11337 that will integrate and propose the addition of sections for the restoration and renewal of historical and monumental buildings.

In short, the level of accuracy related to the BIM environment plays a more meaningful role today than in the past. While the cultural heritage is known, more complete (accurate) representations are present in the database associated with polyhedral shapes analysed below.

Data Enrichment and Organization

The survey acquisition phase should be followed by the implementation of the informative apparatus, the data of which are extremely heterogeneous. The digitalization of all these data will allow the creation of a database associated with the surveyed elements. Therefore, the survey is an integral process of metric surveys, whose final purpose is the representation and documentation of an existing building. It takes into account several kinds of documents, including archival, photographic, previous surveys, dimensioned sketches, material analysis, degradation phenomena analysis and representation (Fig. 5).

We are currently living during a time characterized by numerous technological advancements, especially with regard to architectural survey and design fields. In any case, the statements derived from reference legislation and treaties still remain valid. The introduction of the new BIM methodologies in the construction process also includes surveying and the use of surveys in the field of cultural heritage. In contrast to what has been done in the past, this process allows the creation of a unique database that holds both geometric and non-geometric information, which is implementable and freely accessible.

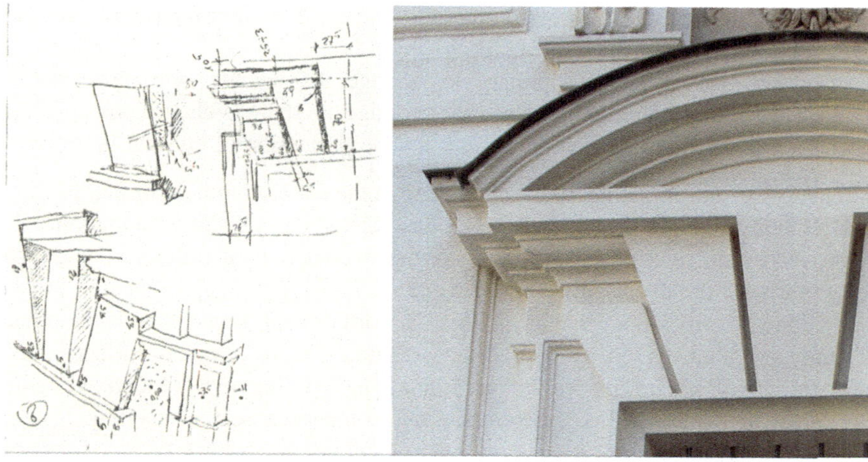

Fig. 5 Dimensioned sketches and related picture

Importantly, an architectural survey is an operation that is not only connected to what one sees, such as the "skin" of the building, but a more complex operation consisting in the retrieval of archival sources, through which it is possible to retrace the building techniques of the time in order to understand the transformations that the artefact has undergone.

The organization of archival sources under a common umbrella that relates them to the main features of a building and to other possible information, such as historical or current pictures, helps the work of historians or restorers. Indeed, they can investigate the reason for the project and can carry out simulations on possible hypotheses. All of this can occur in a unique environment that holds all the information on the object of study, organized according to common taxonomies.

Considering the specific field of restoration, information assets can be integrated by analysing materials and related evaluations that record the graphic and numeric representation of different decay phenomena. This allows the researcher to simultaneously identify the causes, extensions and description of the proposed interventions.

Another relevant topic is the integration of geometric information with the internal stratigraphy of architectural elements. Data can be acquired using sophisticated non-invasive techniques, most of which (e.g., acoustic, electromagnetic) can highlight discontinuity in the walls, historical vault, lacunar and ceilings. Nevertheless, while useful, some of these techniques are very costly, and sometimes the recognition of the element the association of the typological historical element derived from historical manuals could accurately predict the internal composition.

Taking into account the above considerations, it can be assumed that a BIM approach to cultural heritage acts as a mechatronic hinge among all the involved disciplines, for the following reasons:

- It allows the optimization of computing systems, thus improving the efficiency and effectiveness of the entire process.
- It makes use of interdisciplinary knowledge that enables the researcher to organize the hierarchical levels of knowledge to share with the professionals involved, for future interventions and for the protection of cultural heritage.

The latest studies (applications) conducted in the restoration field further develop this relationship by bidirectionally relating graphic design and information. The most well-established procedures provide the explanation of different decay processes through the overlapping of 2D patterns. This type of representation is typically done directly using drawings, which are orthogonal projections. If the object of study is characterized by curved surfaces (e.g., walls, vaulted surfaces), this information will not be an accurate representation. Working on a 3D textured model (i.e., a textured mesh), this issue can be overcome because the mapping is done directly on 3D surfaces, and the computed quantities are correct.

Another issue deals with the difficulty in associating parameters to detailed 2D representations. In the parametric modelling environment, it is possible to use adaptive 3D components that are able to adhere to each surface and, at the same time, to record non-geometric information (e.g., date, author of the surveying, assembled system and degraded finishing material, ID code and description of the proposed interventions).

After data acquisition and management, in terms of processing and data organization, the representation phase can be deeply examined (Fig. 6).

Graphic and Alphanumeric Representation

As is well known, the BIM platform allows the association of multiple data to a single virtual component. A first group controls the graphic content, and international guidelines identify the grade level for managing the graphic representation of building components in orthogonal projection and spatial views, which are congruent with different levels of detail.

A second group concerns alphanumeric data: the international standard identifies the LoD [55] parameter as the degree of reliability of information that can be expected from data contained in the digital model.

Regarding the graphic views associated with different levels of detail, it is possible to base the modelling procedure on dimensioned sketches drawn during the survey, which are consistent with the different representation scales and consistent with the main representation scale used for architectural drawings (Fig. 7).

From a purely graphic point of view, the resolution degrees of the identified models can be divided into the following grades:

- Grade 0 (G0), *Schematic*: at this level, a symbolic or schematic drawing is given in two dimensions and out of scale (e.g., the typical symbols of plan devices).

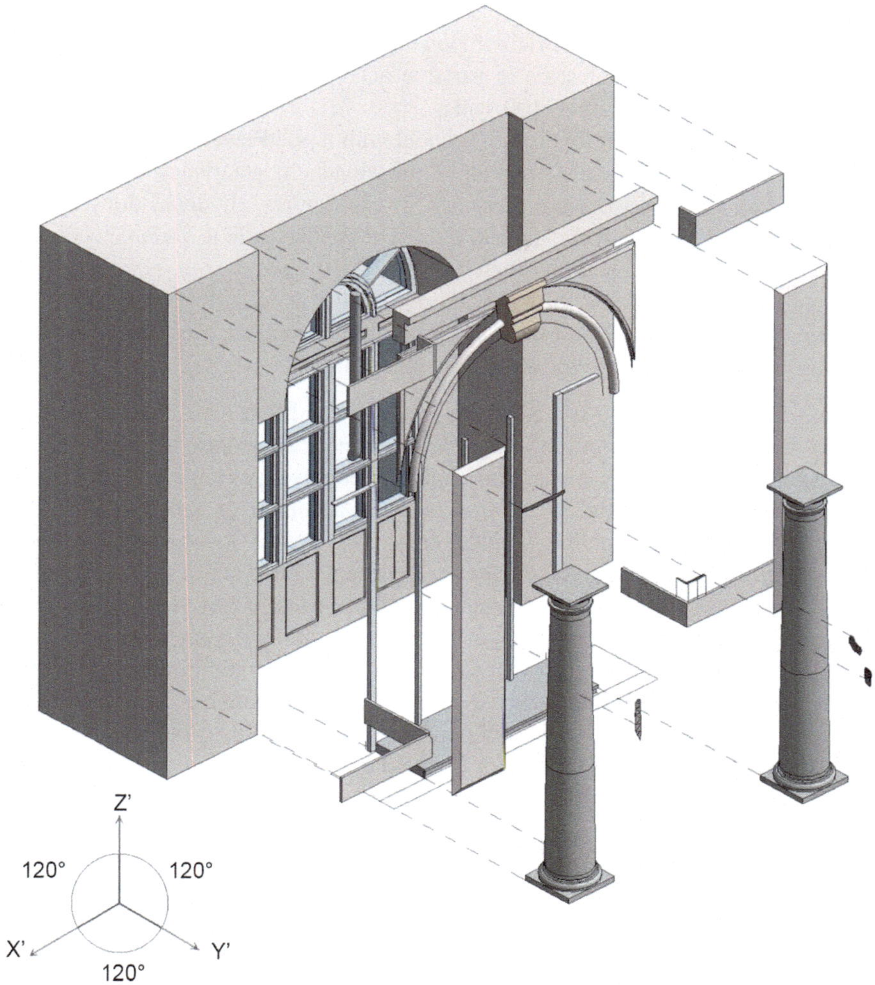

Fig. 6 Semantic subdivision in H-BOM and overlapped information for the restoration

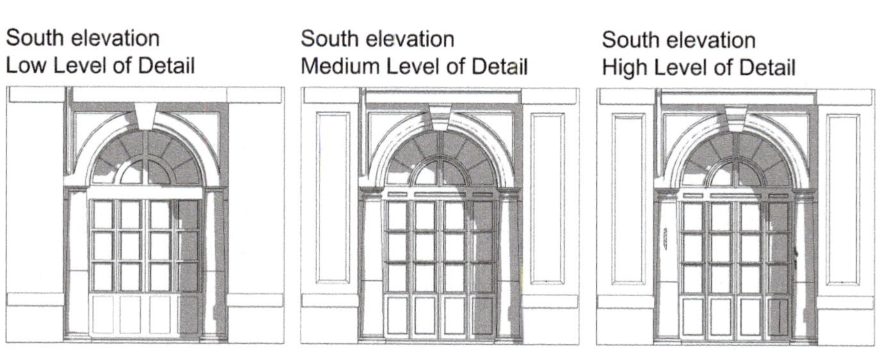

Fig. 7 Grade levels applied to a historical building object model (H-BOM)

- Grade 1 (G1), *Concept*: this is a 3D depiction with the minimum possible detail. There are no associated metric components, and the depiction has approximate dimensions. The model scale is variable according to the represented object, between 1:500 and 1:200/1:100 scale.
- Grade 2 (G2), *Defined*: this is a 3D model with a good level of detail, sufficient to identify its topological, formal and dimensional characteristics. Also, some of the metric characteristics are presented. It can contain 2D detail drawings and can include complete metadata and data. The model scale is variable according to the represented object, between 1:50 and 1:20 scale.
- Grade 3 (G3), *Rendered*: this is a 3D model that is equal to that in grade 2 in terms of technical and informative aspects, but it contains much more accurate graphic features and a photorealistic representation.

From the definitions above, it is easy to understand that the different values of grades are strictly based on the design development and represent increasing graphic detail. In the case of the survey of cultural heritage—particularly regarding the purposes for which the work is intended—the accuracy of acquisition will be defined, as will the so-called graphic error and related iconographic/symbolic graphic choices, depending on the chosen representation scale. As a consequence, for these specific purposes, the schematic level (grade 0) is not very useful, nor is the grade 3 level, which is often not considered within the set of conventional outputs. The choice related to which parts must be represented in the conventional orthogonal projection, rather than modelled in three dimensions, becomes an essential issue. In fact, only 3D components can be linked to the specific informative content that is congruent with the nature and diversity of the data found.

Regarding the LoD specification, such classifications are useful for new building interventions, where the measurement of the LoD is usually based on economic, topological, construction and maintenance information. In the case of interventions on historical buildings, other variables should be included for critical analysis of the richness of the information available; this can be done to measure the reliability degree of the survey. A more complete survey affords a greater opportunity to integrate the various stakeholders who participated in the study.

This procedure includes the retrieval of design archives, state-of-the-art photos, metric survey techniques and degradation surveys. Through the creation of shared parameters (usable for multiple types of components and on several projects), we are able to relate new data to the element detected. The added information can then be incorporated into the model through labels, schedules or thematic views.

Sharing

As for interventions on new construction, H-BIM modelling aims to support the management of a life-cycle process that is of high quality, efficient, safe and in compliance with sustainable development goals. Building information models (and components) can be used throughout the life cycle of the building, starting from the initial design and continuing even during use and facility management (FM) after the construction project is concluded by COBIM2012 [5].

One of the main advantages in using the BIM technology that has not yet been fully explored is the ability to share information among several stakeholders in the design process in order to avoid useless overlapping of information.

The need has therefore arisen to facilitate effective exchanges of information between various software applications by defining standards for sharing data.

Establishing a data archive that gathers all project information in a single integrated project database (IPDB) represents the ideal platform for integration and interaction among the key players in the design phase, thus rationalizing and structuring the process of consulting and subsequently reusing the data needed to describe the building design.

Each data transfer inevitably involves duplication of information, risk of error and thus loss of time. The quest for software interoperability attempts to remove these obstacles in order to enable the individual players in the various stages of the complex design process to share and exchange data automatically. Information exchange is a necessary condition for achieving an exchange of knowledge, because information, to become operational, needs to be interpreted.

Rather than interoperability between software, we choose to discuss semantic interoperability [56], which can be defined as the process that enables software applications or user systems to interpret the meaning of the information exchanged. The ability to manage semantic heterogeneity currently represents a key challenge for information systems integration.

The difficulty is that the meaning changes according to the context and the stage in the process, and various specialist domains have different design requirements, which give rise to different information models. Intuitive and customizable processes thus must be established for handling and editing the data, including its operational meaning [57].

From an information technology (IT) perspective, exchanges are possible both directly and through appropriate formats (proprietary and not). The Industry Foundation Classes (IFC) standard is a class of building data model (BDM); it is specifically developed for the construction industry and is not proprietary. Therefore, it tends to represent a certain set of objects commonly used in the building world.

Horizontal interoperability can be described as the ability to exchange information between platforms that explore the same themes; vertical interoperability, which is certainly more complex, can be achieved when sharing data can be referred to software that integrates their knowledge using a multidisciplinary approach.

The exchanges of data between the source and the receiving software and the interoperability of BIM products at different phases are still limited due to incomplete or ambiguous translations [45]. Most of the applications are still limited to research and academic communities [58].

One-way interfaces are often used to connect the BIM methodology to expert applications, even if it has been considered a real added value to the process, as in some case histories related to the finite element analysis modelling (FEM) approach to complex historical buildings [46, 48].

We also cite the International Framework for Dictionaries (IFD) [59], defined in ISO 12006-3 [37]—then renamed in the building SMART Data Dictionary (bSDD)—as a terminology standard for BIM libraries and ontologies [17]. It is an object-oriented database of multilingual terms that define concepts used in the construction industry and their respective IFC characteristics, such as denotations of objects, parts, attributes, units or values [60].

While information and data exchange is destined to shortly reach a reasonable level of interoperability among machines, the interpretation and understanding of designers remains latent, which ultimately reduces their exchange of knowledge.

Conclusion

It is clear that the current interest in BIM technology has increased the demand for digital representation techniques in the construction industry. Photogrammetry and laser scanning can make a remarkable contribution to the development of interoperable BIM, where the geometric component must be integrated with data surveys. Some reviews [1, 17, 61] have demonstrated satisfactory results for modern buildings, but in the case of cultural heritage buildings, this claim is not well adapted [62]. This essay focuses on enriching the 3D model from the early stages of data collection and segmentation.

This task will require new advanced manual, semi-automatic and fully automated solutions to create tools for as-built BIM creation.

Therefore, it is important to establish methodologies for performance evaluation; currently, no standard evaluation metrics have been established for as-built BIM creation [1].

The term “measurability” has multiple meanings. If the geometric meaning is obviously established and repeatedly mentioned in different documents that fully define the different metric detection operations, the same cannot be said for the ontological meaning, which is intended as a quantitative enhancement of the reliability of a survey.

Future tests will likely lead to the preparation of the intervention protocols replicable on a large scale, evaluating the possibilities of a critical mediation between the Italian public works laws and specification of LoD and the graphic detail (GRADE) by applying them to the themes of architectural heritage to develop proposals to measure their pertinence [63].

This essay has described research related to improving the efficiency and effectiveness of more conventional procedures. The final model synthesizes the information, which is usually fragmented (primarily because it belongs to different representations), and enriches it with new methods of analysis and management of data. This opens up the possibility for new scenarios of “knowledge” [44], particularly regarding the restoration and preservation of built architectural heritage, but also for the dissemination and enhancement of the historical architecture. In this regard, taking into account the statements of the European Charter of Architectural Heritage, the cultural asset is defined as "an irreplaceable expression of the diversity of cultures", and the aim of conservation disciplines is to "preserve the aesthetics and value of the monument" [64]. Indeed, the elements that constitute historical architecture are not only physical artefacts, but hold intangible value, based on highly integrated approaches of programmed preservation [45].

The EU INCEPTION project, funded under H2020 call Reflective 7–Advanced 3D modelling, is related to this. Among the many objectives of this project is the development of an open-standard Semantic Web platform for accessing, processing and sharing interoperable digital models resulting from 3D survey and data capture.

Therefore, one must imagine an evolutionary route that begins with building information modelling an information model that implicitly contains very heterogeneous data. This process is characterized by the following: survey, thematic analysis, maintenance relating to the building, and venturing out towards a possible *building knowledge model* [56, 65]. Therefore, this process is intended to be used to gain knowledge concerning the building structure.

A true cultural step forward must be made in the future, and technology based on human cognitive and communication processes should be used. Intuitive and customizable methods will be important in moving from information to knowledge.

Authors Contribution The paragraphs Introduction and conclusion are written by all authors. Data acquisition is edited by F. Chiabrando. Data processing and integration is edited by F. Chiabrando, M. Lo Turco and C. Santagati. Parametric modelling and semantics is edited by C. Santagati. Measurements of Level of Accuracy is edited by V. Donato. Data Enrichment and classification is edited by M. Lo Turco. Graphic and alphanumeric representation is edited by C. Santagati and V. Donato. Sharing is edited by V. Donato and M. Lo Turco.

References

1. Tang P, Huber D, Akinci B et al (2010) Automatic reconstruction of as-built building information models from laser-scanned point clouds: a review of related techniques. Autom Constr 19:829–843. https://doi.org/10.1016/j.autcon.2010.06.007
2. Eastman C, Teicholz P, Sacks R, Liston K (2011) BIM handbook: a guide to building information modeling for owners, managers, designers, engineers and contractors, 2nd edn. https://doi.org/10.1002/9780470261309
3. AEC (UK) (2012) BIM Protocol, 1–46
4. BSI (2013) PAS 1192-2:2013 Specification for information management for the capital/ delivery phase of construction projects using building information modelling
5. Building Smart Finland (2012) COBIM—common BIM requirements, vol. 1–12
6. Statsbygg (2013) Statsbygg BIM manual version 1.2.1. 1:1–98
7. U.S. General Services Administration's Public Buildings Service: Office of the Chief Architect (2009) GSA building information modeling guide (Series 1–8)
8. Building and Construction Authority (2013) Singapore BIM guide—version 2.0. i-60
9. CRC Construction Innovation (2009) National guidelines for digital modelling

10. AEC (CAN) Committee (2012) AEC (CAN) BIM protocol v1.0—implementing Canadian BIM standards for the architectural, engineering and construction industry based on international collaboration. BIM-standard 54
11. Quattrini R, Malinverni ES, Clini P et al (2015) From TLS to HBIM. High quality semantically-aware 3D modeling of complex architecture. ISPRS—Int Arch Photogramm Remote Sens Spat Inf Sci XL-5/W4:367–374. https://doi.org/10.5194/isprsarchives-XL-5-W4-367-2015
12. Barazzetti L, Fangi G, Remondino F, Scaioni M (2010) Automation in multi-image spherical photogrammetry for 3D architectural reconstructions. Vast 2010 2312. https://doi.org/10.2312/PE/VAST/VAST10S/075-081
13. Chiabrando F, Sammartano G, Spanò A (2016) Historical buildings models and their handling via 3D survey: from points clouds to user-oriented HBIM. ISPRS—Int Arch Photogramm Remote Sens Spat Inf Sci, ISPRS Arch, 633–640
14. Chiabrando F, Spanò A (2013) Points clouds generation using TLS and dense-matching techniques. A test on approachable accuracies of different tools. ISPRS Int Arch Photogramm Remote Sens Spat Inf Sci II:2–6. https://doi.org/10.5194/isprsannals-II-5-W1-67-2013
15. Del Giudice M, Osello A (2013) BIM for cultural heritage. Int Arch Photogramm Remote Sens Spat Inf Sci—ISPRS Arch 40:225–229
16. Thomson C, Boehm J (2015) Automatic geometry generation from point clouds for BIM. Remote Sens 7:11753–11775. https://doi.org/10.3390/rs70911753
17. Volk R, Stengel J, Schultmann F (2014) Building Information Modeling (BIM) for existing buildings—literature review and future needs. Autom Constr 38:109–127. https://doi.org/10.1016/j.autcon.2013.10.023.Abstract
18. Lemmens M (2011) Terrestrial laser scanning. In: Geo-information. Springer, Netherlands, Dordrecht, pp 101–121
19. San Jose Alonso JI, Martínez Rubio J, Fernández Martín JJ, García Fernández J (2011) Comparing time-of-flight and phase-shift. The survey of the royal pantheon in the basilica of San Isidoro (Leon). Arch Photogramm Remote Sens Spat Inf Sci 38:377–385
20. Grussenmeyer P, Landes T, Boegtle T, Ringle K (2008) Comparison methods of terrestrial laser scanning, photogrammetry and tacheometry data for recording of cultural heritage buildings. Int Arch Photogramm Remote Sens Spat Inf Sci 37:213–218
21. Inzerillo L, Santagati C (2013) Using dense stereo matching techniques in survey. Disegnare Idee Immagini 47:82–91
22. Kadobayashi R, Kochi N, Furukawa R (2004) Comparison and evaluation of laser scanning and photogrammetry and their combined use for digital recording of cultural heritage. Int Arch Photogramm Remote Sens Spat Inf Sci (Part B5) 35:401–406
23. Lichti DD, Gordon SJ, Stewart MP et al (2002) Comparison of digital photogrammetry and laser scanning. Int Soc Photogramm Remote Sens 39–44
24. Manferdini AM, Galassi M (2013) Assessments for 3D reconstructions of cultural heritage using digital technologies. Int Arch Photogramm Remote Sens Spat Inf Sci XL-5 W 167–174
25. Roncella R, Re C, Forlani G (2011) Performance evaluation of a structure and motion strategy in architecture and cultural heritage. Int Arch Photogramm Remote Sens Spat Inf Sci (5/W16) 38:285–292
26. Barnea S, Filin S (2008) Keypoint based autonomous registration of terrestrial laser point-clouds. ISPRS J Photogramm Remote Sens 63:19–35. https://doi.org/10.1016/j.isprsjprs.2007.05.005
27. Besl PJ, McKay ND (1992) Method for registration of 3-D shapes. Robot Tentat 1611:586–606. https://doi.org/10.1117/12.57955
28. Lowe DG (1999) Object recognition from local scale-invariant features. Proc Seventh IEEE Int Conf Comput Vis 2:1150–1157. https://doi.org/10.1109/ICCV.1999.790410
29. Vedaldi A (2008) An implementation of SIFT detector and descriptor. Int J 1:1–7. https://doi.org/10.1016/j.applthermaleng.2011.03.022

30. Triggs B, McLauchlan PF, Hartley RI, Fitzgibbon AW (2000) Bundle adjustment—a modern synthesis. In: Proceedings international workshop on vision algorithms Theory Pract, pp 298–372
31. Maoling Q, Songde M, Yi L (2000) Overview of camera calibration for computer vision. ACTA Autom Sin 26:43–55
32. Remondino F, Spera MG, Nocerino E et al (2014) State of the art in high density image matching. Photogramm Rec 29:144–166. https://doi.org/10.1111/phor.12063
33. Benedetti B, Gaiani M, Remondino F (2010) Modelli digitali 3D in archeologia: il caso di Pompei. Strumenti, Edizioni della Normale, Pisa
34. Fantini F, Pini S (2011) Villa Adriana's Serapeum : optimized 3D models for knowledge and distribution of archaeological sites. In: Jerem E, Redő F, Szeverényi V (eds) On the road to reconstructing the past. computer applications and quantitative methods in archaeology (CAA). Proceedings of the 36th international conference, Archeaeolingua, Budapest, 2–6 April 2008, pp 340–345
35. Fantini F (2011) Image based data processing (IBDP): la restituzione attraverso displaced subD a partire da rilevamento laser scanner. In: Il modello in architettura: tra cultura scientifica e rappresentazione, Alinea Editori, Firenze, pp 147–158
36. Guidi G, Angheleddu D (2016) Displacement mapping as a metric tool for optimizing mesh models originated by 3D digitization. J Comput Cult Herit 9(9):1–9, 23. https://doi.org/10.1145/2843947
37. Merlo A, Dalco L, Fantini F (2012) Game engine for cultural heritage: new opportunities in the relation between simplified models and database. In: Proceedings of 2012 18th International Conference Virtual System Multimedia, VSMM 2012 Virtual System Informative Society, pp 623–628
38. Armeni I, Sener O, Zamir A, Jiang H (2016) 3D Semantic parsing of large-scale indoor spaces. CVPR 1534–1543. https://doi.org/10.1109/CVPR.2016.170
39. Apollonio FI, Gaiani M, Sun Z (2013) 3D modeling and data enrichment in digital reconstruction of architectural heritage. ISPRS—Int Arch Photogramm Remote Sens Spat Inf Sci XL-5/W2, 43–48. https://doi.org/10.5194/isprsarchives-XL-5-W2-43-2013
40. Clini P, Quattrini R, Frontoni E, Pierdicca R, Nespeca R (2017) Real/not real : pseudo-holography and augmented reality applications for cultural heritage. In: Cigola M, Ippolito A (eds) Handbook of research on emerging technologies for cultural heritage. IGI Global, pp 201–227
41. De Luca L, Vèron P, Florenzano M (2005) Semantic-based modelling and representation of patrimony buildings. SVE Work Towar Semant Virtual Environ 1–11
42. Fai S, Rafeiro J (2014) Establishing an appropriate level of detail (LoD) for a building information model (BIM)—West Block, Parliament Hill, Ottawa, Canada. ISPRS Ann Photogramm Remote Sens Spat Inf Sci II-5:123–130. https://doi.org/10.5194/isprsannals-II-5-123-2014
43. Biagini C, Donato V (2014) Building object models (BOMs) for the documentation of historical building heritage. EGraFIA 2014 Revis del Futur Previs del pasado V Congr Int Expresión Grá ca en Ing Arquit y Carreras A nes y XI Congr Nac Profesores Expresión Grá ca en Ing Arquit y Carreras A. https://doi.org/10.13140/RG.2.1.1108.2005
44. Lo Turco M, Santagati C (2016) From SfM to semantic-aware BIM objects of architectural elements. EuroMed 2016, Part I, LNCS 10058. Springer, Heidelberg, pp 600–6012
45. Quattrini R, Clini P, Nespeca R, Ruggeri L (2016) Measurement and historical information building: challenges and opportunities in the representation of semantically structured 3D content Misura e historical information building: sfide e opportunità nella rappresentazione di contenuti 3D semanticamente. Disegnarecon 9:11
46. Murphy M, McGovern E, Pavia S (2013) Historic Building Information Modelling—Adding intelligence to laser and image based surveys of European classical architecture. ISPRS J Photogramm Remote Sens 76:89–102. https://doi.org/10.1016/j.isprsjprs.2012.11.006
47. Brumana R, Oreni D, Raimondi A et al (2013) From survey to HBIM for documentation, dissemination and management of built heritage: The case study of St. Maria in Scaria

d'Intelvi, 2013. In: Digital heritage international congress (Digital Heritage), Marseille, pp 497–504. https://doi.org/10.1109/DigitalHeritage.2013.6743789
48. Oreni D, Brumana R, Georgopoulos A, Cuca B (2013) HBIM for conservation and management of built heritage: towards a library of vaults and wooden bean floors. ISPRS Ann Photogramm Remote Sens Spat Inf Sci II-5/W1:215–221. https://doi.org/10.5194/isprsannals-II-5-W1-215-2013
49. Biagini C, Capone P, Donato V, Facchini N (2016) Towards the BIM implementation for historical building restoration sites. Autom Constr 71:74–86. https://doi.org/10.1016/j.autcon.2016.03.003
50. Chiabrando F, Lo Turco M, Santagati C (2017) Digital invasions: from point clouds to historical building object modeling (H-BOM) of a UNESCO WHL site. Int Arch Photogramm Remote Sens Spatial Inf Sci XLII-2-W3:171–178. https://doi.org/10.5194/isprs-archives-XLII-2-W3-171-2017
51. BuildingSmart Finland (2012) COBIM—common BIM requirements 2012 Series 2—modeling of the starting situation
52. Kreider RG, Messner JI (2013) The uses of BIM: classifying and selecting BIM uses. University Park, PA, USA
53. The Computer Integrated Construction Research Group (2009) BIM project execution planning guide, The Pennsylvania State. https://doi.org/10.1017/CBO9781107415324.004
54. U.S. General Services Administration's Public Buildings Service: Office of the Chief Architect (2009) GSA building information modeling guide series: 03—GSA BIM guide for 3D imaging. Natl 3D-4D-BIM Progr
55. BIM forum (2015) Level of development specification
56. Carrara G, Fioravanti A, Loffreda G, Trento A (2009) An ontology-based knowledge representation model for cross-disciplinary building design. In: 27th eCAADe Conf Proc, pp 367–374
57. Lo Turco M (2015) Il BIM e la rappresentazione infografica nel processo edilizio. Dieci anni di ricerche e applicazioni - BIM and infographic representation in the construction process. A decade of research and applications, Aracne, Roma
58. Abanda FH, Tah JHM, Keivani R (2013) Trends in built environment semantic web applications: where are we today? Expert Syst Appl 40:5563–5577. https://doi.org/10.1016/j.eswa.2013.04.027
59. Building SMART (2008) IFD library white paper
60. ISO (2012) BS ISO 12006–3: 2007—Building construction—organization of information about construction works—part 3: framework for object-oriented information. ISO 3:32
61. De Luca L (2011) La fotomodellazione architettonica—Rilievo, modellazione, rappresentazione di edi ci a partire da fotografie. Dario Flaccovio Editore
62. Hichri N, Stefani C, De Luca L, et al (2013) From point cloud to BIM: a survey of existing approaches. ISPRS—Int Arch Photogramm Remote Sens Spat Inf Sci XL-5/W2:343–348. https://doi.org/10.5194/isprsarchives-XL-5-W2-343-2013
63. Lo Turco M, Bocconcino MM (2016) Collaborative environments for knowledge sharing: first building information modeling applications for public works Ambienti collaborativi per la condivisione della conoscenza: prime applicazioni di building information modeling in ambito pubblico. Disegnarecon 9:10
64. ICOMOS (1975) European charter of architectural heritage. http://www.icomos.org/en/charters-and-texts/179-articles-en-francais/ressources/charters-and-standards/170-european-charter-of-the-architectural-heritage. Accessed 14 Jun 2016
65. Simeone D, Cursi S, Toldo I, Carrara G (2014) Bim and knowledge management for building heritage. Acadia 2014:681–690

3D Survey Systems and Digital Simulations for Structural Monitoring of Rooms at the Uffizi Museum in Florence

Sandro Parrinello and Sara Porzilli

Abstract This article presents research activities conducted on several rooms as part of the extension project of the Grandi Uffizi Museum in Florence (Italy). The research addresses survey methods for monitoring the static performance and evaluating the structural plastic deformations of vaulted systems and architectural elements, mostly subjected to structural reinforcements. In order to achieve positive results, the most appropriate 2D and 3D graphic representation methods have been identified, to enable the realization of detailed, technical documents. Laser scanner survey activities have been executed along with photo-modeling and extensive photographic documentation, crucial for the operations of photogrammetry and photomapping reconstructions. Post-production and data processing steps have produced substantial documentation of graphic materials through the development of floor plans, detailed measurements of series of sections, photomap reconstructions and 3D simulations. The thrust of the innovative research deals with the development of 3D computational models by implementing and refining reverse engineering processes for the simulating the static performance and plastic deformations that overlap the various stages of each investigation. These monitoring techniques have made it possible to determine the effect of the consolidation interventions operated, through a progressive implementation of the point cloud. The opportunity to follow the construction site from 2010 until now has contributed significantly to the enhancement and refinement of these detection and representation techniques, enabling the development of new operational methods with higher metric reliability to support the activities that such a sophisticated construction site as the Grandi Uffizi in Florence represents.

Keywords Laser scanner · Reverse engineering · Digital representation Structural monitoring · Morphological analysis of strain

S. Parrinello (✉) · S. Porzilli
Department of Civil Engineering and Architecture, University of Pavia, Pavia, Italy
e-mail: sandro.parrinello@unipv.it

S. Porzilli
e-mail: arch.saraporzilli@globeit.it

E. Ottaviano et al. (eds.), *Mechatronics for Cultural Heritage and Civil Engineering*, Intelligent Systems, Control and Automation: Science and Engineering 92,
https://doi.org/10.1007/978-3-319-68646-2_9

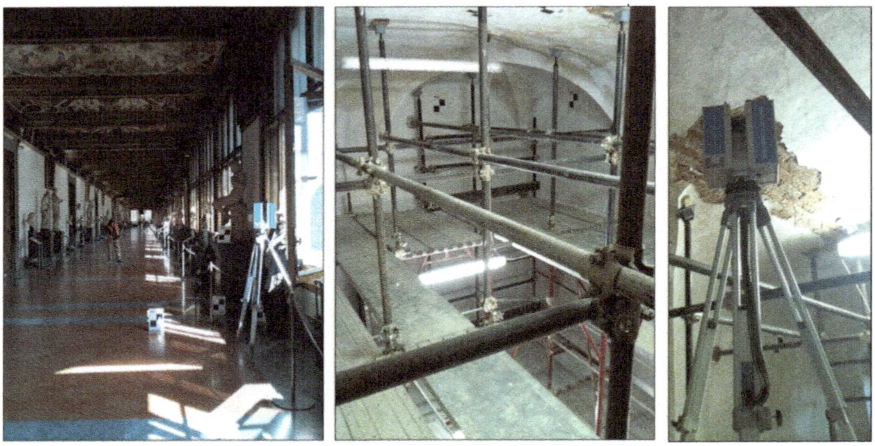

Fig. 1 Laser scanner survey activities in the museum complex of the Uffizi

Introduction

This contribution presents the key results obtained from digital surveys conducted on several rooms of the Uffizi Museum in Florence (Fig. 1), primarily as a result of the significant rearrangement of the exhibition spaces over the past few years. A detailed structural analysis provides us an understanding of the static situation and allows us to expand the consolidation project whenever necessary [2].

The research project is carried out in partnership with the institutions (Fig. 2) primarily responsible for the construction works related to the extension of the exhibition areas within the Uffizi Museum.[1] Because of this fruitful collaboration, which started in 2010 and is still active, it has been possible to carry out several analyses and investigations in support of the technicians and specialists involved in the construction work. From a scientific point of view, this experience has enabled the development of important experiments for the advancement of innovative 3D digital survey methods specifically devoted to the structural monitoring of historical architecture. The historical, artistic and architectural context in which this research

[1]The activities were conducted through a specific agreement with the Department of Civil Engineering and Architecture of the University of Pavia. Project Scientific Coordinator: Prof. Arch. Sandro Parrinello. Technical coordinator of the project: PhD Arch. Sara Porzilli. Promoting Institutions for the survey and documentation activities: "CCC Consorzio Cooperative, Costruzioni, Imprese esecutrici CMSA Soc. Cop.", "MiBACT Soprintendenza Belle Arti e Paesaggio per le province di Firenze, Pistoia e Prato" (MiBACT Superintendency of Fine Arts Academy and Landscape for the provinces of Florence, Pistoia and Prato), "Soprintendenza per la Tutela e valorizzazione beni architettonici, paesaggistici, archeologici, storico-artistici ed etnoantropologici per le province di Firenze, Prato e Pistoia" (Superintendence for the Protection and Valorization of architectural, landscape, archaeological, historical, artistic and ethno-anthropological Heritage to the provinces of Florence, Prato and Pistoia).

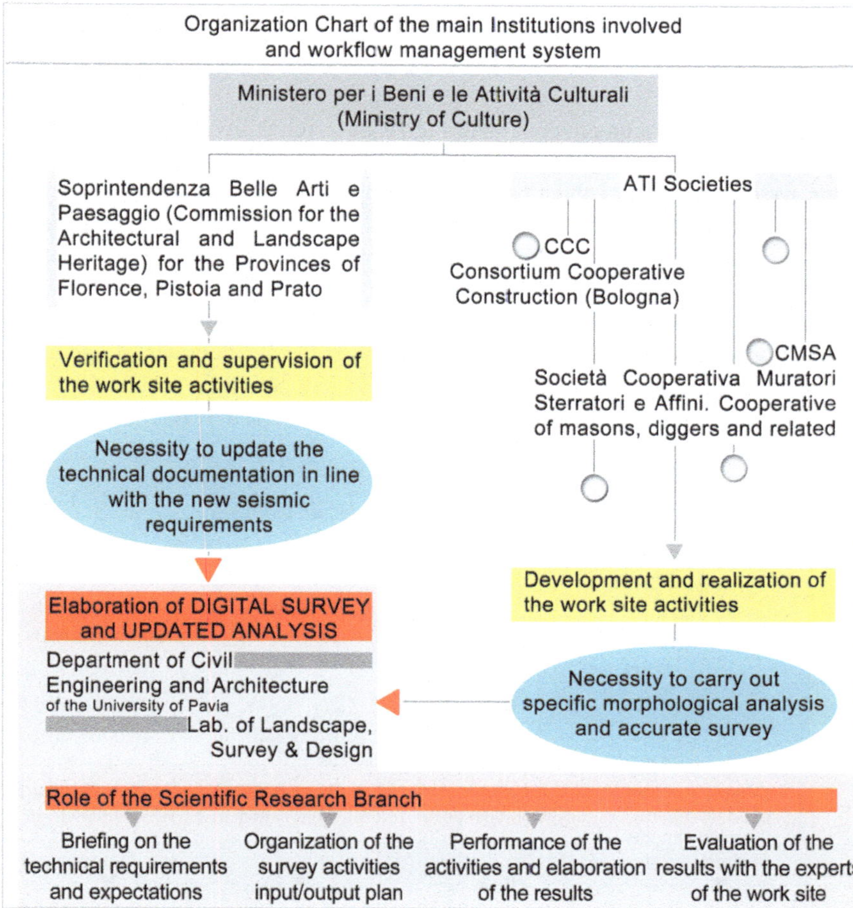

Fig. 2 Scheme showing the institutions primarily involved in the promotion and execution of the process. Survey activities carried out by the research team have produced extensive documentation used by several authorities and for different purposes

has been executed represents a key aspect that has made this such a fascinating and compelling experience, engendering professional passion. The Uffizi Gallery is part of the Italian cultural heritage, not only for its priceless paintings and sculptures, but also for the significance of the building itself, which, together with the city of Florence, is a designated UNESCO World Heritage sites [5]. In this context, monitoring through laser scanner surveys has been carried out with the goal of obtaining detailed morphological analysis with higher metric reliability. This activity has enabled technicians and experts to design more efficient intervention strategies based on detailed documentation. This includes diagnostic analysis, structural inspections of static structures, planning of improved safety measuring activities for each room, restoration and structural consolidation, design of plant facilities, technical corridors and platforms for maintenance and inspection activities.

Research Approach and Development of Operational Methodologies

In order to obtain documentation with high metric reliability, it was necessary to carry out careful planning of all the survey phases (Fig. 3). Due to the complexity of the rooms, as well as their partial accessibility and usability, the survey and photogrammetric data acquisition have relied on reconnoiter surveys and analysis to establish the most strategic locations and methods for instrument placement (Fig. 4).

During the preliminary phase, it was important to define the most appropriate methodologies for achieving the objectives, through the development of operational schemes on which to affix notes and planned on site activities, to identify all the different data typologies obtained and to understand the achievable results [3].

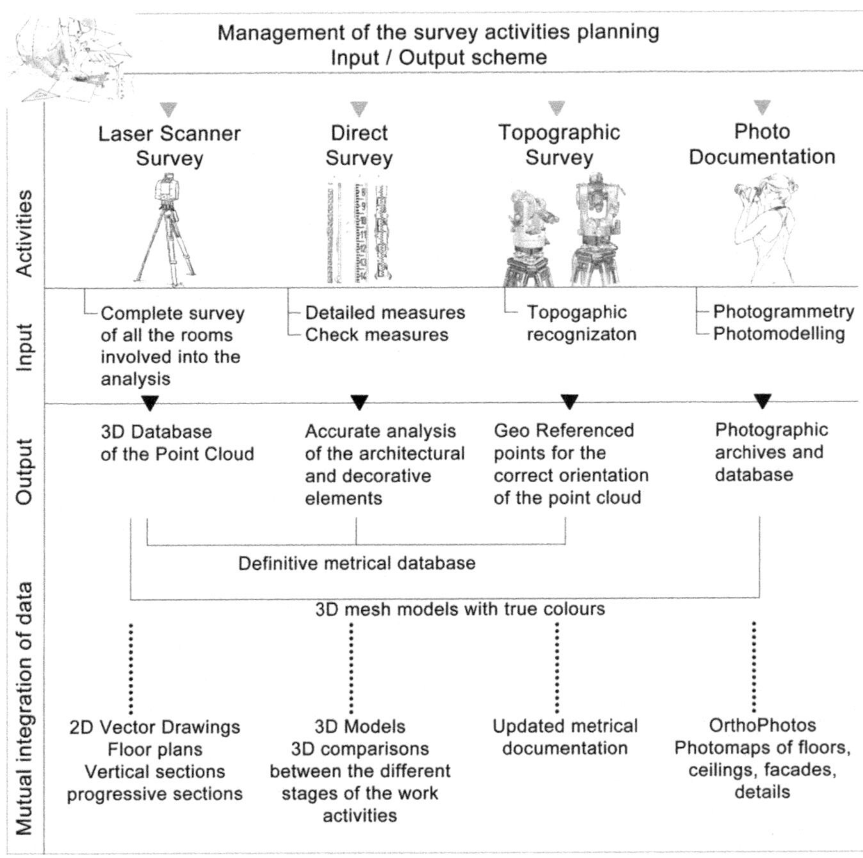

Fig. 3 An integrated research project requires detailed organization of the different activities carried out. By using schemes and diagrams it is possible to develop suitable solutions for achieving the best results from each work operation

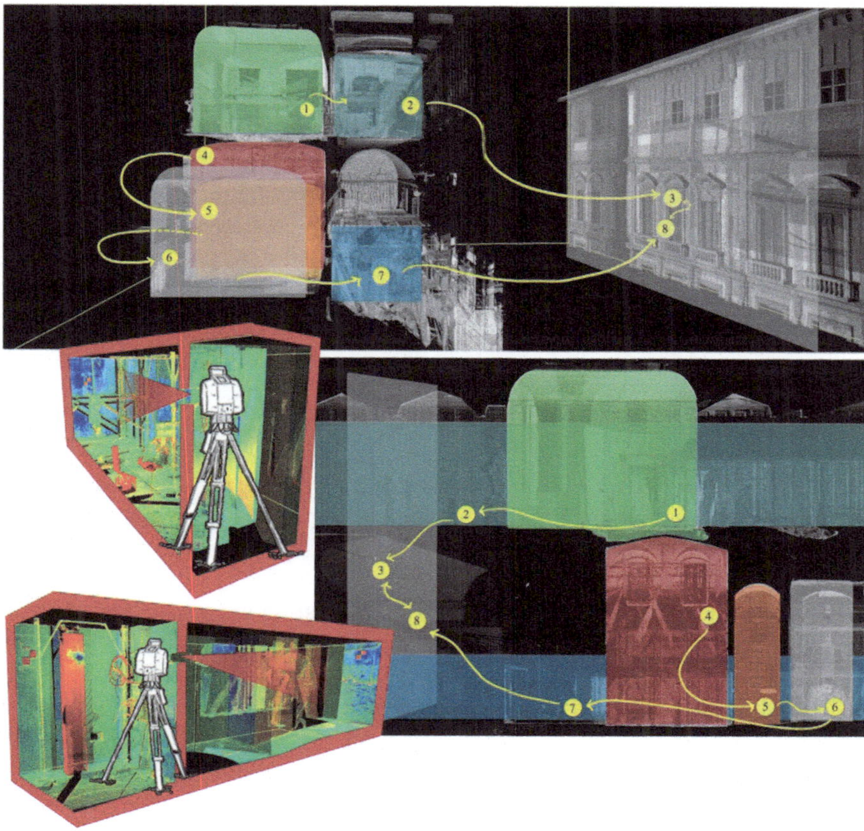

Fig. 4 Laser detection scheme for defining the instrumental path

The second fundamental aspect was the identification of the most appropriate schedule through which to carry out the periodic monitoring scans for obtaining the most relevant data. The elaboration of this information has made it possible to increase the process facilities on the condition of the architectural elements using comparisons and elaborated overlaying data documents.

Analysis and survey activities produce a large volume of updated documentation, characterized by the different data typologies:

- Metric databases (point clouds) obtained from laser scanner survey;
- Vector bidimensional drawings obtained from the elaboration of the point clouds—in particular, general plans, vertical and horizontal sections, progressive sections, comparisons of the data obtained during the different detection phases with dimensioning of the height differences;
- Three-dimensional models obtained through the elaboration of the point clouds with overlapping of the different steps to understand the minimum movements

of the structures and for the static assessment of floors, vaults and wooden structures.

Alongside the technical-operational aspect, this experience has also involved a theoretical and an academic approach for the advancement of the new integrated digital survey systems and for increasing the 2D and 3D post-production methods in order to obtain the newest and most up-to-date procedures for systematic analysis. The fruitful synergy between technicians and professional experts and managers from the university has added value to the research experience, thus increasing our positive results.

The Integrated Survey Project

The most complex aspect of this research has been the ability to perform a laser scanner survey from which all the architectural structures were described completely, obtaining a full survey and investigation of all the extrados and intrados environments and surfaces of all the vaulted rooms, wooden structures, and horizontal and vertical architectural elements (Fig. 5). During the preliminary planning of the survey, it was important to set the laser scanner stations and all the survey activities. The purpose of this phase was not only to identify the best means of understanding the morphological complexity of the building, but also to perform the activity without disruption to the museum. Survey activities covered additional areas around the main survey topic in order to elaborate the triangulation of the points and the closing of polygonal paths, which were necessary steps for the final evaluation and for error compensation (Fig. 6). The survey generated a complete and accurate description in all parts of the masonry floors, the vaulted ceilings, the roofs, the wooden structures with trusses, and the trellis coatings.

Similar to that for the walls and the vertical elements, all aspects of the metric description were complete due to the detection of the architectural elements in their entirety (Fig. 7). The system of laser scanner stations with the polygonal reference used for connecting rooms located on different levels defined a complex structure of relations between the target system and the detected areas (Fig. 8).

Certain rooms of the museum were covered by the investigation so that it was possible to obtain the necessary amount of data for verification of survey correctness and metric reliability. For these reasons, this integrated survey project was initiated, with the use of various detection methods simultaneously, which differed not only in the instruments used but also the results achieved. Three-dimensional laser scanning was the methodology most widely adopted for the detection of the environments under investigation. The laser scans were always carried out at a high resolution in order to simultaneously acquire both the constructive details, deformations and chromatic alterations (deduced from the reflectance value of each different surface), and the targets located at considerable distances (Figs. 9 and 10).

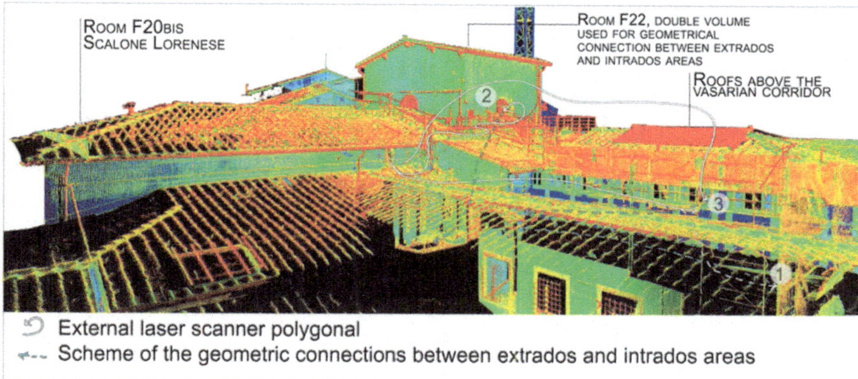

Fig. 5 Graphic representations related to the different types of polygonal designs for the laser scanning survey. The most complex aspect of the survey activities was the need to connect extrados and intrados rooms with the highest metric reliability

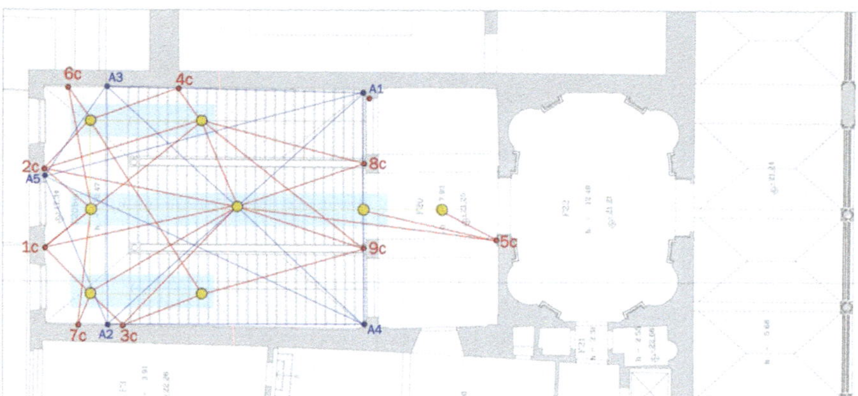

Fig. 6 Reference system with fixed targets for the registration of point clouds

A final aspect, of no less importance, was the ability to manage the instruments (e.g., initiating the scans, monitoring the scanning stages) using wireless remote control. This was particularly important for the management of the scan activities in partially stable environments, such as ladders, scaffoldings and trellis coatings. Oscillation and vibrations produced during instrument installation are detrimental to scan quality. Therefore, under these circumstances, scanning was controlled via a tablet connected to the laser scanner instrument once these disturbances had ceased.

High definition is needed not only to facilitate the redrawing of the architectural elements and = the many decorative elements present, but also to enable a very high level of metric accuracy for measuring the horizontal and vertical bearing structures and their deformations (Fig. 11). For these structures, it facilitated the

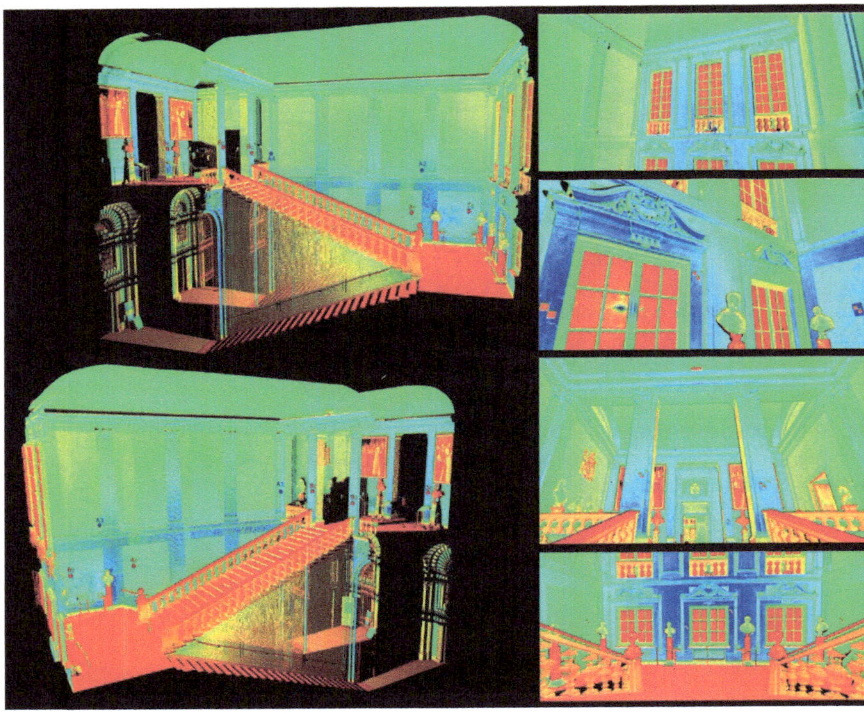

Fig. 7 Views of point clouds of the museum access staircase; the survey was conducted for study of the vault

development of a monitoring system for verification of the stability of complex structures with increased deformations (Figs. 12 and 13).

In some cases, the morphological complexity of the environment required the integration of data obtained from laser scanner survey by performing photo-modeling analysis. From this activity, we developed three-dimensional models of portions of environments investigated directly, thanks to intensive and detailed photographic campaigns using specific software dedicated to this type of function (Fig. 14).

In order to perform periodical monitoring at the worksite, it was vital to place specific "cornerstones points", consisting of targets strongly anchored to the architectural structures, as this made it possible to report each measurement in the general coordinate system adopted for the whole 3D laser scanner survey (Fig. 15).

Verification of the monitoring was then defined by comparing the three-dimensional databases produced by the laser survey in the same reference system. Each group of points with spatial coordinates belonging to the same acquisition campaign was associated with a specific surface. This made it possible to assess more clearly the detachment or the interpenetration that, within a certain tolerance scheme, appeared between the different models and the original surface of the first

Fig. 8 Section of the point cloud where it is possible to see the cutting action of a wall placed over a vaulted ceiling; the survey was conducted to analyze the micro displacement of the ceiling once relieved of the wall weight

detection. The positioning of the pillars constituted the most delicate phase for all the detected environments. Through this activity, it was possible to validate and define the success of the morphological survey, as it allowed us to execute overlaps and three-dimensional comparisons that otherwise would not have been possible to quantify. The network of points constituted by the system of targets represents the geometric platform on which all the stages of monitoring are fixed, and it gives form to a digital archive of metric information that relates to the same reference

Fig. 9 Surveys carried out in the Botticelli room for the insertion of the new suspended ceiling

system and has the same orientation (UCS). In this experience, different point clouds obtained were progressively overlapped as successive layers, thereby allowing a comparative reading to be conducted entirely within the digital simulation (Figs. 16, 17 and 18). Intensive photographic campaigns were carried out in parallel with the laser scanner survey activities. These followed an operational logic process that generally focused on particular architectural and decorative details. The intensive photographic campaigns were necessary for the realization of the photo plane of walls, floors and ceilings. This activity was also important because it allowed archive documentation of several rooms to be updated during the working phases.

In many instances it was possible to create a photographic historical memory of how rooms and spaces appeared before the redesign assessment, and in other situations, it was possible to document architectural features that are no longer visible. This was the case, for example, with the "Sala della Niobe" (Niobe Room), where the floors were reported as the previous ancient rustic asset characterized by ancient lighting systems and vaults in the masonry, and then after certain fundamental structural engineering interventions, they were again covered by the original restored floor (Figs. 19 and 20).

Fig. 10 Detail of the monumental wooden trusses of the Botticelli room

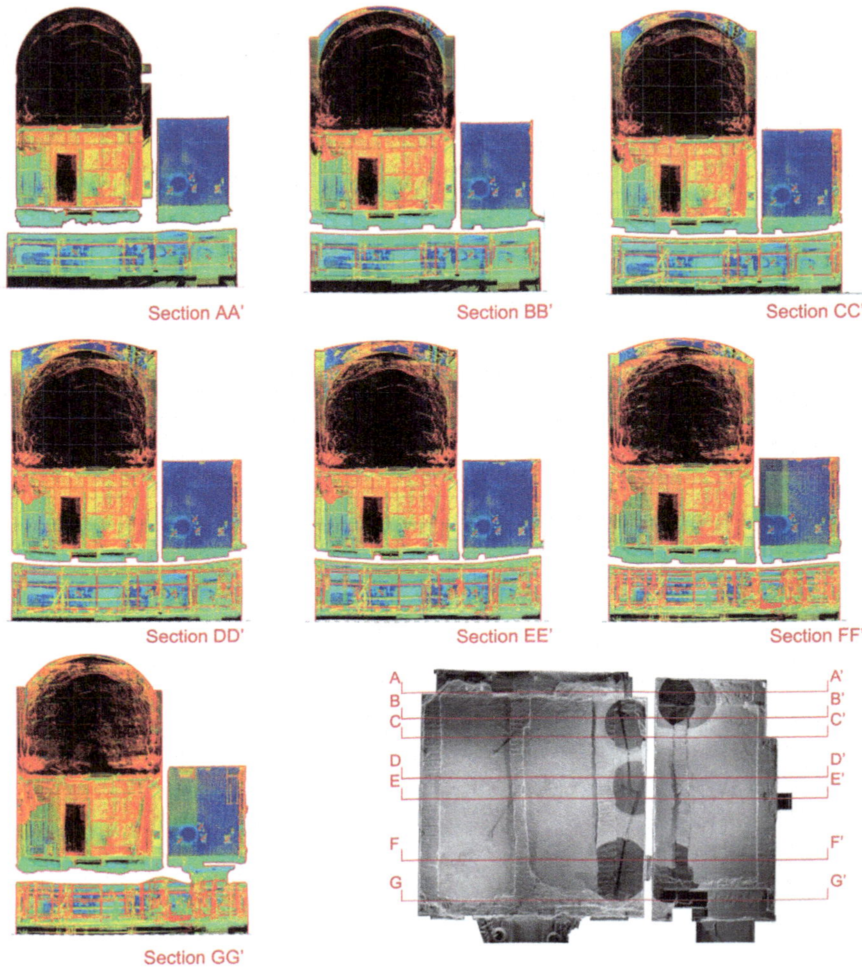

Fig. 11 Analysis of the deformations of a vaulted ceiling due to the weight of a wall

Data Storage Management for the Post-production Phase

Each survey activity has produced a vast amount of data which needed to be carefully scrutinized, evaluated and classified. The laser scanner activities have produced a metric database containing all the partial 3D scans performed (Fig. 21). An essential component of the processing stage was registering[2] and performing all

[2]The "registration" operation represents the technical procedure that enables the combination of different partial point clouds from the same survey activities into one general and complete point cloud. This operation is performed due to the use of common targets surveyed from different scan positions. The specific software can elaborate geometric calculations and algorithms that are aided

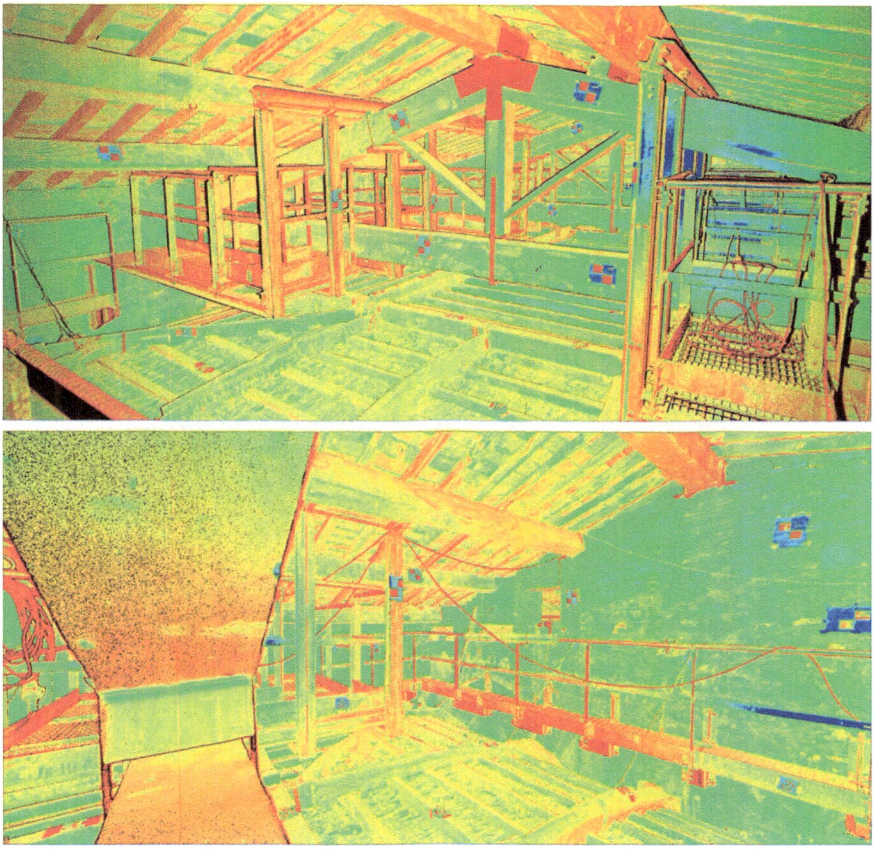

Fig. 12 Point cloud under the roof, above the museum access staircase. It is possible to see the wooden trusses and the top surface of the vault with the wooden structure

tests to verify their reliability. The photographic campaigns are ordered according to an archiving system based on folders marked by reference codes. From this, one can identify each singular architectural element, including the walls, vaults, trusses, staircases and statues (Fig. 22). Due to the use of a code system for storing the information, it is now possible to use the database according to different channels of access that are related to the various operators involved during data processing.

The walls of the documented rooms were defined with respect to their orientation with the cardinal directions N-S-W-E (north, south, west, east). For photo campaigns involving the floors and roofing systems, however, we created descriptive drawings enriched with specific geometric divisions into numbered quadrants. In this way, each individual part is assigned its own related folder. In

---

by these specific points. Geometric recognition produces a global unique point cloud of the whole survey database.

Fig. 13 Path of the laser stations for the survey under the roof, above the museum access staircase

general, all folders of the photo documentation are named with the same codes used for technical drawings or 3D models. Therefore, it is now easy to search for a specific image within a precise section of the architecture documented. To store

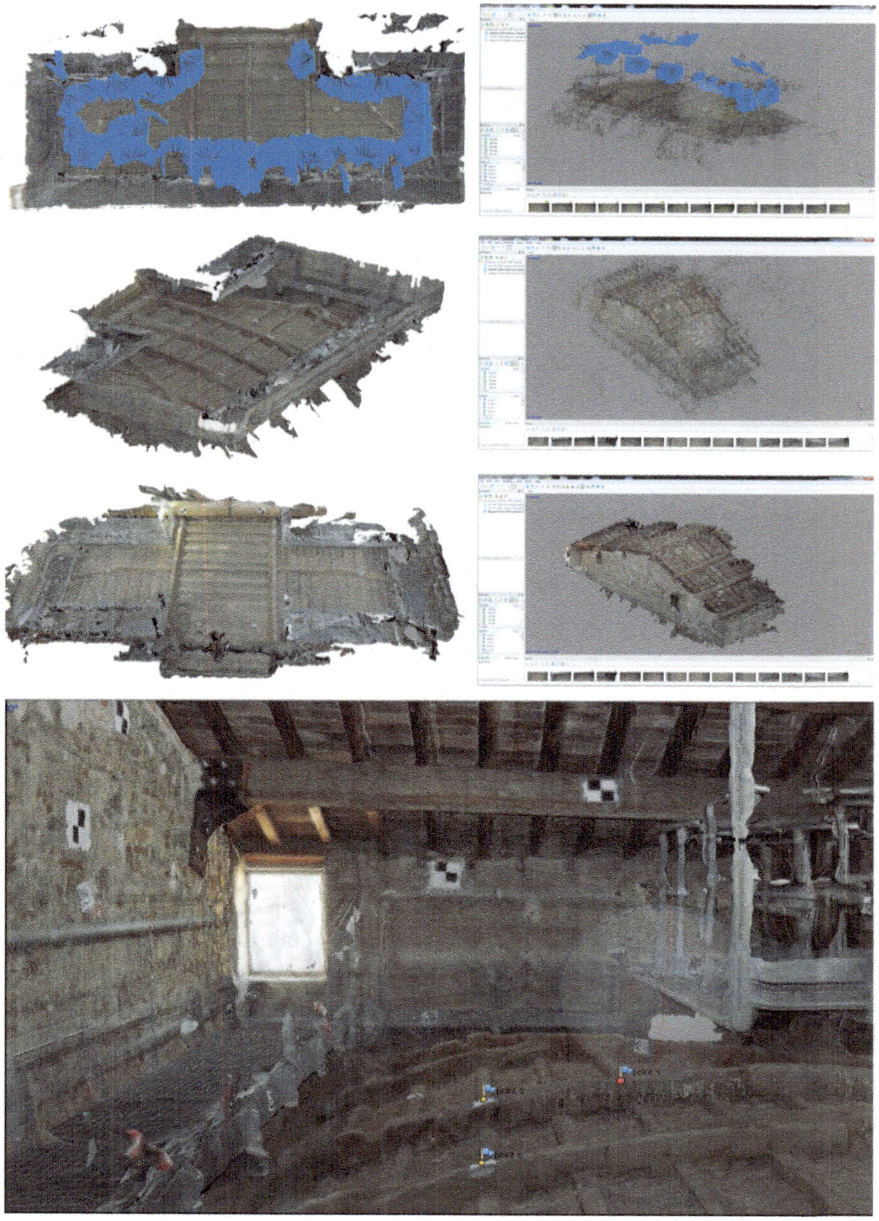

Fig. 14 Photogrammetric survey by PhotoScan software

photos related to the wooden structures, trusses and beams, we have used reference grids where "xi" identifies x-axis and "yi" y-axis. In this way, it was possible to locate precisely each element. For example, an entire truss is identified with an "xi"

Fig. 15 View of the target system for the registration of the 3D point cloud

or = "yi" code depending on the warping direction, while a node or a joint is assigned an "xiyi" code [1].

After cataloging the metric information and processing data, it is possible to start the post production step. This step is devoted to the interpretation and representation of data by developing cross-comparisons for critical evaluations, among other techniques.

Computer Data Processing

The registration procedure and the testing elaboration on the point clouds constitute the first critical steps of data processing. Through implementation of the target system explained above, it was possible to merge several partial 3D scans to obtain a complete three-dimensional model of entire rooms and additional spaces. Testing is the final delicate phase, where the level of detail and accuracy of the registration phase are verified and quantified. In this phase, we strategically elaborate certain sections to ensure the absence of erroneous overlapping and roto-translations between partial ScanWorlds.[3] We then assess the level of detail exhibited by all parts of the point cloud in order to define the overall quality of the work done. The same check is performed at the end of the registration process, where complete sections are elaborated in both directions. With the Cyclone software package it is possible to use the "slice" function to obtain section profiles featured on the cutting plane selected. This option gives the surveyor an important digital instrument that enables verification of the quality of the registration process made. The final step is devoted to cleaning the point cloud from all elements of disturbance considered

[3]We define "ScanWorld" as the partial point cloud obtained from a single scan position.

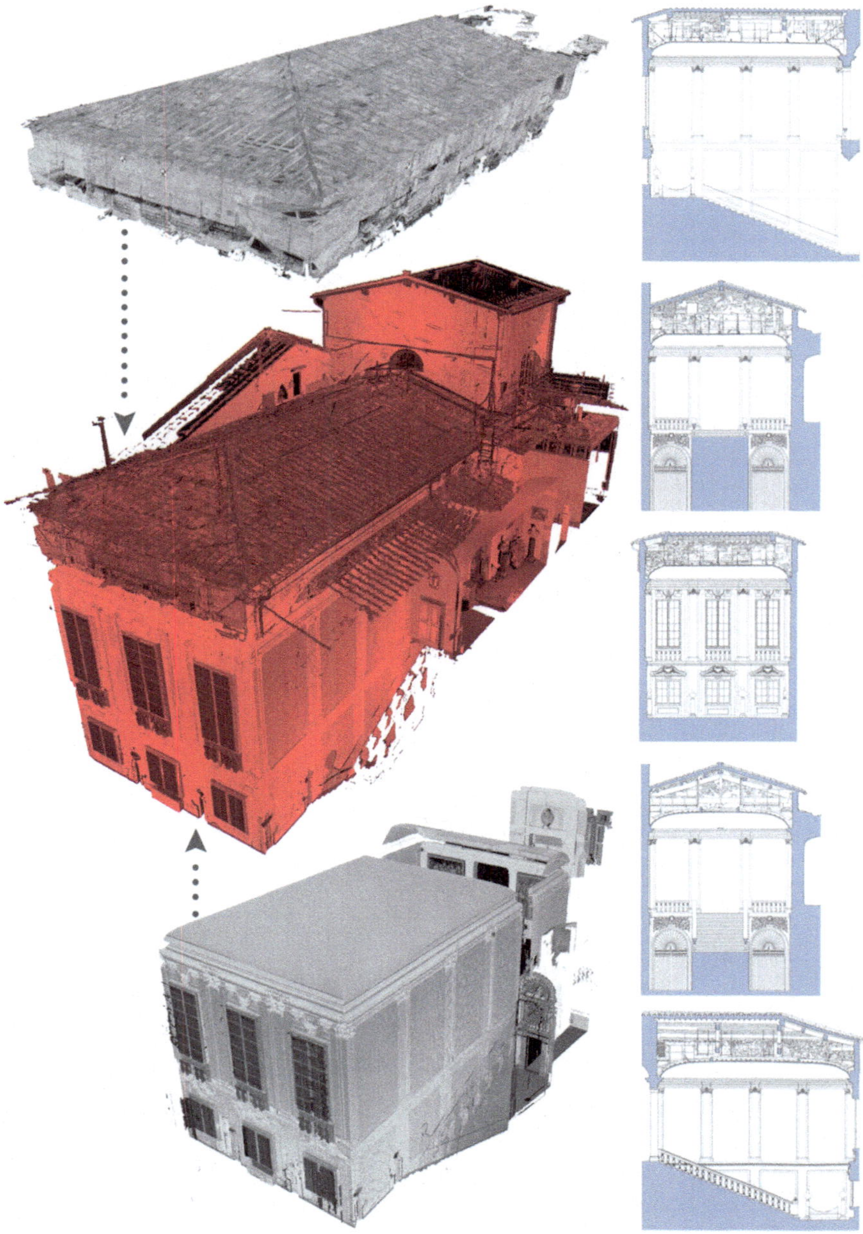

Fig. 16 Union of the extrados and intrados areas in the same reference system

"noise". This "noise" might disrupt measurement, the ability to understand the environment and elaboration of the technical drawings. The "noise" can be bundles or agglomerates of points generated by the instrument during the survey process.

Fig. 17 Section of the staircase for verification of geometrical alignment

The main causes include the presence of surfaces that create reflections; dust in the air; and natural or artificial light that disturbs the pulsed laser light of the scanner, creating agglomerations of points and interference. In addition, operators and

Fig. 18 Point cloud slice of the vault and cross sections

people in general present during the survey activities can create significant deflections. Furniture, supplies, temporary structures, such as scaffolding and platforms related to the construction site, and permanent instruments and equipment

Fig. 19 Details of the graphic restitution: point cloud, wire frame drawing, photomap reconstruction

can emerge as disruptive elements. During post-production these disruptive elements are removed from the final point cloud.

Interpretation and Representation of Data

During the first phase of elaboration of data acquired, we developed all the two-dimensional vector drawings. Specifically, we developed intrados and extrados floor plans with metric dimensions (by using a reference horizontal plane or level previously agreed upon with supervision of works), longitudinal sections, cross sections, and progressive sections performed with a minimum constant dimensional

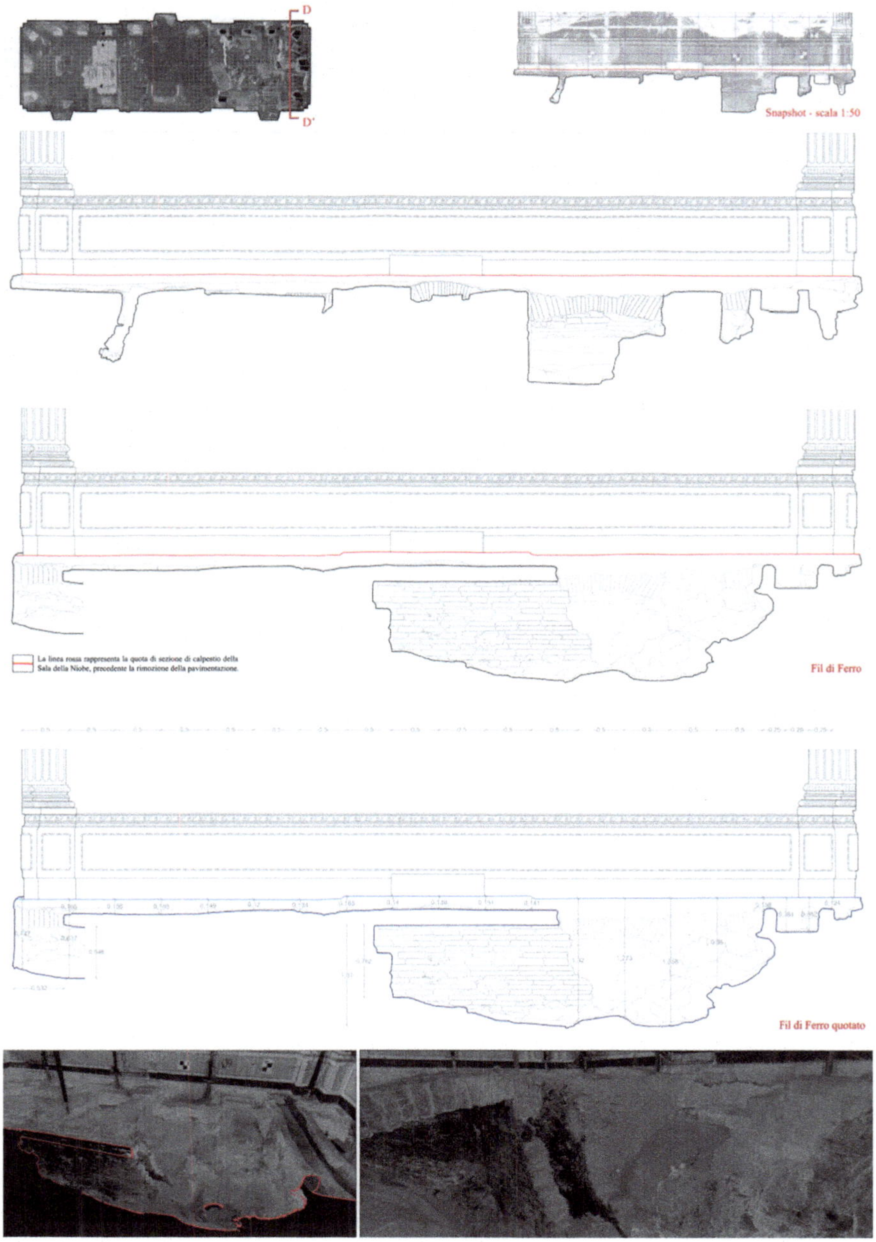

Fig. 20 Detail of the drawing of the excavation under the floor

interval of 5–10 cm. All CAD drawings were produced with a level of detail equal to the metric scale 1:10–1:5. Following the elaboration of the vector drawings, ortho-photo planes were processed for the representation of each wall and façade [6].

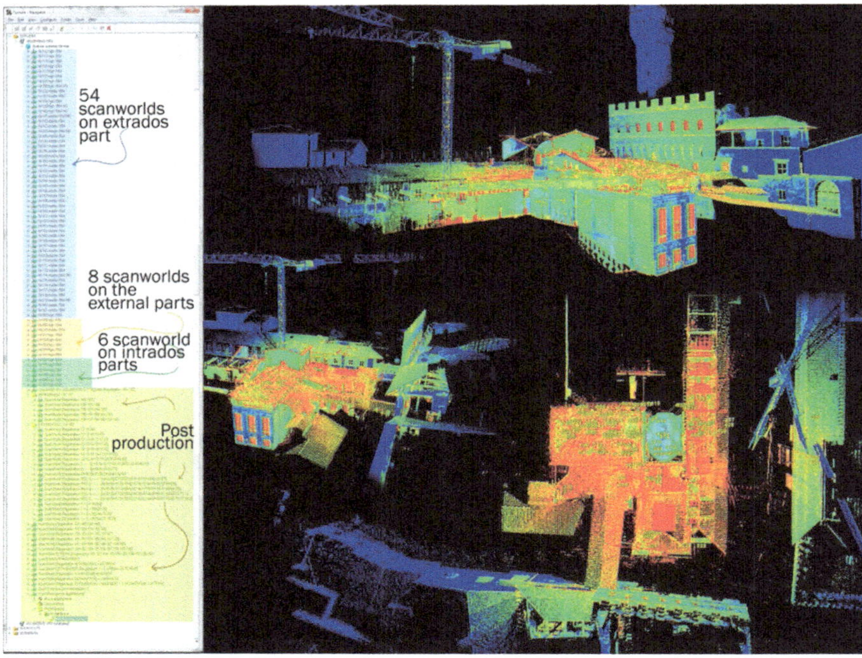

Fig. 21 The database organization of the laser scans uses a large system of reference for the general alignment and a close-range system for the alignment between the individual portions of the building

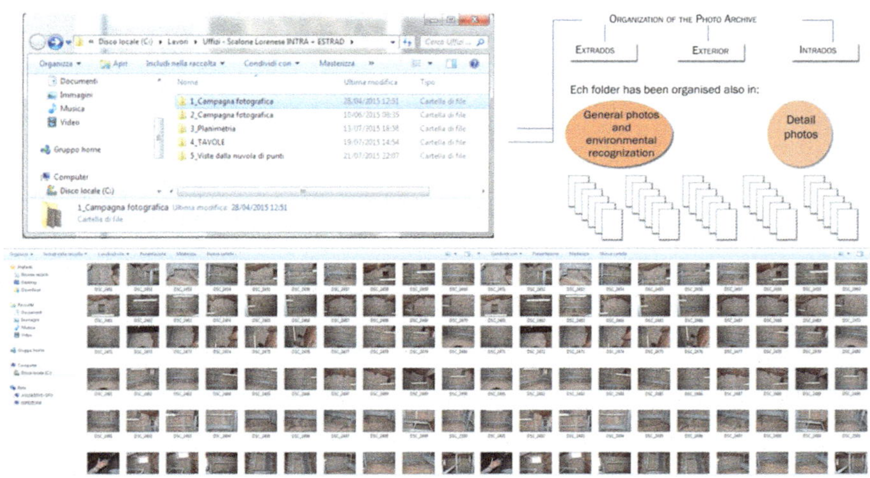

Fig. 22 Organization of the photo documentation

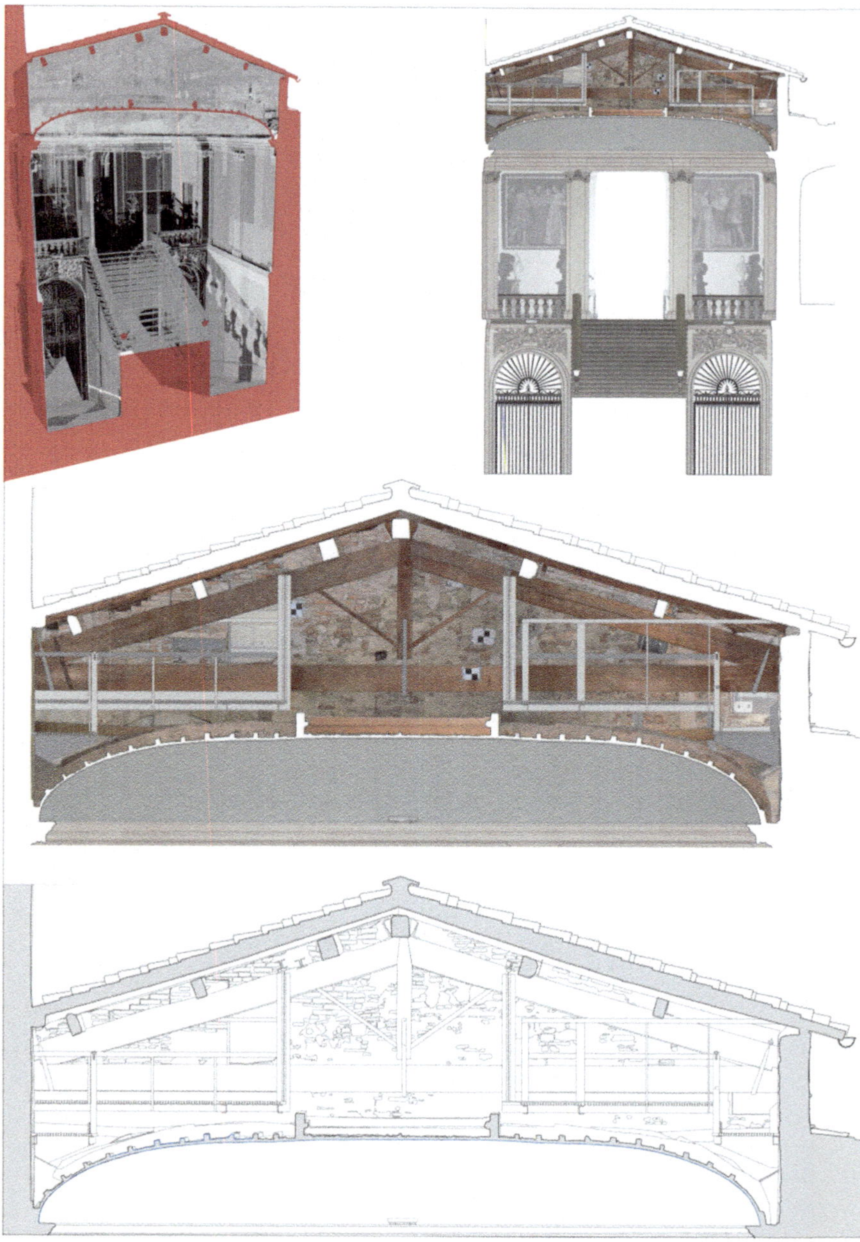

Fig. 23 Main steps during the post-production phase. The point cloud allows technicians to elaborate specific sections in order to proceed with detailed drawings. This figure shows a trasversal section which includes both extrados and intrados parts. A comparison between wire frame drawings and photomapping elaborations demonstrate the high resolution and quality of details obtained

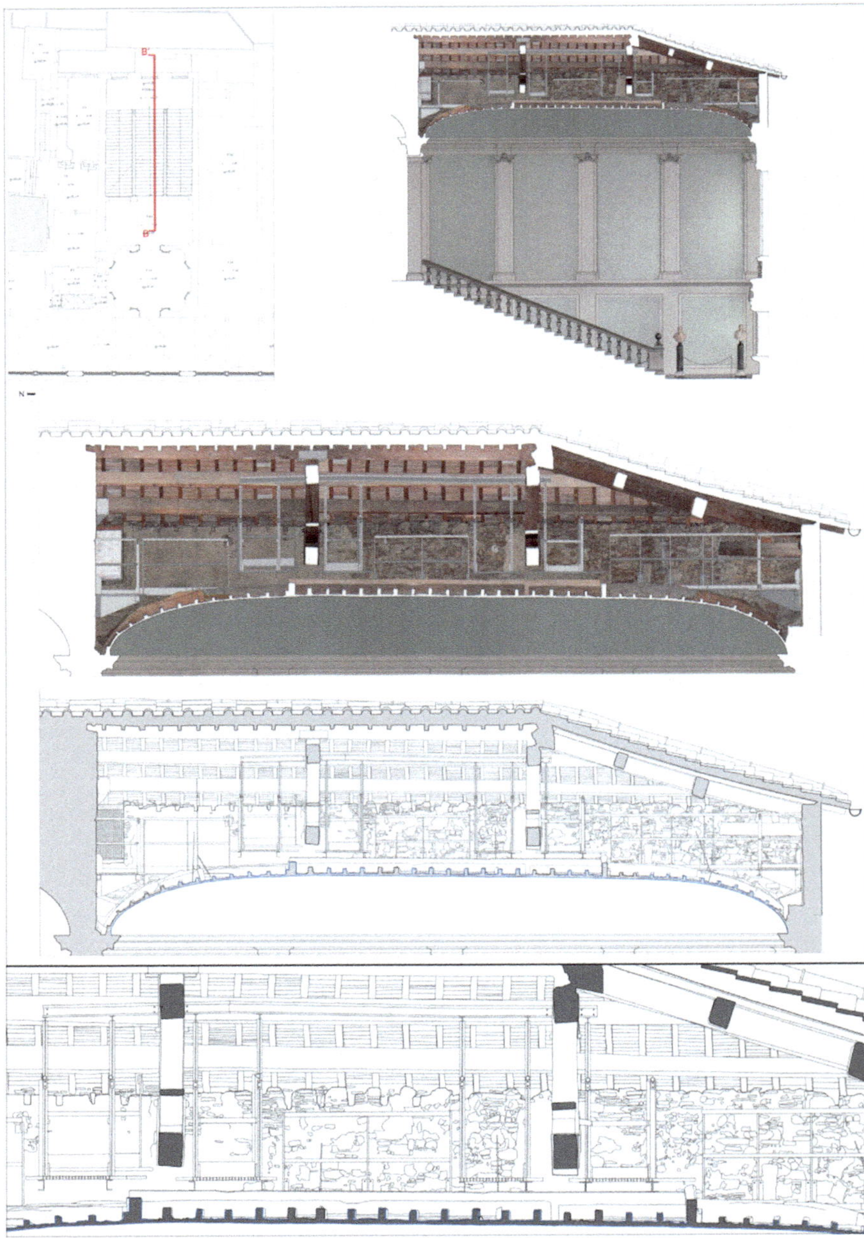

Fig. 24 Longitudinal section for the understanding of the timber structure of the main roof and for the assessment of the ceiling vault, main object of this specific investigation. Comparison between wire frame drawings and photomapping elaborations

This is necessary in order to obtain material for the realistic description of the rooms and an updated recognition of each surface detected. The processing of ortho-photo planes is a more accurate way to perform the necessary deterioration mapping analysis for restoration and consolidation projects (Figs. 23 and 24).

Images were acquired at very high resolution in "RAW + JPEG fine" mode in order to manage the balance and color correction during the post-production phase.

The high quality of the photo documentation used for the elaboration of the ortho-photos guaranteed great resolution for graphics printed up to a scale of 1:20–1:10. In some cases, the investigations carried out in the room of the Niobe being exemplary, processing the photo representations of the structurally ancient floor slab structures while the floor and underlying materials were removed presented an interesting opportunity to admire the original floor slab and its complex system of vaults. We have completed a laser scanner survey of this precious and unique architectural feature, and by doing so we have revealed an important descriptive graphic material that would not otherwise be visible without removing the upper structural layers.

Digital Models for the Simulation of Static Assessment and Virtual Applications for the Conclusive Survey Results

Following the development of two-dimensional vector drawings and vectors on CAD software, we proceeded to develop three-dimensional models for each survey phase directly from the point clouds (Fig. 25) using dedicated software capable of transforming the point cloud database in mesh models [4]. This phase involved the most innovative experimentation, and this has made it possible to determine the optimal methods for facilitating the activities of a worksite as complex as the Uffizi Museum. We have documented all operating phases by analyzing the morphology of the structures using virtual simulation during the period concentrated on consolidating works. During the post-production phase, including the practical survey activities, the results were integrated to enhance the level of detail and the metric reliability of the research. For each phase of monitoring and analysis, 3D models were produced directly from the point cloud database. This intricate operation was performed using *Rapidform* software (Fig. 26). This software has several commands and toolbars that enable the necessary corrections to the processed three-dimensional models. These corrections preserve the metric reliability of the point cloud, but at the same time they generate mesh models in which walls, vaults and floors are represented by continuous closed surfaces [7].

Where the data provided by the point cloud were found to be insufficient, we combined the information with direct measures, either by comparison with the intensive photo documentation or with the support of the photo-modeling activities. After the predetermined phases of monitoring were performed, the individual 3D models related to a single room or a group of rooms, but descriptive of different

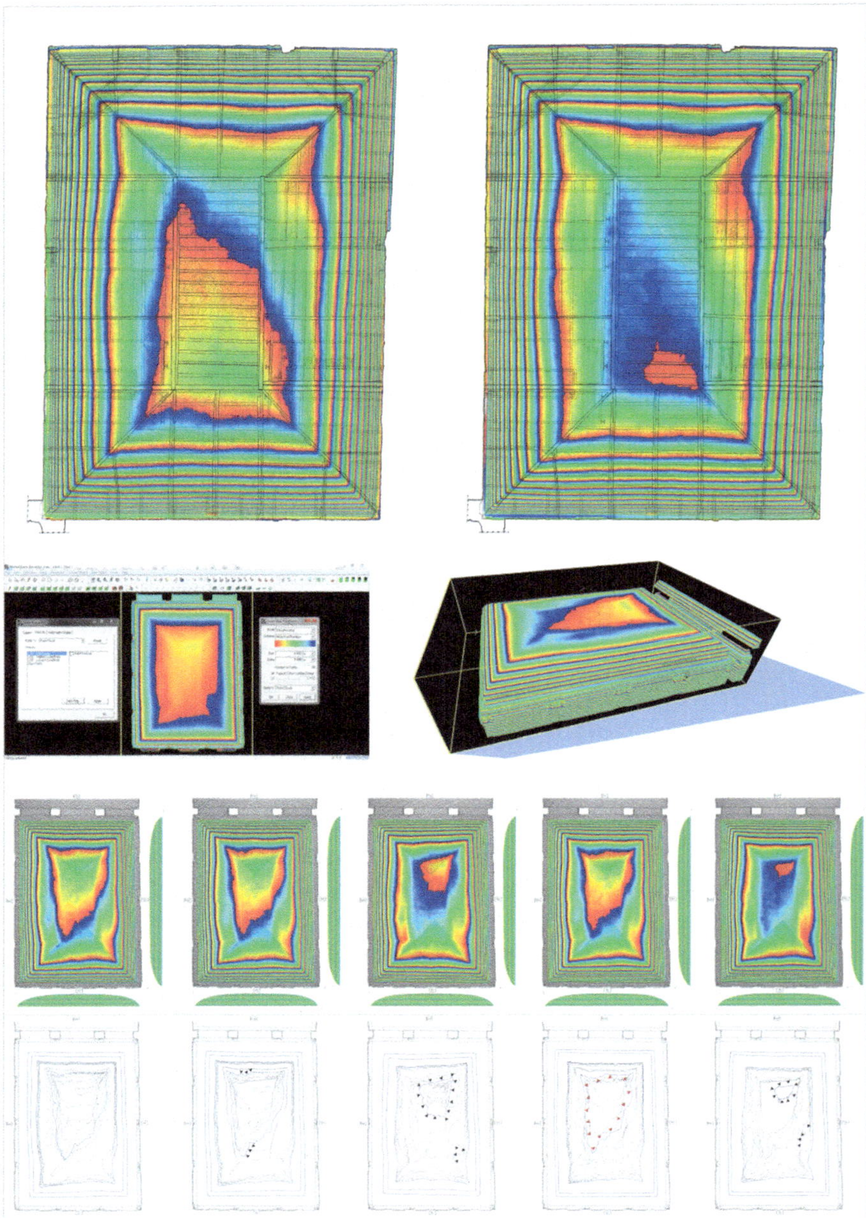

Fig. 25 Elevation map calculated on each point cloud of the studied vault. The geometry of colors shows a deformation and a collapse. Drawings show the results of the five surveys of the vault conducted over two years

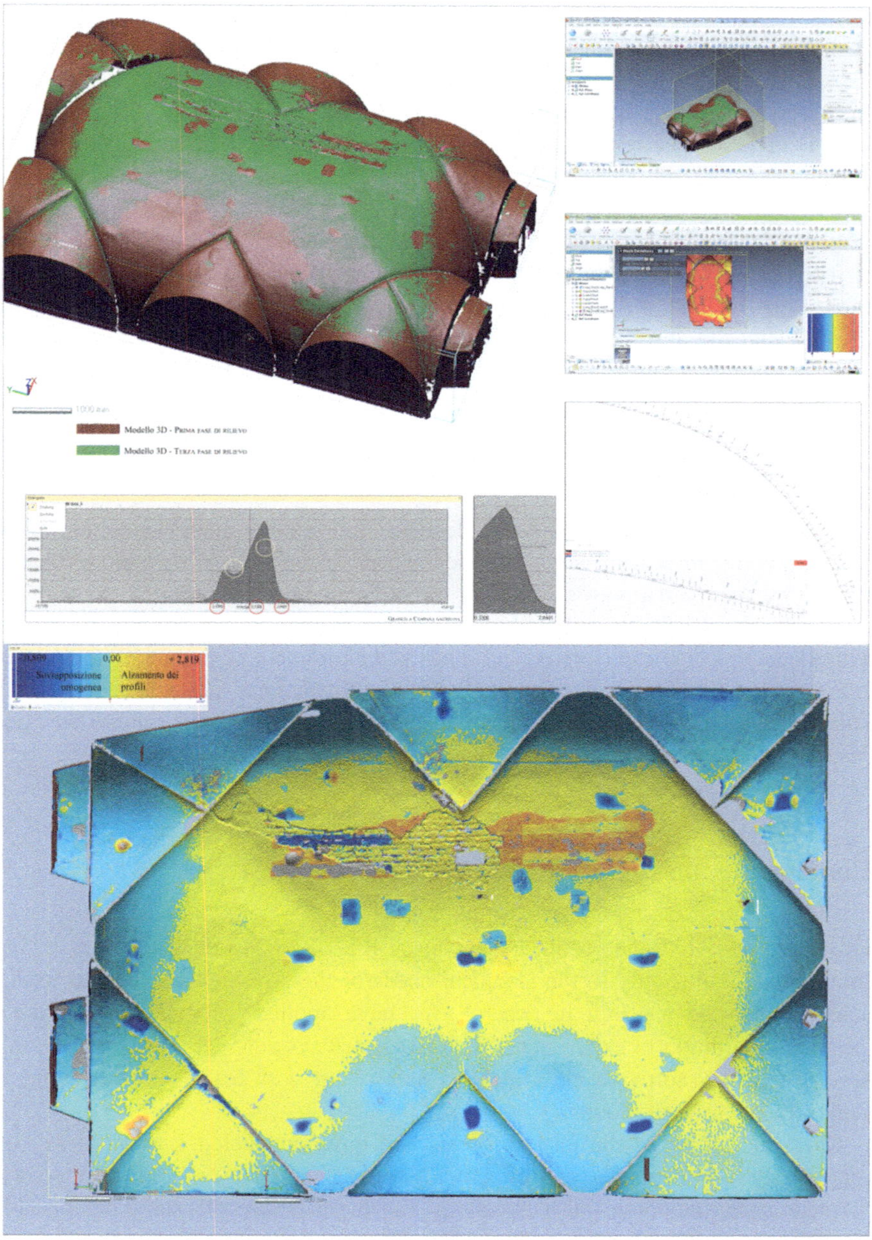

Fig. 26 Mesh model for verification of the most deformed areas through the overlapping of different 3D models

work phases, were overlapped by use of the same reference system (UCS) to identify potential deviations between surfaces. These deviations are identified by the software through the geometric analysis algorithm by the "deviation parameters" registered along the three directions X, Y and Z.

The same surface represented and documented with a different surface development demonstrates that in the period between successive laser scanner campaigns, the structure suffered sensitive movements. The *Rapidform* program and related software can quantify the value of this data element. The level of detail and metric reliability ensures the ability to switch between processing parameters. These parameters include the tolerance for instrumental error exhibited by the laser scanner, and the absolute error verified at the end of the point cloud registration process, during alignment of the partial point clouds. The software not only allows the direct calculation of various deformations of the processed point cloud, but it also offers, during the final synthesis and comparison step, the opportunity to represent this metric information along with statistics obtained through graphics and Gaussian curves.

The results of calculations have identified the maximum deviation measured over the entire three-dimensional model, where the maximum deviation value is defined by the overlapping of different three-dimensional models.

Thanks to these charts and models, it has been extremely easy to share the results with the technicians and experts, and it has had an immediate impact on the critical interpretation of numerical percentages of the dimensional deviations registered. This data has provided engineering managers critical elements to consider when designing the most appropriate intervention strategies [11].

Conclusions on the Research Experience

This research, which has been under way at the Uffizi Museum since 2010, has contributed significantly to the technical needs of the construction site and to this field of scientific research in general. We have applied technical laser scanner methods in addition to experiments with integrated systems. At the same time, we have tested new software elevating the state of art with respect to the dynamic management of point clouds.

Over the course of the past six years, we have conducted a wide survey on approximately ten rooms, each of which is characterized by unique architectural characteristics, including the morphology and the presence of usable or unusable floor slabs, precious vaulted roofs, and wooden truss systems (as was the case for the Botticelli Room or for the extrados room above the main access of the Scalone Lorenese stairs leading to the Vasari Gallery) [9].

Survey procedures have also identified different intervention strategies for planning and executing the activities based on the location of the rooms studied and the ease or difficulty in accessing the connecting areas. Throughout these studies, we have explored operational methodologies for data acquisition and post-production information handling. This has enabled the exploration of new techniques and has significantly reduced the magnitude of procedural operational errors.

In all point clouds produced and registered by different software programs, a consistently high level of metric reliability was maintained, with minimal dimensional error. This was confirmed via parallel tests carried out by technical engineers at the worksite whose responsibility is monitoring the reinforcements and consolidation solutions introduced in various rooms that we analyzed.

The photographic documentation was expanded, in the case of both orthophoto image processing and the realization of 3D models through photo-modeling activities. The use of professional digital cameras and their associated software in post-production allowed us to obtain highly realistic mapped models that depict the natural colors of the plastered or painted walls of the wooden and masonry structures documented. The photo campaigns have also enabled documentation of areas that are no longer visible. For instance, the floor structure in brick vaults of the Niobe room is presently covered with the original restored floor. The massive structure of wooden trusses present in the Botticelli room defines another example: it is presently hidden by the ceiling system required for placement of the air conditioning designed for the renovated room.

It has been necessary to enhance vector drawings by the use of more suitable graphic tools and to interpret the symbol system to which we transfer the metric information on the two-dimensional drawings. This has increased the readability of comparisons between the same architectural sections that are composed of the sectional profiles related to the different time periods. The investigations have produced transverse and longitudinal sections, with scales of 1:20–1:25, accompanied by detailed drawings with scales of 1:10–1:5 for representation of specific portions with obvious deformations. Along these lines, we have integrated enhanced processing techniques for the progressive sections, and this has made it possible to obtain a series of parallel sections characterized by minimal range. With these kinds of technical drawings it is possible to check on the morphology of the architectural structures in detail. This procedure has given rise to valuable results during analysis of Room F61 and its adjacent rooms. This room was characterized by the presence of a vaulted structural floor which contained a "false wall" that compressed the stability of the vault itself, thereby compromising the static assessment of the whole structure. In this case, the progressive sections have enabled technicians to identify the points at which the wall has further impaired the vaulted structure, as well as parts of the vault that have already started to show signs of structural failure. These technical drawings have made it possible to act quickly on the damaged wall and reinforce the vault at specific points [8].

In the framework of this research project, the implementation of reverse engineering activities has strengthened the technique of 3D modeling directly from the point cloud. By increasing the potential of the tested software, we have strengthened the operational techniques that allow one to overlap 3D models derived from each phase of the work and elaborate calculations of the deviations for evaluation of the dimensional differences [10]. The integration of different detection techniques has also aided the post-production phase, specifically through the integration of the results from new methods of technical documentation, which is critical for supporting a site as complex as the Uffizi Museum.

In this research project, we have implemented and updated specific documentation for a monumental system, and we have produced analytical methods for the prevention of seismic risk by monitoring the structural conditions of different construction systems. Considering the numerous threats to our cultural heritage today, this represents an invaluable procedure. Digital technologies provide experts a precious opportunity to know and work on the architectural heritage with the delicacy required for preservation of heritage. It is currently essential to improve innovative systems by way of constant experimentation.

Restoration and structural consolidation provide a unique challenge for the scientific field that derives from the infinite complexity of historical architecture. Drawing—especially the architectural drawing germane to this study—remains the main method for governing technology and controlling digital processes which, in the representation of the image, create results arising due to the application of complex and advanced technology, equipment and methods. The representation of vaulted systems, monumental halls, stairways and wooden structures at the Uffizi Museum has given rise to precious updated historical documents. It has been possible to represent the condition of one of the most historical Florentine buildings, a building that constitutes one of the most important contemporary restoration sites in Florence. From an academic point of view, this experience has also led to improved methodologies for the study of historical architecture. This experience should serve as a starting point for further application of three-dimensional data banks for the preservation, management and understanding of architecture.

Credits Sandro Parrinello wrote the introduction and chapters: 1, Research approach and development of operational methodologies; 2.2, Interpretation and representation of data; 3, Conclusions on the research experience.

Sara Porzilli wrote chapters: 1.1, The integrated survey project; 2, Data storage management for the post-production phase; 2.1, Computer data processing; 2.3, Digital models for the simulation of static assessment and virtual applications for the conclusive survey results.

The reference list is limited to main works for further reading and to author's main experiences. Citation of references has not included in the text since space limits.

References

1. Barthes SR (1980) AAVV. In: La documentazione dei beni architettonici e ambientali. Strumenti, indagini, esperienze. La Camera Chiara. Nota sulla fotografia. Piccola Biblioteca Einaudi, Torino, Italia
2. Bertocci S, Puma P, Balzani M (2007) Banca dati 3D per la valorizzazione, il restauro e la conservazione dell'opera michelangiolesca a Firenze. In: Ferrara CFR (ed) Bologna. Zanini, Italia
3. Bertocci S, Bini M (2012) Manuale di rilievo architettonico e urbano. Città Studi edizioni, Torino, Italia

4. Bertocci S, Parrinello S, Vital R (2013) Masada notebook. Report of the research project, vol I. Edifir, Firenze, Italia
5. Bertocci S, Parrinello S (2015) Digital Survey and Documentation of the Archaelogical and Architectural Sites. UNESCO World Heritage List. Edifir, Firenze, Italia
6. Bini M (2000) Tecniche grafiche e rappresentazione degli elementi dell'architettura. Alinea, Firenze, Italia
7. De Carlo L (a cura di) (2007) Informatica e fondamenti scientifici della rappresentazione. Collana Architettura, Urbanistica, Ambiente. Gangemi Editore, Roma, Italia
8. Docci M (2005) Metodologie innovative integrate per il rilevamento dell'architettura e dell'ambiente. Gangemi Editore, Roma, Italia
9. Giordano G (1997) Tecnica delle costruzioni in legno. Caratteristiche, qualificazione e normazione dei legnami da costruzione. Progettazione e controllo delle strutture lignee tradizionali. Applicazione dei moderni metodi di calcolo alle nuove tipologie costruttive. Hoepli Editore, Quarta edizione, Milano
10. Guidi G, Russo M, Beraldin JA (2010) Acquisizione 3D e modellazione poligonale. McGraw-Hill, Milano, Italia
11. Migliari R (2004) Disegno come modello. Kappa, Roma, Italia

GIS-Based Study of Land Subsidence in the City of Bologna

Rose Line Spacagna and Giuseppe Modoni

Abstract Geographic information systems (GIS) are very powerful tools capable of performing spatial analyses over large amounts of the territory representing large amounts of data. They consist of hardware, software, and human activities that facilitate collection, analysis, and storage of data, in different forms, related to the features of the territory. In the present chapter, the potential of the GIS tool is explored in a study of the effects of subsidence that occurred during the previous decades in the city of Bologna, where a valuable historic and cultural heritage made up of an intricate system of old masonry buildings, churches and monuments, is exposed and at risk. Data from previous geological, hydrogeological and geotechnical studies was collected and georeferenced. Then, topographical investigations and periodic records of the groundwater level in the network wells were superposed to establish a relation between causes and effects. The analysis reveals the spatial and temporal distribution of settlements in the city center and in the surrounding area. The evolution of the subsidence phenomenon has been possible, and the use of GIS has revealed the role of different factors and their mutual correlation. By combining all information, it can be seen that the area is severely affected by the distribution of total and differential settlements and that a particularly critical situation emerges in the city center. Additionally, the causes of the phenomenon could be established with great precision, enabling the identification of appropriate countermeasures.

Keywords Geographic information systems · Subsidence · Geostatistics

R. L. Spacagna (✉) · G. Modoni
Università di Cassino e del Lazio Meridionale, Frosinone, Italy
e-mail: rlspacagna@unicas.it

E. Ottaviano et al. (eds.), *Mechatronics for Cultural Heritage and Civil Engineering*,
Intelligent Systems, Control and Automation: Science and Engineering 92,
https://doi.org/10.1007/978-3-319-68646-2_10

Land Subsidence Induced by Groundwater Withdrawal

The Earth's surface is the result of the combined action of exogenous and endogenous morphogenetic processes that lead to modification and adaptation of the chemical, physical and biological environment. Through their activities, humans interact with these processes and affect the balance of nature by acting as an additional "geological agent", capable of triggering new mechanisms that sometimes produce more conspicuous and incisive effects than natural agents.

For instance, in the recent past, due to the high population density in big cities and the consequences of choices based solely on socioeconomic criteria, a pronounced and quantifiable degradation of groundwater resources has occurred. In fact, an indiscriminate exploitation of aquifers has occurred in conjunction with an increased use of chemical products, including pesticides and fertilizers and an increased production of civil and industrial wastes. These activities have placed pressure on vulnerable territories.

Therefore, in the context of geological risks, the decline of the quality and quantity of groundwater resources has become increasingly relevant. Defining and identifying feasible methodologies to assess this risk has become a major concern from both a scientific and technological standpoint.

One of the most drastic effects resulting from the over-exploitation of aquifers is land subsidence, which arises from the depletion of flat areas. The efficiency and functionality of structures and infrastructures present in these areas may be compromised in relatively short times, and this is a particularly dangerous situation. This phenomenon, observed in many areas globally, is only the surface manifestation of a large number of modifications that occur within the subsoil. The deformation of the subsurface is generally gradual and protracted and is therefore difficult to perceive. Therefore, there is increased need for prompt and efficient monitoring tools.

In general, the deformation of the earth's surface arises from several natural causes, including tectonic processes, isostatic movements and chemical-physical transformations of sediments resulting from lithostatic load or oscillation of the groundwater table. The importance of subsidence increases significantly when these geological processes are accelerated by engineering activities. The extent of urbanization and industrialization in an area "sensitive" to subsidence may enhance the importance of this phenomenon.

The importance of subsidence in densely populated areas has been documented in the literature. London [32], Mexico City [15, 34], Houston [13], Santa Clara Valley [21], and cities in Japan [33], Greece [27], China [24] and Spain [30] provide remarkable examples. In all of these examples, the settlement varies as a function of the subsoil composition and the level of groundwater exploitation from a few millimeters to more than 12 cm per year. In some cases, the equilibrium alteration induced by subsidence affects areas with numerous historical and cultural sites and monuments. The ancient Italian cities are among the most remarkable examples of contemporary civilization, not only for the buildings and churches that constitute authentic architectural masterpieces, but also for their ability to preserve and pass to new generations the memory and the culture of the past.

The study of subsidence is multidisciplinary. Additionally, the information available does not cover the entire area under study, nor the reference time interval. In this context, it is hard to describe this complex geotechnical phenomenon without making some erroneous assessments. However, the use of data processing in conjunction with powerful software and the implementation of simple models can facilitate definition of the comprehensive framework.

Geographical information systems are powerful platforms capable of managing and analyzing the entirety of this heterogeneous information, thereby enabling identification of the causes and effects of the subsidence phenomenon to be identified.

Geographic Information Systems (GIS)

The current evolution of decision-making models has led to the development of efficient instruments capable of providing more accurate analysis within a much shorter time. In this context, public authorities and, more recently, private companies have implemented geographic analysis to provide more efficient solutions to problems in various sectors.

Most of the activities relate to space and geography. The information, directly or indirectly connected to the earth's surface, is called geographic information (GI). However, GI does not only refers to the description and position of entities on the territory and their representation with geographical maps; it also includes the spatial relationships between objects and events on the territory that affect human activities.

The GI stores data in digital form and processes it by computer. The information can be collected from different sources (e.g. satellites, GPS, aerophotogrammetry, maps, topography) and catalogued at different scales of representation. While these systems greatly expand the capacity for knowledge, they must also establish connections between data sets to enable efficient use of the acquired information.

The development of applications based on GI has expanded considerably for within the GIS framework that use highly advanced software and hardware platforms. The GIS is generally considered one of the most powerful tools in the field of information technology. A GIS is a complex system of hardware, software and human activities that enables the capture, analysis, storage and return of data related to the territory in both graphic and alphanumeric form. By integrating different types of data (vectors, images, text, etc.), GIS-based technology enables the acquisition and analysis of GI within different thematic areas with unrestricted resolution. This flexibility expands potential applications and dynamically and interactively generates answers to the problems at hand.

The importance of GIS stems from often inaccurate and poorly maintained geospatial data, out-of-date maps and statistics, and a lack of data sharing and retrieval services. Therefore, the creation of GIS provides several benefits. For instance, geospatial data and information is maintained and updated in a standard format, and is easier to locate, analyze and present. This facilitates rapid and informed decision making.

All these aspects have permitted the utilization of spatial data for analysis and planning in many fields, including civil engineering. Some of the analyses that can be performed by GIS are:

- Soil analysis
- Environment analysis
- Risk assessment
- Management of cultural heritage
- Land management for emergencies.

GIS Structure

GIS combines a powerful visualization environment with an analytical and modeling structure. To support this setting, platforms must be organized according to three modalities: geodatabase, geovisualization and geoprocessing (Fig. 1).

The *geodatabase* is a spatial database comprising data sets with geographic information. Specifically, it is a geographical information model organized with thematic layers and spatial representations that is used for accessing and managing GIS data. Its main utility is the organization of data, which can be useful for understanding complex scenarios (e.g. environmental impact or risk connected with the use of natural resources or damage of infrastructures).

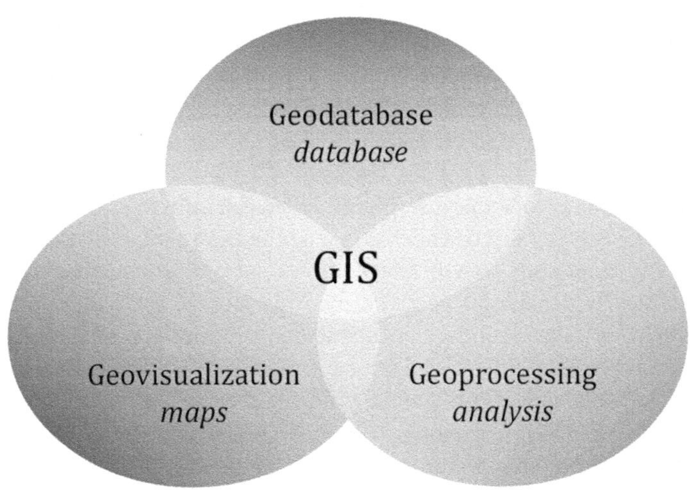

Fig. 1 GIS structure

The *geoprocessing* comprises operational tools for geographic analysis and processes capable of generating derived data sets. This toolset is used to operate on GIS information objects, including data sets and cartographic elements of printed maps. Altogether, these comprehensive commands and data objects form the basis for the geoprocessing framework.

The *geovisualization* constructs complex geographical representations that display the elements and their spatial relationships on the earth's surface. This includes maps, interactive maps, three-dimensional scenes, charts, tables and schematic views of the network relationships. Interactive maps provide a user interface at many levels.

The creation of the data set is a critical step, and it is closely related to the ability to georeference the data. It primarily relies on knowledge of the data's original coordinates (longitude, latitude, altitude or by other geocode systems).

The information is organized by the software in the form of thematic layers called shapefiles (Fig. 2). Several geographical features are adequately expressed by a precise range of geometric elements: points (dimensionless elements used for the simple location within an aerial representation); lines and polylines (one-dimensional elements used for the localization of linear elements); polygons (two-dimensional elements used for the representation of geographic elements characterized by coverage of a particular soil surface). Therefore, each layer summarizes a geometric feature; the properties of each element forming part of a layer are collected in the geodatabase.

Finally, due to the remarkable potential the GIS tool holds, spatial analyses can be performed by adopting specific statistical and modeling methods.

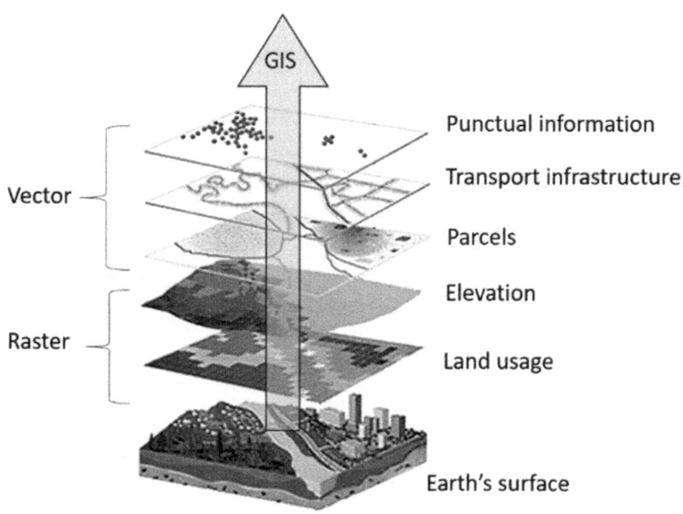

Fig. 2 GIS thematic layers

Geostatistical Tool Analysis

The management of territory and heritage conservation relies on cartographic representation to track the evolution of natural phenomena. Sampling data, which provides a basis for knowledge, allows for the description and representation of phenomena over the whole area through interpolation of information exploiting the natural structure of the data distribution. Every natural phenomenon (the stratigraphy of subsoil, dispersion of a pollutant in groundwater, etc.) is characterized by its structure, which depends on the origin of the phenomenon itself.

The discipline of geostatistics, developed since the early sixties, defines the dimension and extension of mineral deposits. The term geostatistics was adopted for the first time in 1962 by Georges Matheron, unanimously considered the founder of the discipline. His work focused on the development of geostatistics and its mathematical and statistical aspects. Matheron is also the founder of the Centre de Géostatistique de l'Ecole des Mines de Paris, based in Fontainebleau, France, which is currently the reference center for the study of this discipline. Geostatistics takes into hinges on the so-called "theory of regionalized variables", which accounts for the spatial dependence between variables of interest [14].

Natural Phenomenon and Regionalized Variables

The method begins from a regionalized phenomenon, which emerges from a space defined as a geographical (one, two or three dimensions) or temporal space. The structural characteristics of the phenomenon are sought from the measured data, but are sometimes revealed by additional qualitative information that improves understanding of the phenomenon.

Even the collection of large amounts of raw data would not provide exhaustive knowledge of regionalized phenomena. In fact, it does not allow to fully understand the processes that lead to the physical phenomenon and its evolution. Supposing the regionalized phenomenon is completely known within a domain D, it can be described satisfactorily by one or more functions defined on D. The variables of these functions are called "regionalized variables" because they are related to their location in the field (for example, the thickness of a geological formation for evaluation of natural resources [7], or the concentration of pollutants in the groundwater [26]).

Instrument for Modeling the Structure of Data

The value assumed by the regionalized variable in a point is not independent from a value measured at a different location. Specifically, values measured at neighboring points are more closely related than values measured at distant points. This spatial correlation constitutes the structure of the regionalized phenomenon. It is described by the spatial laws associated with random functions that represent the behavior of the phenomenon.

In particular, the linear geostatistics define the spatial structure by two moments. The moment of order 1 (expected value) does not provide any structural information. The moment of order 2 (covariance and variogram) is defined according to two different points, and it describes the interaction between them. The two moments provide an elementary description of the spatial structure of the regionalized phenomenon.

The interpretation of the variogram is performed using a series of experimental data. The regionalized variable z(x) is considered to be a realization of the random function Z(x). The variogram is written as follows:

$$\widehat{\gamma}(h) = \frac{1}{2|N(h)|} \sum_{N(h)} \left[z(x_\alpha) - z(x_\beta)\right]^2$$

$N(h) = \{(\alpha, \beta): x_\alpha - x_\beta = h\}$ and $|N(h)|$ number of pair.

Calculation of the experimental variogram relies on considering the difference in the regionalized variable values at two locations separated by a distance h (Fig. 3a). The requisite procedure is as follows:

- Plot the squared increments between each couple of measured data as a function of the distance (h) between sampling points (Fig. 3b); the obtained graph is the variogram cloud.
- Define a number of intervals with an amplitude (lags, number of lags).
- Compute the average values falling within the intervals; this defines the experimental variogram (Fig. 3b).

In this way, the expected value for two points, separated by a distance equal to h, can be expressed by the following expression:

$$\gamma(h) = \frac{1}{2} E\left\{[Z(x+h) - Z(x)]^2\right\}$$

The experimental variogram is a discrete series of points, it is not defined for all distances h, and it cannot be used directly. Therefore, it is necessary to infer a theoretical variogram, as similar as possible to the experimental variogram. This theoretical variogram contains all the structural features of the regionalized variables describing the phenomenon. This type of modeling is called structural

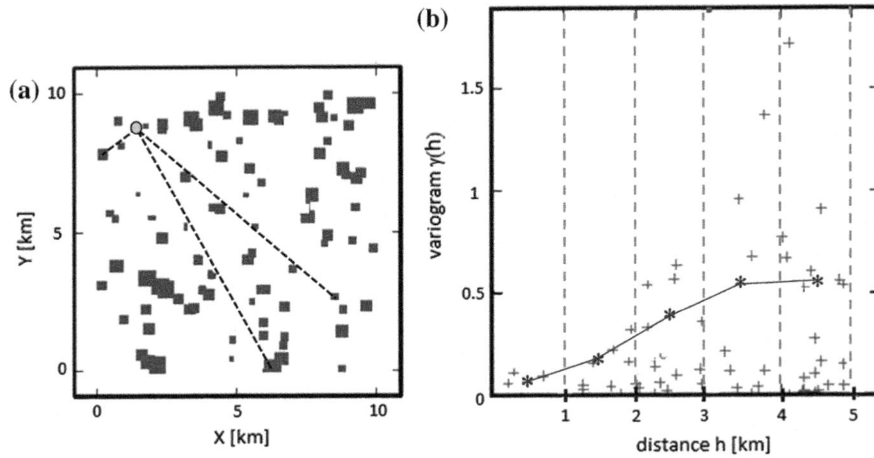

Fig. 3 Example of an experimental variogram calculation. **a** map showing the location of the measurements (the dimension of the point indicate the entity of the value of the variable); **b** variogram cloud (cross), lag intervals (dashed lines) and experimental variogram (line)

analysis or variographic analysis, and it is a fundamental stage in geostatistics. A poor model can produce inaccurate, unreliable results.

The main characteristics of the variogram include:

- the behaviour at the origin, which indicates the degree of regularity of regionalization;
- the presence or absence of the sill; the presence of the sill is symptomatic of second order stationarity. In this case, the covariance function is deduced from the variogram by the following relationship:

$$\gamma(h) = C(0) - C(h)$$

The variogram γ(h) describes the link between two values of the variable measured at a distance equal to h. In general, the difference grows with h, indicating that the variability increases with distance, until approaching a limit value γ(∞) called "sill" *C*. In this case, the random function is stationary of order 2, and the sill *C* and the variance are equal. The distance within which the equality occurs is called "range" *a* (Fig. 4). Two values Z(x) and Z(x + h) are related if the length of the vector h is less than the distance *a*. The range leads to the notion of the "area of influence" of a value. Beyond *a*, the variogram assumes a constant value equal to the sill *C*, and the variables Z(x) and Z(x + h) are no longer related (independent).

Several variogram models appear in the literature [7]. Table 1 lists the most common models for the study of regionalized variables.

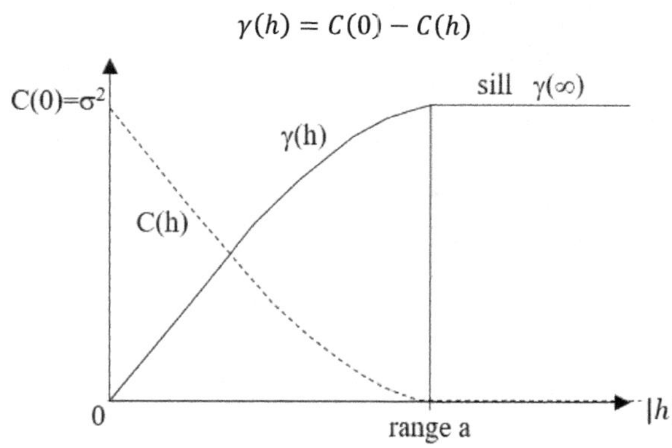

Fig. 4 Covariance C(h) and Variogram (h)

Table 1 Main variogram model

Spherical	Exponential
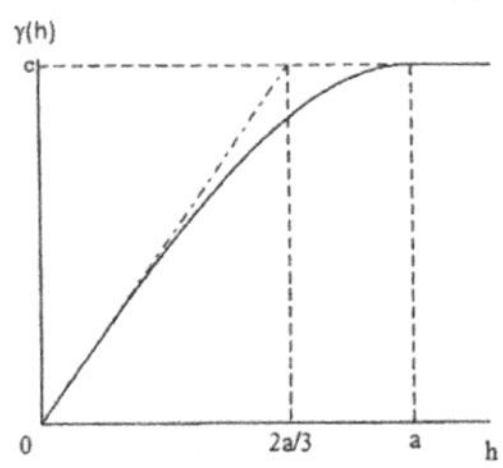	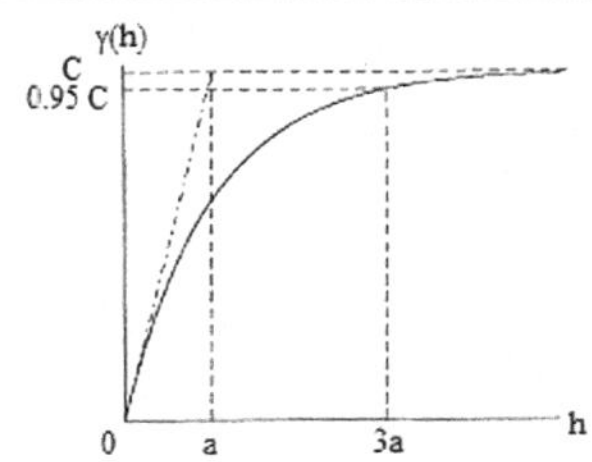
$\gamma(h) = \begin{cases} C\left(\frac{3}{2}\frac{h}{s} - \frac{1}{2}\frac{h^3}{a^3}\right) & 0 \leq h \leq a \\ C & h > a \end{cases}$	$\gamma(h) = C\left(1 - \exp\left(-\frac{h}{a}\right)\right)$

The spherical model is the most common model. The polygonal expression increases to a certain distance after which the value stabilizes. The tangent at the origin intersects the sill *C* abscissa 2a/3.

In analogy to the spherical model, the exponential model exhibits linear behavior for small values of h. The exponential model approaches the sill *C* asymptotically. At constant sill, it is observed that the exponential model, relative to the spherical model, grows faster initially, but this increase decelerates as the same value of *C* is approached. However, these models are very similar. The difference lies in understanding how fast the model diminishes the value within the range of influence.

The nugget effect represents small-scale changes and/or measurement errors. This discontinuity at the origin is derived from applications in mineralogy (mines gold) and has been proposed by Matheron [14]. The nugget effect model translates a phenomenon of absence of correlation between neighboring values of the variables.

Estimation

The estimate of a variable is carried out using a limited number of points (sampling points). The Geostatistics program provides estimators that make this, along with quantification of the estimations reliability, possible.

The global estimation covers the entire area necessary for characterizing the variable that describes the phenomenon. In general, the objective is to evaluate the value at a non-sampled point, using a weighted linear combination of the measured values at the sampled points. The process should take into account the distance between sampled points and points where estimation of the variable is required; close points are given more weight than those further away. The weighting should also account for the variability of regionalized structures, specifically the degree of regularity and the spatial arrangement of the samples.

Analysis of the spatial dependence of the data allows quantification of the correlation between values at different sites, and it determines the size of the "zone of influence" for observations. The continuity and regularity of the regionalized variable is evaluated through easily interpreted instruments. The study of different directions allows any anisotropy to be found, which indicates that regionalization is more structured in these directions. The interpretation of a regionalized phenomenon leads to see if the makes it possible to determine if processes that constitute the phenomenon have evolved at the same scale.

The linear interpolator use in Geostatistical analysis is called Kriging. This interpolator offers several advantages over conventional interpolation techniques:

- it provides an accurate estimation, without distorting the value, while taking into account the geometrical nature of the data (number and configuration of data) and structural information contained in the variogram;
- It quantitatively addresses the accuracy of the estimation by way of the estimation variance.

The punctual kriging estimates the value of the variable in each node of regular grid covering the whole area. The estimation of the variable Z at the point x_0 is given by the following equation:

$$Z^*(x_0) = \sum_{\alpha=1}^{n} \lambda_\alpha Z(x_\alpha)$$

The expected value of the error estimation is written:

$$E\left[Z^*(x_0) - Z(x_0)\right] = \sum_{\alpha=1}^{n} \lambda_\alpha E\left[Z^*(x_0)\right] - E[Z(x_0)] = m\left(\sum_{\alpha=1}^{n} \lambda_\alpha - 1\right)$$

In the case of the punctual ordinary kriging, where m is unknown, ensuring the absence of error distortion imposes the following condition, called universality condition:

$$\sum_{\alpha=1}^{n} \lambda_\alpha = 1$$

The variance of estimation-error $Var[Z^*(x_0) - Z(x_0)]$ needs to be minimized to adhere with the required of the absence of dual-distortion, i.e. $\sum_{\alpha=1}^{n} \lambda_\alpha = 1$.

The weights λ_α of kriging are evaluated considering:

- the distances between the points to be estimated and observed points;
- the geometric configuration of the observed points;
- the spatial structure of the regionalization described by the variogram γ.

In general, the weights assigned to the measurement points are much higher as they are close to the points where the estimation is required.

The weight and variance estimation relies only on the kriging layout and the variogram model, and not the values of the data. Therefore, the accuracy of the estimation can be evaluated if the configuration of the measuring points and the variogram model is known. However, in highly variable areas, with pronounced outliers, the accuracy decreases compared to areas of low variability.

If the measured points between them are all distinct, the kriging system is regular and, therefore, it provides a unique solution. The punctual Kriging is an exact interpolator. The estimation in a measurement point returns the measured value and the variance of kriging in this point is zero. The maps of the values estimated by kriging are always characterized by differences between neighboring values that are less obvious than in the real system (minor fluctuations). In fact, the estimation cannot reproduce the details of real situations that do not appear in the observations provided by the sampling. If it is necessary to reproduce the variability of regionalization, simulation techniques must be applied.

The Case Study: Subsidence Over the City of Bologna

The city of Bologna was settled before 1000 Bc in the southern part of the Po valley, at the foot of the Italian Apennines. Owing to the presence of monuments and architectural masterpieces, it is a nexus of culture and art in Italy. From the time

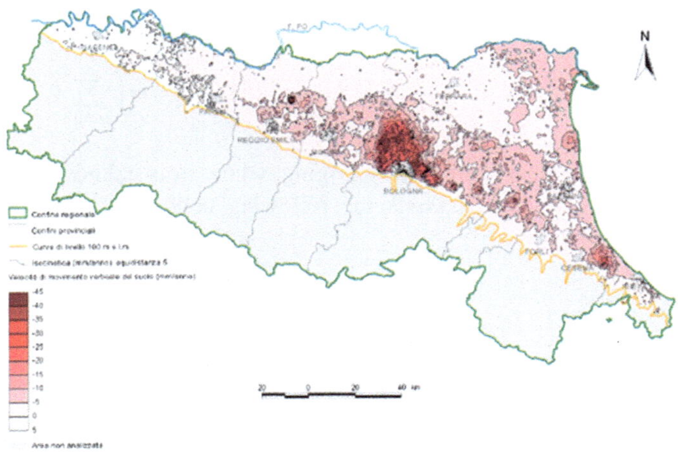

Fig. 5 Map of settlements in the period 1992–2000 [22]

it was founded, the city has been affected by a slow deformation of the ground, induced by the self-weight consolidation of the thick alluvial deposits. However, settlement proceeded with a rate on the order of a few millimeters per year and was not of serious concern. After 1950, in analogy to occurrences in other cities of northern Italy (e.g. Milano, Venezia, Ravenna), a rapid deformation of the ground ensued within the town and the surrounding areas, and in some cases this proceeded with rates of dozens of centimeters per year [6].

The ground deformation recorded throughout the Emilia Romagna region in the period 1992–2000. Figure 5 shows that the phenomenon is prominent near the city of Bologna.

Damage to several buildings in the historic center were noted at the time, especially the buildings of via Zamboni, a major urban road located in the northeast side of the historic center (Fig. 6). Displacements surveyed along this road between 1979 and 1983 [1] give rise to the profile evolution reported in Fig. 7.

With the goal of discovering the causes of this unexpected phenomenon, many studies were promoted to survey the ground deformation in the area by means of increasingly more precise geometrical levelling networks [2, 3, 11, 18–20]. Finally, a satellite recording campaign was undertaken in 2005 to monitor the vertical movement of the territory [28].

A mechanical explanation for the phenomenon, founded on the quantification of the different variables, was provided by Darini et al. [8, 9] as well as Modoni et al. [16, 17]. These authors merged measurements of different variables with results of numerical analyses to create a comprehensive database where information of a distinct nature can be combined.

Here attention is confined to the analysis of settlements. For this investigation, an area of 272.25 km^2 has been delimited (Fig. 8). This area includes the city center and the upper plain area, where a grid of points has been ideally superposed. The

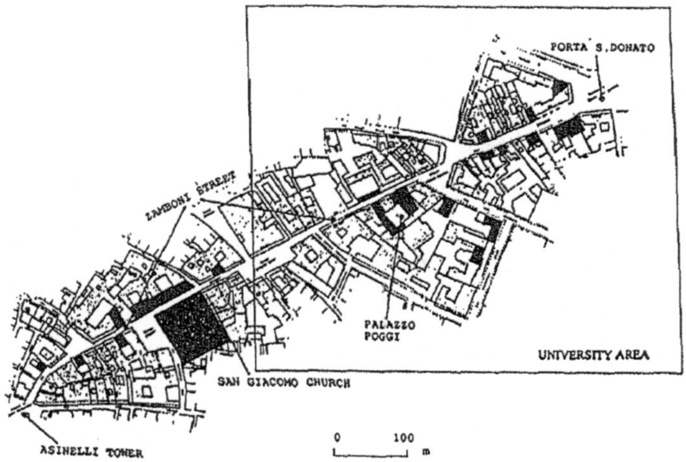

Fig. 6 Damaged buildings recorded along via Zamboni [5]

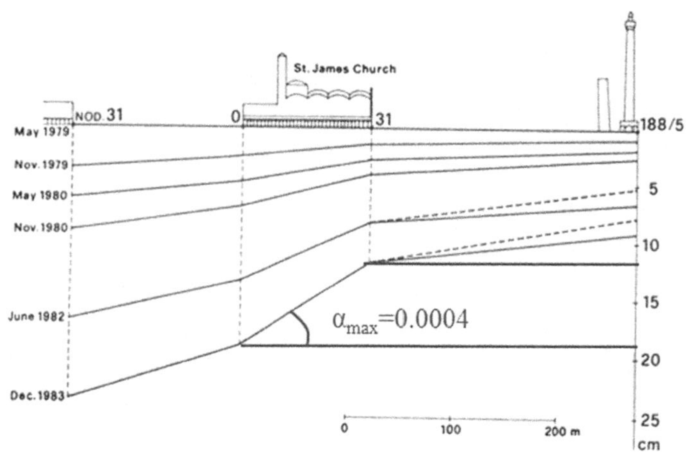

Fig. 7 Settlements recorded along via Zamboni [1]

variables have been computed in each node of this grid using a geostatistical interpolation method [7].

From a geological viewpoint, the province of Bologna extends over an area of about 3702.5 km^2 and can be subdivided into distinct geomorphologies: *flood plains* consisting of alluvial materials transported by rivers and deposited in the plain mostly during flooding; *foothill plains*, which constitute a morphological connection between the hill and the alluvial plain, characterized by decreasing topographical gradients, and containing coarser materials deposited by the two major rivers (Reno and Savena); *hills*, characterized by a parallel to dendritic

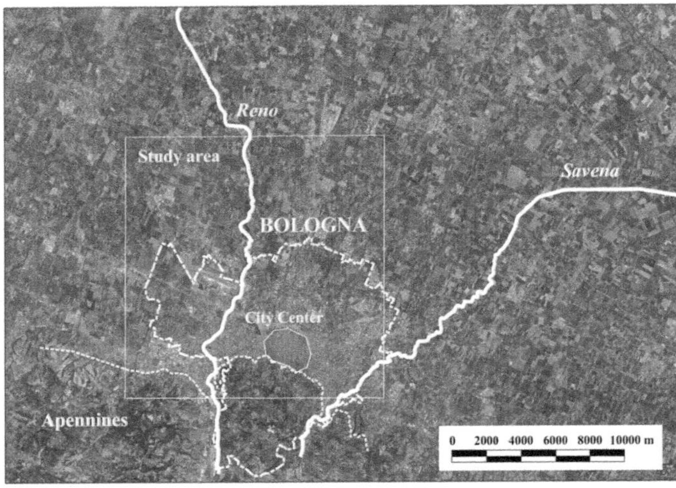

Fig. 8 Plain view of the studied area

drainage system; *mountains*, consisting of geological units formed from chaotic complexes of clay with sub-horizontal flaps of more resistant lithological units.

The hydrogeological layout [23] shows a recurrent alternation of permeable (aquifers) and semi-permeable or impermeable strata (aquitards). Here three large aquifer groups (A, B and C) can be distinguished that separate each other by thick and widely extended fine-grained strata (Fig. 9). The most important of the studied phenomena is the upper group A, about 300 m thick and dominated by the presence of the two alluvial fans of the Reno and Savena rivers which are the most important groundwater supplies [10, 12]. Coarse-grained materials of marine origin can be found in the lower strata of aquifer group B and in Group C. The groundwater circulation in the upper layers is mostly fed by the dispersion of the two rivers, while a secondary and more limited contribution from remote sources can be observed in the lowland area.

The withdrawal of water from wells occurs mostly from the upper aquifer unit A.

Ground Deformation

Due to arising from tectonic movement and sediment compaction, the area of Bologna is affected by a natural long-term subsidence progressing with rates ranging from 1.5 to 2 mm/year. Some benchmarks measured in the period 1897–1902 and 1947–1957 [18, 19] reveal settlements ranging from 10.7 to 20.9 cm (Fig. 10).

This subsidence became increased after 1950, as evidenced by the geometrical levelling carried out between 1943–1950 and some repetitions performed by the

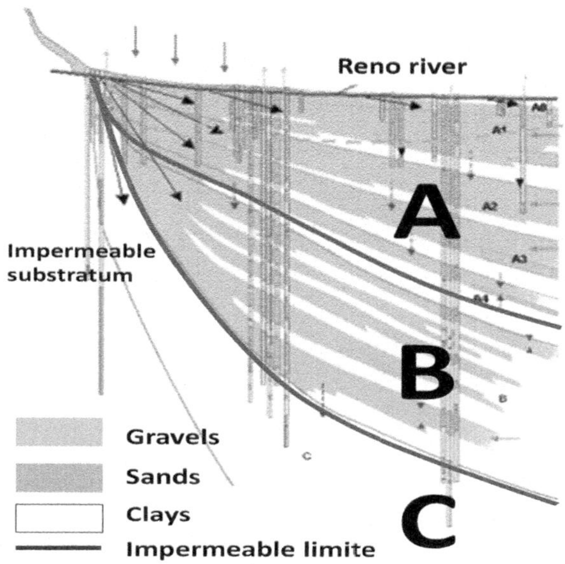

Fig. 9 Longitudinal section taken along the Reno river [23]

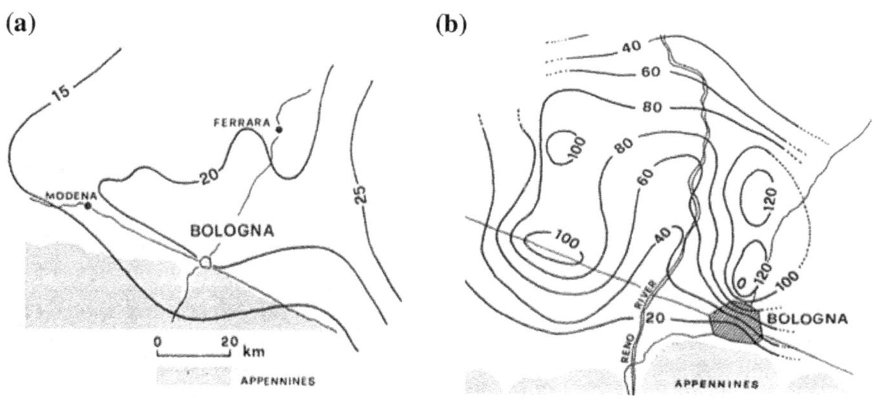

Fig. 10 Contour lines of equal subsidence (cm) in the period 1897–1957 (**a**) and in the period 1943/1950–1970/1973 (**b**) [19]

Italian Geographical Military Institute (IGMI) during the period 1970–1973 (Fig. 10). The map extracted from Pieri and Russo [19] shows concentrated settlements in the areas surrounding the Reno river (with settlements greater than 1 m), whereas smaller settlements are noticed along the river.

Thereafter, the *Autorità di Bacino del Fiume Reno* (Fig. 11a) undertook a campaign to during the period 1950–1997 with the goal of extensively monitoring the subsidence in *Emilia Romagna*. In the area of Bologna, this database includes 200 benchmarks collected from three different geometrical leveling networks

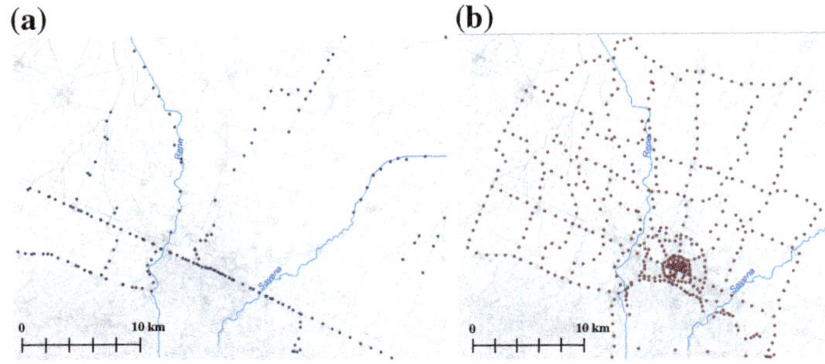

Fig. 11 Monitoring network of the *Autorità di Bacino Reno* (period 1950–1997) (**a**) and of the *ARPA Emilia Romagna* (period 1983–1999) (**b**)

previously installed by various public authorities (*Autorità di Bacino del fiume Reno, Istituto Geografico Militare Italiano, Ministry of Public Works*).

The second database (Fig. 11b) was included by the *ARPA Emilia Romagna* and the municipality of Bologna in a larger database comprising 527 benchmarks for monitoring the subsidence during the period between 1983 and 1999. During this period more comprehensive and refined analysis could be performed in the area due to the much higher spatial density of this network compared with the previous one (in the city center benchmarks were located at mutual distances shorter than 250 m).

Analysis of Settlements

All data have been combined in the GIS to observe the progression of settlements in the area studied. The measurements taken at singular points have been interpolated with a series of geostatistical analyses. Specifically, setting the subsidence for the period before 1943–1950 at zero, a sequence of contour lines have been created to reproduce the computed settlements during the different periods (Fig. 13).

As an example, Fig. 12 shows the experimental and theoretical variograms for the settlements computed in the period 1983–1987. The theoretical variogram was then used to interpolate the data.

The subsidence (Fig. 13) is repetitive for all plots; a larger concentration of settlements is recurrently observed around the alluvial fan of the Reno river. This result can be explained by the fact that the largest amount of groundwater was withdrawn near the Reno river [31] to intercept the fresh water coming from the mountains. Although the decline of the water head is maximal in the alluvial fan, where the most active wells are located, higher subsidence does not occur in this zone due to the prominence of coarse-grained materials in the subsoil strata. The observed settlements occur in the surrounding area, where fine-grained soil is

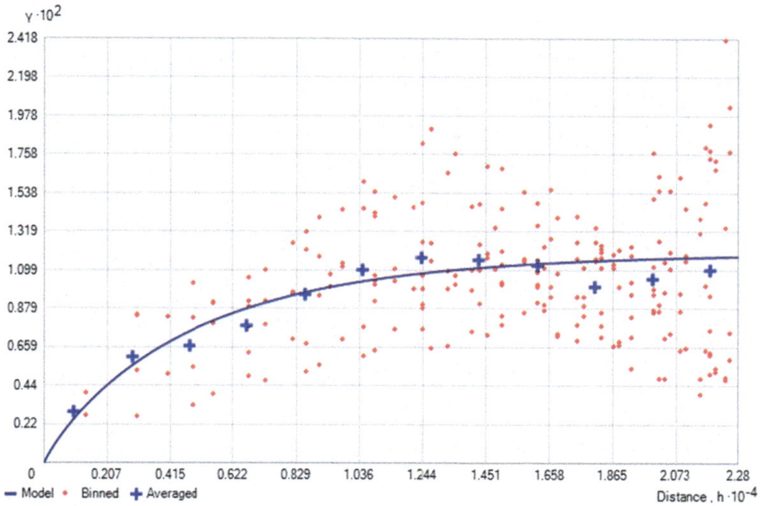

Fig. 12 Experimental and theoretical variograms (settlements in the period 1983–1987)

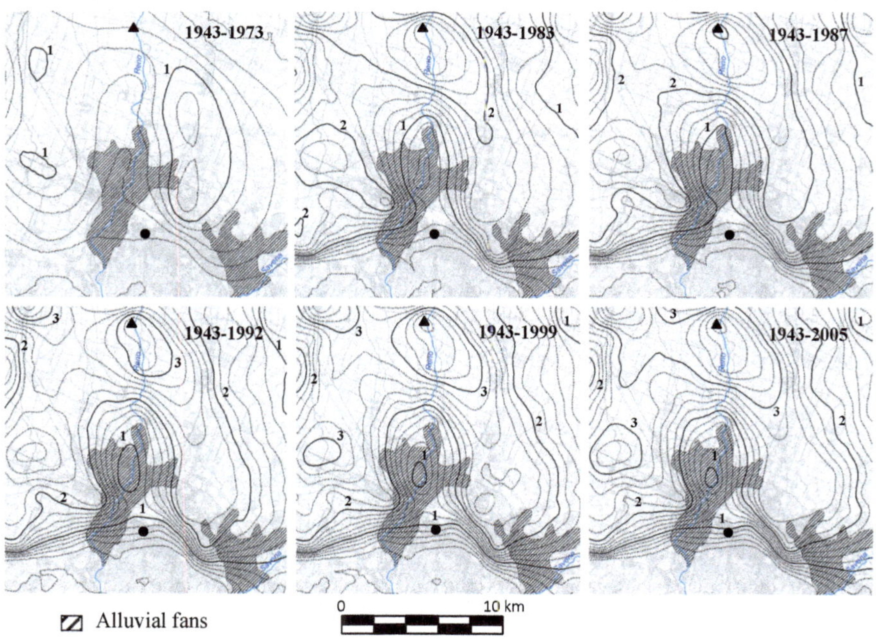

Fig. 13 Settlements measured for different time intervals between 1943 and 2005

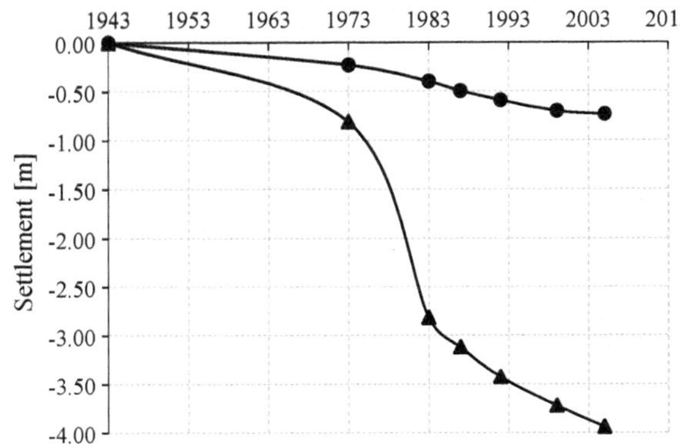

Fig. 14 Time histories of settlements for two positions of the city and surrounding area

prevalent. Settlement progression as a function of time for two points representing different locations is plotted in Fig. 14.

The two curves show a maximum rate during the period between 1970 and 1980, i.e. when groundwater was more heavily extracted. The curve representative of the situation in the city center is less pronounced, while larger settlements occurred in the surrounding area, where the rates are still significantly high, even after 2005.

As an exemplary application of GIS, the present study focuses on possible damage to buildings. As described in the literature, damage to buildings, ranging from the formation of small fissures to total collapse, can correlate with angular distortion at the foundation level [25] or tensile strain that develops in the upper structure [4]. The extent of damage depends on the type of structure (open framed structures show more limited effects under higher deformation in comparison to load bearing walls) and on the deformation mode (sagging or hogging). Irrespective of these criteria, it is generally acknowledged that angular distortions larger than $0.5\text{–}1 \times 10^{-3}$ produce cracks in continuous walls [4, 29].

It is known that the angular distortion is proportional to the derivative of rotation, consequently the latter quantity has been computed herein to forecast the damage to buildings.

In particular, the analysis has been restricted to the historic center of the city. It superposes an ideal grid of points (nodes are positioned at a mutual distance $\Delta = 60$ m) and computes settlements at each node (i, j) with the previously described geostatistical method. Rotation in each node $\alpha_{i,j}$ is then calculated with the following relation:

$$\alpha_{i,j} = \sqrt{\left(\frac{w_{i+1,j} - w_{i-1,j}}{2\Delta}\right)^2 + \left(\frac{w_{i,j+1} - w_{i,j-1}}{2\Delta}\right)^2}$$

Calculation of these values within the studied area (Fig. 15a) shows that the largest derivatives of rotation occur in the. Focusing on Via Zamboni, where noticeable cracks were recorded (Figs. 6 and 7), it is seen that the curvature of the deformed ground (given by the gradient of rotation) is maximum along the direction

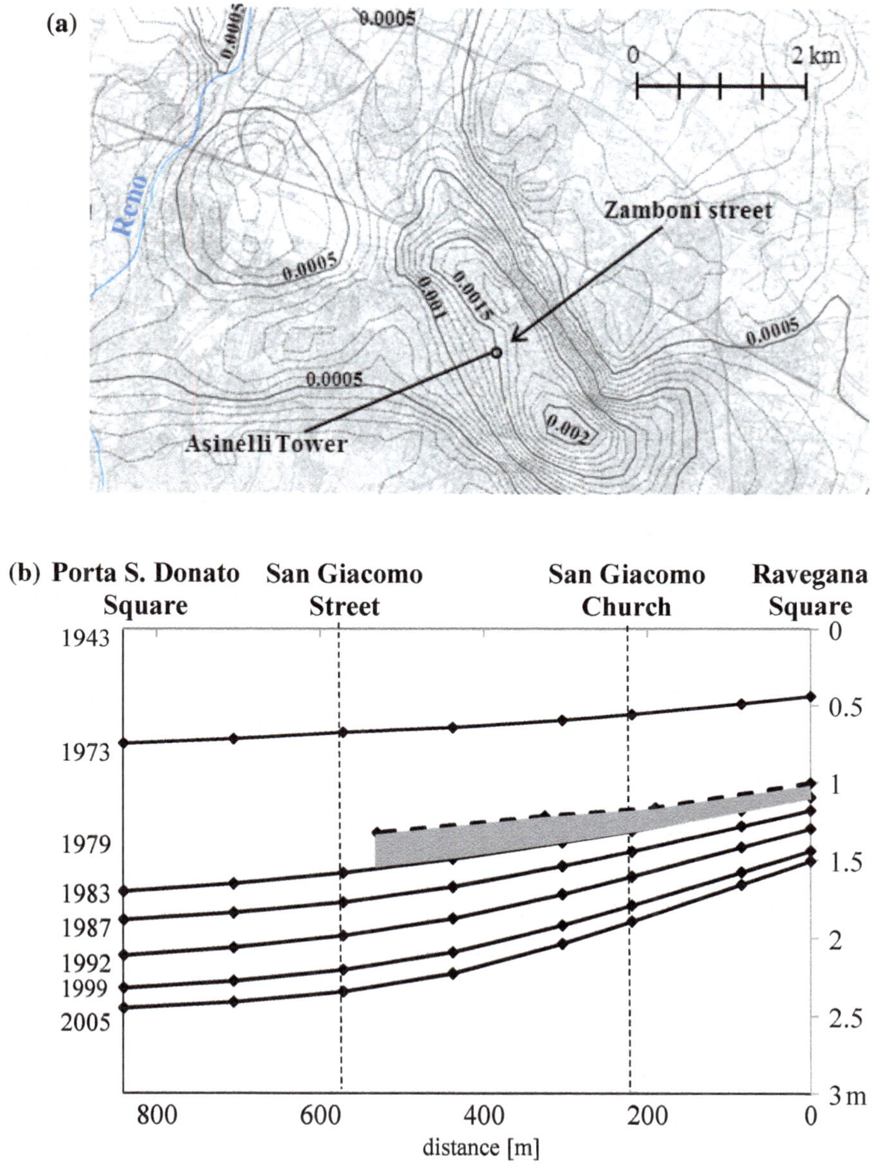

Fig. 15 Map of rotations computed for the period 1943–2005 in the historical center of Bologna (**a**) and settlement profiles along Zamboni road (**b**)

of the road. In fact, the maximum rotation computed in this area for the period 1943–2005 is equal to about 0.002, nearly five times larger than the value (0.0004) calculated by Alessi [1] for the period May 1979 to Dec 1983 (Fig. 7). It is worth noting that the settlement profiles traced along this road with the present analysis (Fig. 15b) are fully consistent with those (reported in the same plot with a shaded area) derived by Alessi [1].

Conclusions

The GIS is one of the most powerful information technology tools, capable of performing spatial analyses over large areas. Large and diverse sets of information can be collected, catalogued and processed. In the application of GIS presented here, the analysis of data has been carried out using geostatistical tools that allow interpolation of data recorded at distinct sampling points over the whole area. An enlightening example of the combined use of GIS and geostatistical analyses has been carried out to assess the settlements induced in the area of Bologna by intensive groundwater removal. The phenomenon, first noticed during the second part of the last century, has been quantified by combining information reported in the literature with results of topographical investigations. Settlements reconstructed for the period from 1943 to 2005 show a recurring trend, specifically, larger settlements of, up to 4 m occur in the area surrounding the alluvial fan of the Reno river. Fortunately, these settlements occur in the countryside. Nevertheless, large settlements have been recorded in the city center as well, and these are associated with relatively large rotations. This pattern is consistent with damage observed on the buildings of via Zamboni. It is noteworthy that the rotation computed for the period 1943–2005 is about five times larger than the one observed by previous authors during the period 1979–1983. The evolution of subsidence with time shows that the condition for the historic center will stabilize, while relatively fast rates are still observed in the outer part of the city.

Finally, it is worth noting that the computed distribution of settlement is consistent with the leading causes of the phenomenon as well as the mechanical properties of the subsoil. In fact, the largest settlements have been recorded where the decline of the water table matches the compressibility of fine-grained deposits.

References

1. Alessi R (1985). La subsidenza nel centro storico della città di Bologna. Il grado di dissesto dei fabbricati nella zona di via Zamboni. Inarcos, no 456. Bologna (Italy)
2. Arca S, Beretta GP (1985) Prima sintesi geodetico-geologica sui movimenti verticali del suolo nell'Italia Settentrionale (1897–1957). Bollettino di Geodesia e scienze affini, no 2
3. Bondesan M, Gatti M, Russo P (1997) Movimenti verticali del suolo nella Pianura Padana orientale desumibili dai dati I.G.M. fino a tutto il 1990. Bollettino di Geodesia e scienze affini, no 2
4. Burland JB, Wroth CP (1974). Settlement of buildings and associated damage. In: Proceedings of Conference on Settlement of Structures, Cambridge, UK, 1974, 611–654

5. Capra A, Folloni G (1991) Subsidence controls in the town of bologna. In: Proceedings of the fourth international symposium on land subsidence. IAHS Publication no 200
6. Carminati E, Martinelli G (2002) Subsidence rates in the Po Plain, northern Italy: the relative impact of natural and anthropogenic causation. Eng Geol 66:241–255
7. Chilès JP, Delfiner P (1999) Geostatistics: modeling spatial uncertainty. Wiley, New York
8. Darini G, Modoni G, Saroli M, Croce P (2008) Land subsidence induced by groundwater extraction: the case of Bologna, Proceedings of IEMSs 2008: International congress on environmental modelling and software, Barcelona 10–13 Luglio, pp 1386–1393, ISBN 978-84-7653-074-0
9. Darini G (2007) Land subsidence due to groundwater withdrawal: the case of Bologna. PhD Thesis. Università degli Studi di Cassino
10. Elmi C, Bergonzoni A, Massa T, Montaletti V, Barattella PL, Ronchi A (1984) Il territorio di pianura del Comune di Bologna: aspetti geologici e geotecnica. Giorn Geol 46(2):127–152
11. Folloni G, Radiconi F, Russo P (1996) La subsidenza del territorio bolognese dal 1983 al 1983. Inarcos, n. 571. Bologna (Italy)
12. Fava A, Farina M, Marcaccio M (2005) Le caratteristiche degli acquiferi della Regione Emilia-Romagna. Report 2003. RER-APAT. Grafiche Pioppi Settembre 2005
13. Lockwood MG (1954) Ground subsides in Houston areas, Civ Eng 24(6), 48–50
14. Matheron G (1965) Les variables régionalisées et leur estimation: une application de la théorie des fonctions aléatoires aux sciences de la nature. Masson, Paris, France
15. Marsal RJ, Mazari M (1962) El subsuelo de la Ciudad de Mexico, The subsoil of Mexico city, Universidad Autonoma de Mexico, Facultad de Ingenieria, 2nd edn p 614
16. Modoni G, Darini G, Spacagna RL, Saroli M, Russo G, Croce P (2013a) Spatial-temporal analysis of the subsidence in the city of Bologna. In: Proceedings of the Conference Geotechnical Engineering for the Preservation of Monuments and Historic Sites, May 2013, Napoli, Taylor & Francis Group, ISBN 978-1-138-00055-1
17. Modoni G, Darini G, Spacagna RL, Saroli M, Russo G, Croce P (2013b) Spatial analysis of subsidence induced by groundwater withdrawal. Eng Geol 167, Dec 2013, pp 59–71. ISSN 0013–7952. https://doi.org/10.1016/j.enggeo.2013.10.014
18. Pieri L, Russo P (1977) Studio del fenomeno di abbassamento del suolo in atto nella zona di Bologna. Bollettino di Geodesia e scienze affini, n. 3
19. Pieri L, Russo P (1984) The survey of soil vertical movements in the region of Bologna. In: Proceedings of the Third International Symposium on Land Subsidence. Venezia (Italy)
20. Pieri L, Russo P (1985). Situazione attuale delle ricerche sull'abbassamento del suolo nel territorio bolognese. Inarcos, 456
21. Poland JF (1958). Land subsidence due to groundwater development. ASCE J Irrigat Drain Div 84(1774), 11
22. Arpae Emilia-Romagna. https://www.arpae.it
23. Regione Emilia Romagna and ENI-AGIP (1998) Riserve idriche sotterranee della Regione Emilia-Romagna. S.EL.CA. (Firenze)
24. Shi X, Fang R, Wu J, Xu H, Sun Y, Yu J (2012) Sustainable development and utilization of groundwater resources considering land subsidence in Suzhou. China Eng Geol 124(4):77–89
25. Skempton AW, Mac Donald DH (1956) The allowable settlements of buildings, In: ICE Proceedings: engineering divisions, vol 5, Issue 6, 01 Dec 1956, pp 727–768
26. Spacagna R-L, De Fouquet C, Russo G (2013) Geostatistical analysis of groundwater nitrates distribution in the Plaine d'Alsace, Coupled Phenomena in Environmental Geotechnics: From Theoretical and Experimental Research to Practical Applications—Proceedings of the International Symposium, ISSMGE TC 215
27. Stiros SC (2001) Subsidence of the Thessaloniki (northern Greece) coastal plain, 1960–1999. Eng Geol 61(4):243–256
28. Stramondo S, Saroli M, Tolomei C, Moro M, Doumaz F, Pesci A, Loddo F, Baldi P, Boschi E (2007). Surface movements in Bologna (Po Plain-Italy) detected by multitemporal DInSAR. Remote Sensing of Environment, XX

29. Sowers GF (1962) Shallow foundations. Foundation Engineering, Leonards GA (ed), McGraw-Hill Book Co., New York, NY, USA pp 525–632
30. Tessitore S, Fernández-Merodo JA, Herrera G, Tomás R, Ramondini M, Sanabria M, Duro J, Mulas J, Calcaterra D (2016) Comparison of water-level, extensometric, DInSAR and simulation data for quantification of subsidence in Murcia City (SE Spain). Hydrogeol J ISSN 1431–2174, pp. 1–21. https://doi.org/10.1007/s10040-015-1349-8
31. Vassena C (2003) Sviluppo e applicazione di modelli di flusso delle acque sotterranee nella conoide alluvionale del fiume Reno, PhD Thesis, Università degli Studi di Milano
32. Wilson G, Grace H (1942) The settlement of London due to underdrainage of the London clay. J Inst Civil Eng. London 19:100–127
33. Yamamoto S (1995) Recent trend of land subsidence in Japan. In: Land Subsidence, (eds) Barends FBJ, Brouwer FJJ, Schroder FH. IAHS, Publication n. 234:487–492
34. Zeevaert L (1983) Foundation Engineering for difficult subsoil conditions. Van Nostrand Reinhold Company Inc. (USA)

Infrared Thermography of Walls in Residential Buildings in Historic Workers' Housing Estates in Upper Silesia

Magdalena Żmudzińska-Nowak, Paweł Krause and Magdalena Krause

Abstract Thermal diagnostics of historic buildings in most cases requires the application of non-destructive testing methods. They are needed to carry out historical analysis of walls in historic buildings as well as to assess their level of thermal insulation. One of the most commonly used methods of measurement is infrared thermography. It allows visualisation, recording and interpretation of temperature distribution on the surface of tested walls. Diverse thermal imaging methods can be used, including passive and active thermography. In this article selected results of infrared thermography of external walls of historic workers' residential buildings localised in Upper Silesia are presented. The scope of the tests was to diagnose the condition of the buildings and to formulate guidelines for their protection consistent with recommendations of the conservator. Numerical analysis was performed with Therm 7.4 software, based on the finite-element method.

Introduction

Recent trends in parallel manipulators show a growing interest in modular re-configurable parallel robots. This class can be referred to as either classical parallel manipulators or cable-based systems. In this context, issues related to modular re-configurable parallel robots are presented for both classical and cable-based manipulators.

Tests performed on historic buildings in many cases require the application of non-destructive methods. Analysis of the hygrothermal performance of external walls of buildings with historic façades defines one example. The drive to improve the energy efficiency of buildings often requires modifications that increase the thermal resistance of external walls. In the case of historic buildings, it can be done by applying thermal insulation from the inside. It is commonly known that internal insulation impacts on the possible condensation of water vapour in the wall structure.

M. Żmudzińska-Nowak (✉) · P. Krause · M. Krause
Silesian University of Technology, Gliwice, Poland
e-mail: magdalena.zmudzinska@gmail.com

E. Ottaviano et al. (eds.), *Mechatronics for Cultural Heritage and Civil Engineering*, Intelligent Systems, Control and Automation: Science and Engineering 92,
https://doi.org/10.1007/978-3-319-68646-2_11

In view of this, detailed hygrothermal diagnostics need to be performed before adopting particular designs for improved thermal insulation of walls. The diagnostics should evaluate thermal insulation, airtightness and the level of moisture in external walls. The existing structures are often buildings that were constructed many years ago, when regulations concerning thermal insulation differed or were even non-existent. The level of thermal insulation of buildings is very diverse. In many cases, defining the level of thermal insulation is quite difficult due to the absence of design documents or data concerning the structure of particular walls. Non-destructive testing by hygrothermal analysis usually involves testing the distribution of the thermal field on the wall surface, the distribution of the surface moisture of the wall and airtightness of the building envelope. Reliable assessment of a building's thermal insulation level must rely on more than macroscopic studies or point openings. The development of measuring techniques allows us to perform many tests on building walls without interfering with their material structure, and thus without damaging the tested walls. Devices used to evaluate the physical properties of historic buildings can be divided into three main groups:

- devices used for the assessment of the level of thermal insulation, in particular for studying the temperature of the outer and inner surface—pyrometers and thermographic cameras;
- devices used to assess the level of moisture in construction materials, moisture on the surface—contact hygrometers;
- devices for the assessment of airtightness—blower door systems.

The simplest device used for measuring the temperature on the surface of a structural element is a pyrometer. It is an instrument used for contactless temperature measurement. It works on the basis of measurement and analysis of thermal radiation emitted by the tested bodies. The test provides a temperature measurement only for one conventional point of measurement. Thus, using such an instrument to evaluate temperature distribution of the whole surface of a wall is difficult, and for bigger surfaces it is impossible. The device is rarely used for broader thermal diagnostics in the building industry. It is mainly used as an auxiliary device. One of the fastest methods of measurement used in thermal diagnostics is thermography. Thermography is a method which involves visualisation, recording and interpretation of temperature distribution on the surface of tested walls. In construction, the most often used type of thermography is passive thermography. It consists of identifying temperature distribution on the boundary surfaces of a wall without controlled external interference and thermal stimulation of their thermodynamic state. In other words, that the thermal envelope of a building is tested in the conditions of regular use, without additional thermal stimulation. The method is involves measuring the intensity of the infrared radiation emitted by the tested object and then transforming it into a visible image. In the case of thermal imaging, qualitative thermography is most commonly used. The proper performance and analysis of thermographic tests allow the thermal quality of walls in the tested building to be defined, and it enables determination of material solutions applied in other periods of the building's life (e.g. remade brickwork).

Thermography in the building industry has been described in domestic and international literature by e.g. Fouad and Richter [1], Nikzad, et al. [9], Fox, et al. [2] and Nowak [10]. The application of infrared measurements in passive thermography is described in selected works: Grinzato [3], Riegert [11] and Kominsky et al. [5]. Heinrich et al. [4] used active thermography to diagnose walls in the basement area. Löther et al. [8] tested murals using active thermography.

Quantitative thermography makes it possible to indirectly determine the thermal insulation characteristics of a wall on the basis of infrared measurements. Such characteristics are described, for example, in the work Biot-Zahlen zur Kennzeichnung von Temperaturfeldern in Außenbauteilen—Krause [7].

The purpose of this article is to determine the state of thermal insulation of historic workers' residential buildings. The buildings, commonly referred to as *Familoki* (from German *Familien-Block*) were constructed for coal mine and steel works employees by the enterprise owners. Historic building complexes and workers' housing estates of interesting urban arrangement are characteristic features of the cultural landscape of Upper Silesia and they define the region's historical identity. As the Upper-Silesian industry collapsed, degradation of historic residential complexes began. Many such buildings have already been demolished and significantly reconstructed, while others are neglected or not maintained properly. In the last few years, particularly valuable complexes of workers' homes have been entered in the register of historic monuments or covered by protection by inclusion in the so-called conservation protected zone.

Infrared Thermography

With regard to parallel manipulators, a re-configurable system consists of a set of modules,—which may be identified as actuators, passive joints, rigid links (connectors), mobile platforms and end-effectors—that can be rapidly assembled into a complete robot with various configurations. A modular parallel robot configuration can thus be rapidly constructed and its workspace can be modified by changing the leg attachment points, the joint types, and link lengths for a diversity of task requirements. Because of the re-configurability, a modular parallel robot system has unlimited configurations. The subject of tests described herein are multi-family residential buildings (*Familoki*) constituting a workers' colony complex at 1 Maja Street in Chwałowice, Rybnik, in Fig. 1. The complex is a part of an estate built between 1910 and 1916 for the workers of the Donnersmarck coal mine (today's Chwałowice Coal Mine). The housing estate area is under the conservator's protection.

The tested objects are two-storey buildings of brick constructed in a traditional technology on a rectangular plan, covered with a gable roof with the ridge parallel to the street. The buildings are symmetrical, and they have two staircases with

Fig. 1 Modules for parallel robots: **a** prismatic (P); **b** revolute (R); **c** universal (U); **d** spherical (S)

entrances in the ridge wall. All the buildings have basements and functional attics. The walls of the basement and of the overground storeys are made of solid bricks being 51 and 38 thick, accordingly. The ceiling above the unheated basement is a vault supported on steel beams. The ceiling between the storeys is made of wooden beams of 20 × 20 cm with spacing of 60–80 cm. The ceiling above the uppermost storey is made of beams with a sound boarding.

Thermography was performed to provide a qualitative assessment of the level of thermal insulation of external walls. The principle of diagnosing thermal insulation involves establishing whether the temperature distribution is correct and the identification of thermal defects (e.g. altered material structure of the wall). The tests were carried out in winter using a FLIR ThermaCAM infrared camera. The camera was equipped with a 20° × 20° lens, which at a distance of 5 m creates a field of view of 1.6 × 1.6 m. The value of emissivity of the tested surface was adopted as 0.92 on average.

Air temperature was measured with a GANN HTR 300 research set with an OT 100 probe for continuous air temperature measurements—sampling every 0.1 s. The instrument's range of temperature measurement is from −10 to 90 °C. As reliable data, mean values measured at the level of 1.50 m above the ground were given. The outside and inside air temperature was measured prior to and during the tests. The outside (ambient) temperature ranged from −4.5 to −5 °C. The temperature of the air inside the building was 16–20 °C. Infrared measurements were taken at a minimum difference in temperature of 20.5 K. The maximum difference in temperature was 25 K. Temperature distribution for the tested surface was examined visually and then significant thermal images were saved directly from the screen and recorded for further quantitative computer-aided analysis. The tests were performed in accordance with PN-EN 13187 "Thermal performance of buildings. Qualitative detection of thermal irregularities in building envelopes.

Infrared method". During the tests there was no wind (wind velocity below 2 m/s). No significant differences in air pressure were observed at opposite surfaces of the building walls. The tests resulted in images presenting visualisation of temperature differences on tested surfaces of the external walls. The tests were carried out from the inside and from the outside. The energy of thermal radiation emitted from the surface of a body is a function of the temperature of the surface of that body. The heat flux flowing through the wall encounters specific thermal resistance from the thermal insulation layer. Variable properties of the thermal insulation layer in the wall gives rise to an uneven temperature distribution on the wall surface which, as a consequence, affects the changeable nature of infrared radiation emission. The energy is transformed inside the infrared camera into electric impulses which create a thermal image of the tested area on the screen, allowing conclusions to be drawn with respect to the quality of thermal insulation of the wall.

The analysis of the results involves isolating points with different levels of image brightness which corresponds with variable thermal insulation of the wall. Thermal

Label	Value
SP01	0,2°C
SP02	-3,5°C
SP03	-1,8°C
SP04	-3,6°C
SP05	-2,7°C

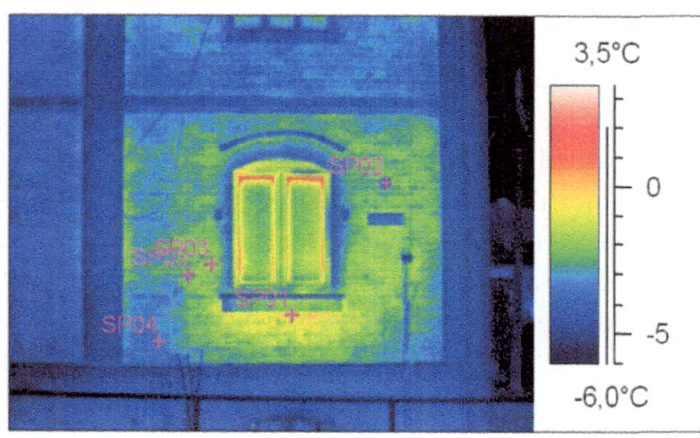

Fig. 2 Thermogram of the external wall with temperatures in points SP01–SP05

Label	Value
AR01 : max	2,4°C
AR01 : min	-2,3°C
AR02 : max	0,7°C
AR02 : min	-1,4°C
AR03 : max	0,3°C
AR03 : min	-1,2°C
AR04 : max	-1,6°C
AR04 : min	-3,3°C
AR05 : max	-0,3°C
AR05 : min	-2,3°C

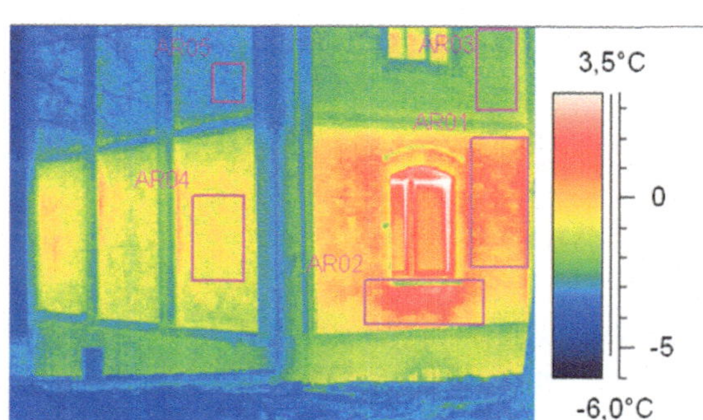

Fig. 3 Thermogram of the external wall with the minimum and maximum temperatures in fields AR01–AR05

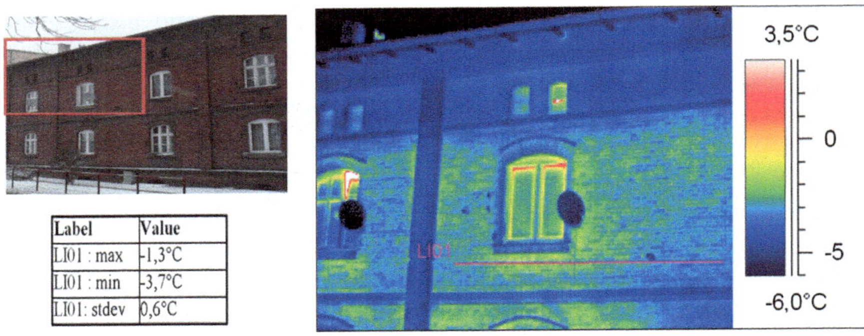

Label	Value
LI01 : max	-1,3°C
LI01 : min	-3,7°C
LI01: stdev	0,6°C

Fig. 4 Thermogram of the external wall with the minimum and maximum temperature along the LI01 line

anomalies which cannot be explained on the basis of geometrical analysis, wall structure analysis or other factors that could have impact upon measurement result, were treated as thermal defects. Coloured thermograms present temperature distribution in the form of coloured isothermal lines, where each line represents a certain temperature range shown on the right-hand side of the printout on the colour scale. The analysis of temperature distribution obtained from the thermogram can be performed on the basis of diverse data. It can be presented in the form of point temperature on the wall surface. An example of a thermogram with a set of temperatures in five selected points is presented in Fig. 2. Another form of analysis involves presenting the maximum and minimum temperatures in the analysed area. These data representation methods enable more accurate comparison of differences between the maximum and minimum surface temperature. In many cases the analysis is complemented by showing the mean value of temperature for the discussed area. The result of infrared thermography using the temperature field analysis for a selected area is presented in Fig. 3. In some cases, linear analysis of temperature distribution is an essential issue. It is important when assessing thermal uniformity of a wall, which allows identification of structural changes that occurred during the buildings' life (reconstruction, bricking up and overhauls). The results of tests carried out on external walls are presented in the following thermograms in Figs. 2, 3 and 4.

Result Analysis

When analysing temperature distribution on the surface of external walls in the tested buildings from the outside, we can state that there are some thermal anomalies concerning irregularities in uniformity of thermal insulation of external walls. In Fig. 2, the maximum temperature in selected points on the outer surface of the wall is 0.2 °C (SP01 temperature). The minimum temperature of the outer wall for a single residential room is −3.6 °C (SP04 temperature). The maximum difference in temperature is 3.8 K. It can be caused, for example, by the location of an individual heat source (individual heating solutions in particular flats) or material defects in the ceramic bricks or mortar. The thermogram presented in Fig. 3 allowed for the specification of the minimum and maximum temperatures in five fields AR01–AR05. For a single room, three temperature fields are presented, i.e. AR01, AR02 and AR04. The maximum temperature of the front wall is 2.4 °C (Field AR01). The minimum temperature on the outer surface of the wall is −2.3 °C. The maximum difference in temperature in the AR01 field is 4.7 K. In the same room the biggest difference in temperature on the surface of the gable wall, which has identical structure to the front wall, is 1.7 K. The analysis of thermal

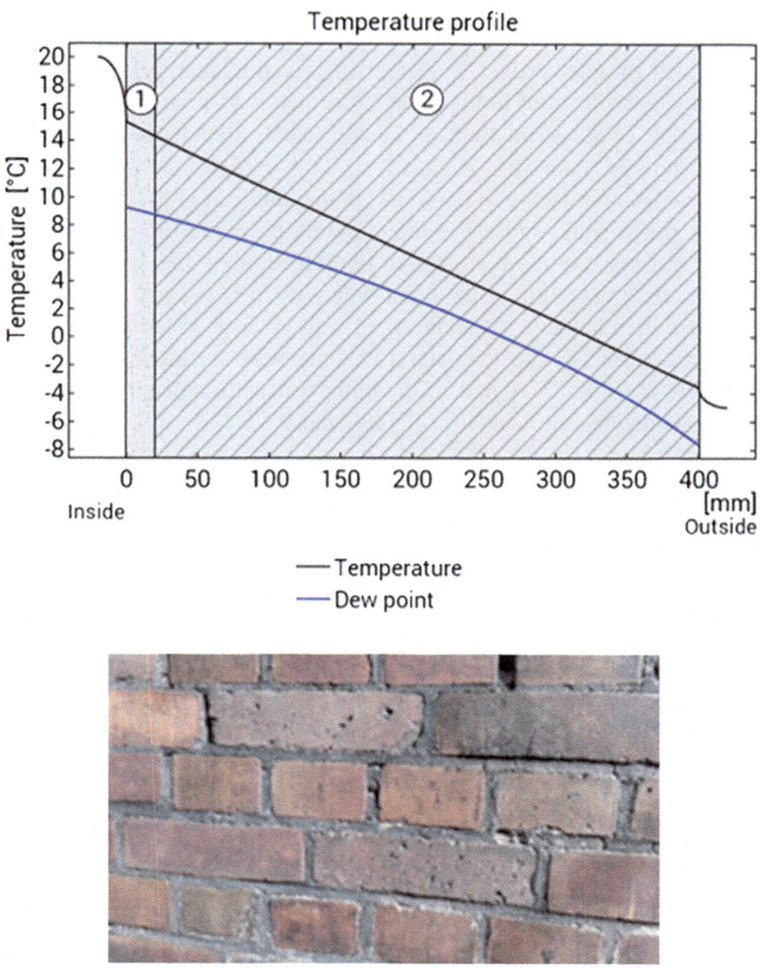

Fig. 5 Temperature profile in the 38-cm-thick external wall

insulation uniformity in the horizontal plane of the front wall of the last storey showed maximum difference in temperature of 2.4 K.

In order to evaluate the state of thermal insulation of the buildings, additional calculations of thermal insulation of external walls were performed. The external walls were made of solid ceramic brick, 38÷51 cm thick, joined by lime mortar. The heat transfer coefficient U for a wall of 38 cm, without taking thermal bridges into account, is 1.48 W/m^2K, while for a wall of 51 cm it is 1.18 W/m^2K. If the temperature inside the room is 20 °C, and assuming that the temperature outside is about −5 °C, the calculated temperature on the outer surface of a 38-cm-thick wall is −3.6 °C (Fig. 5). A change in thickness of the outer wall to 51 cm decreases the temperature on the outer surface of the wall to −3.8 °C. In the calculations the assumed that the thermal conduction coefficient for the ceramic brick was $\lambda = 0.60$ W/mK, while for the mortar it was $\lambda = 1.00$ W/mK. Thermal conductivity of cement and lime plaster was at the level of $\lambda = 0.82$ W/mK.

Evaluation of the technical condition of external walls revealed local defects in the façade layer of outer walls. The defects involved a local lack of horizontal and vertical joints in the form of lime mortar (or cement-and-lime mortar that was used in later periods). To assess the impact of those defects upon temperature distribution on the wall surface, numerical calculations were carried out with the Therm 7.4 software. Such an approach was presented in the publication by Krause [6]. For the 51-cm-thick outer wall, under the assumptions presented above, the calculated temperature for the point of structural imperfection of the wall was obtained. The decrease in temperature at the point where there is a 6-cm-deep decrement in the joint is 0.7 K relative to an undamaged wall. When the temperature inside the building rises to 24 °C, the temperature of the wall surface at the point of the damaged joint is −2.8 °C. A part of the outer wall with defects and the results of numerical calculations are shown in Figs. 6 and 7.

Fig. 6 Fragment of the external wall with damaged vertical and horizontal joints

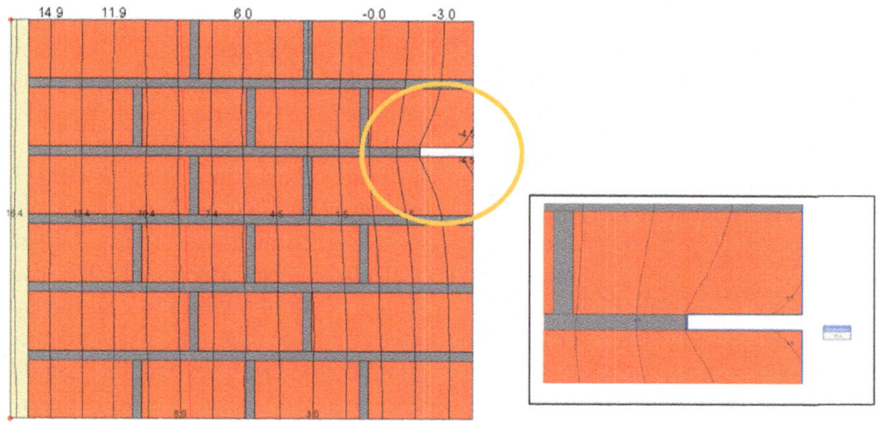

Fig. 7 Computational model for the 51-cm-thick wall. The temperature at the point of damaged joint is −3.1 °C

Fig. 8 Linear temperature distribution along the LI01 line in Fig. 4

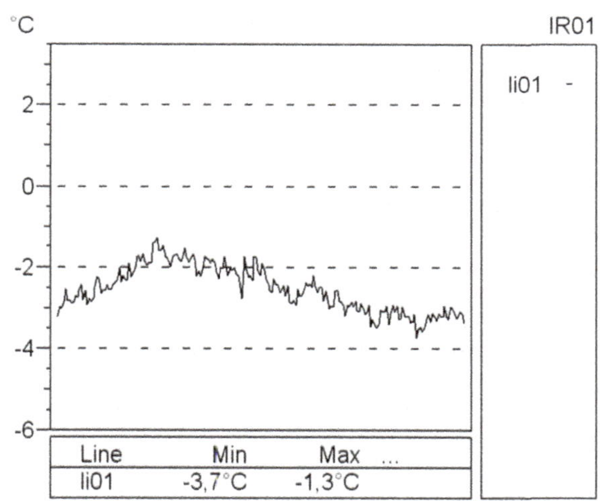

On analysing temperature distribution along the horizontal LI01 line in Fig. 8, we can state that the maximum difference in temperature is 2.4 K. The highest line temperature occurs right next to window frames and it is additionally correlated with the location of a panel radiator under the window. At the distance of more than 1 m from the window frames we observe temperature diversification of no more than 1 K.

Conclusion

The combination of diagnostic methods using infrared measurements for the analysis of historic buildings makes it possible to obtain satisfactory results. The applied tests allow detection of areas in the building walls which include building materials characterised by diverse thermal properties and also areas related with local wall defects. The infrared thermography performed on external walls of historic buildings of the workers' housing estate in Chwałowice, Rybnik, enables the assessment of the state of their thermal insulation. Test results show large differences in temperature on the outer surface of the wall. The maximum difference in temperature for a single room at a specified inside temperature is 5.7 K (Fig. 3). Due to the fact that some fragments of external walls were damaged since construction of the building, numerical analysis was performed with Therm 7.4 software, based on the finite-element method. Calculations performed under the given assumptions showed that joint damage reaching a depth of 6 cm causes an increase in temperature on the inner surface of the wall by no more than 1 K. The differences measured by infrared thermography are fivefold that value. They can be related to the diverse influences of local heat sources, e.g. tile stoves (local heating) or panel radiators (central heating). Uncontrolled air infiltration through the outer wall is another factor that may result in an increase in temperature relative to expected calculated values. In some cases there are perturbations related to the lack of complete airtightness with respect to air flowing through external walls. Infrared testing enables the detection of areas in the walls which were constructed in a different period than the rest of the building. In addition, the application of infrared testing enables the state of thermal insulation to be evaluated for the purpose of minimising consumption of energy for heating purposes and negative consequences related to, for example, water vapour condensation in thermal bridge locations. The combined application of infrared testing with techniques such as heat transfer coefficient measurement, airtightness measurement using the blower door method, and wall structure moisture measurements, creates compelling opportunities for further research.

The diagnostics performed and analysis of the results enables the selection of appropriate technology and actions consistent with the historic monument conservation guidelines.

References

1. Fouad NA, Richter T (2006) Leitfaden Thermografie im Bauwesen. Fraunhofer IRB Verlag, Stuttgart
2. Fox M, Goodhew S, De Wilde P (2016) Building defect detection: external versus internal thermography. Build Environ 105
3. Grinzato E (2012) IR thermography applied to the cultural heritage conservation. In: 18th world conference on nondestructive testing, Durban, South Africa
4. Heinrich H, Dahlem K-H, Edinger M, Vogel J (2001) Einsatz aktiver Thermographie zur Detektion von Einschlüssen in Baukonstruktionen und im Erdreich. DGZfP, Thermografie-Kolloquium, Stuttgart
5. Kominsky JR, Luckino JS, Marin TF (1998) Passive infrared thermography—a qualitative method for detecting moisture anomalies in building envelopes
6. Krause P. (2017) Obliczenia numeryczne wybranych mostków termicznych w ścianach z betonu komórkowego. Cement Wapno Beton 5/2017.

7. Krause H, Krause P (2002) Biot-Zahlen zur Kennzeichnung von Temperaturfeldern in Außenbauteilen. Bauphysik 4/2002
8. Löther T, Meinhardt-Degen J, Weise S, Franzen C (2007) Hohlstellendetektion an mittelalterlichen Wandmalereien mittels aktiver IR-Thermographie. Thermografie-Kolloquium 2007. Institut für Diagnostik und Konservierung an Denkmalen in Sachsen und Sachsen-Anhalt. Dresden
9. Nikzad S, Kari BM, Tahmasebi F (2011) The application of thermal imaging as a nondestructive test in historic buildings. In: XII International Conference DBMC, Porto
10. Nowak H (2012) Zastosowanie badań termowizyjnych w budownictwie. Oficyna Wydawnicza Politechniki Wrocławskiej, Wrocław
11. Riegert G (2007) Induktions-Lockin-Thermographie—ein neues Verfahren zur zerstorungsfreien Prufung. Dissertation, Stuttgart

Part III
Civil Structural Health Monitoring

Structural Health Monitoring Systems for Smart Heritage and Infrastructures in Spain

F. Javier Baeza, Salvador Ivorra, David Bru and F. Borja Varona

Abstract The development of information and communication technologies (ICT) and robotics is currently demonstrating its potential impact on different fields of application. With regard to cultural heritage, and architectural and engineering heritage in particular, these new technologies are changing the possibilities for structural capacity assessment and health monitoring (SHM). The objective of smart heritage can be achieved thanks to properly designed SHM systems, which when connected to an automated diagnostic system can even self-evaluate retrofitting needs. This chapter includes a brief summary of the SHM technologies applied for cultural heritage management in Spain during the early 2000s.

Introduction

The concept of smart cities (SC) has been proposed as a new urban development strategy in which, through the use of information and communication technologies (ICT), the city itself can adapt to actively improve the quality of life of its citizens and visitors. Spanish standard UNE 178104:2015 [2] defines the main objectives for an SC as an improvement in sustainability, efficiency and the quality of life of its inhabitants, which will be achieved through the use of technology. Actuation guidelines focus on improved management of administration and services, decision-making based on real-time data, and enhanced city resilience (defined as the capacity to meet and overcome unexpected challenges). The early detection of needs was also noted as a key aspect of an SC in a recent study conducted by Deloitte for the Spanish government [15]. The concept of smart heritage thus appears as a definition of an architectural heritage connected to a structural health monitoring (SHM) system (measurement + data management + damage evaluation tool), which is capable of detecting its own service condition, even responding automatically if an alarm threshold is triggered.

F. J. Baeza (✉) · S. Ivorra · D. Bru · F. B. Varona
Department of Civil Engineering, University of Alicante, Alicante, Spain
e-mail: fj.baeza@ua.es

E. Ottaviano et al. (eds.), *Mechatronics for Cultural Heritage and Civil Engineering*, Intelligent Systems, Control and Automation: Science and Engineering 92,
https://doi.org/10.1007/978-3-319-68646-2_12

All three aforementioned actuation lines can be directly linked to architectural heritage. The management of historic buildings usually lacks the information needed to optimize the funds available for maintenance tasks. Moreover, structural retrofitting is sometimes proposed only after catastrophic events, such as earthquakes [11]. In this regard, various preventive management methodologies have recently appeared, based on expert systems [45] or fuzzy logic [31], which aid in determining where and when an intervention should be made. These decision-making algorithms could even be applied in the construction stage of infrastructures [29]. In addition, open data from ongoing monitoring systems could be used to improve vulnerability studies—for example, risk management and building performance using geographic information system (GIS) tools applied to Barcelona [6]. Therefore, ideally, smart heritage would help public administrators by offering real-time data and analysis tools to improve service life management and crisis intervention, in which economic investments (maintenance and structural retrofitting) would be optimized through early damage detection and preventive conservation.

The future success of these proposals will depend on the fundamental role played by technological development, both ICT and automated devices (like robots or drones). Civil engineering and building monitoring have traditionally required complicated and expensive wired systems, which in most cases interfered with the normal use of the facility. Wireless communication is simplifying the experimental setup necessary for SHM purposes. However, some issues must be confronted for successful implementation, such as power supply (batteries vs. power harvesters), signal interference or data loss [44].

The SHM system installed in the Humber Bridge is a good example of an automated SHM system that is not only capable of measuring certain variables, but can also automatically diagnose, and consequently take the necessary actions to correct, undesired situations. The objective of this system was the prevention of steel corrosion on the anchorage chambers of the suspension bridge [57]. For this purpose, temperature and relative humidity sensors were distributed along the chambers, and dehumidification units were connected directly to the SHM system. If hygrothermal conditions reached a threshold that could trigger the onset of corrosion, dehumidifiers would be activated to prevent corrosion on the unprotected steel wires.

Traditional monitoring techniques, on the other hand, are also being improved through the implementation of robotics. For example, unmanned aerial vehicles (UAV) have become popular for visual infrastructure inspections [36]. UAV monitoring reduces work safety issues and inspection uncertainties due to human error, and can easily reach almost every part of any structure. Early applications of UAV in engineering include visual inspection and damage detection via digital image correlation (DIC) and thermography, and dynamic characterization with equipped radar devices.

In summary, automation and ICT could have a huge impact in the field of civil engineering and architectural heritage management in the coming years. In this chapter, an overall vision of structural health monitoring in Spain is presented, considering all monitoring stages, including analysis, design, implementation and management.

SHM Systems, Damage Assessment and Structural Analysis

Heritage monitoring comprises a wide variety of fields—works of art, built heritage and written documents. Over the past decade, for example, the Santa María la Real Foundation has developed their Monitoring Heritage System (MHS) as a tool for integrated management of all types of heritage [28]. Initially conceived as simply a monitoring tool for more sustainable management, implementation has recently begun on an expert system for preventive conservation based on pattern recognition [12]. In this chapter, however, only issues related to structural monitoring of built heritage will be considered.

A successful monitoring system usually comprises three or four stages:

- *Preliminary investigation and planning*, in which the objectives of SHM should be defined. A pathology report would help to determine the best techniques for properly evaluating the structure. At this point, a visual inspection is required. In efforts to avoid human errors, automated vision-based techniques are being implemented today to detect surface deterioration [33, 36], cracking patterns [30], geometric characterization of masonry elements [37], or lateral drift or deflections that can jeopardize the stability of the structure [46].
  Another key aspect of SHM planning is the time frame desired for monitoring: permanent monitoring requirements differ from single-day (or a few days) measurement setups. The former aims to characterize long-term behavior of the building (e.g. thermal strains between day and night or between seasons, ongoing settlement due to non-stabilized ground movements, control during construction or retrofitting work). Measurement duration should be at least long enough to register a complete cycle of variable actions— temperature, for example [48]. Hence, the irreversible or cumulative component can be separated from cyclical counterparts. Temporary, or short-term, monitoring is usually undertaken to evaluate the structural performance or state under certain circumstances. Examples include loading tests on bridges just after construction to assess normalized requirements, or dynamic testing to determine the mechanical parameters of a structure (which can later correlate to material degradation).
- *Installation and actuation*. Once the objectives have been defined and the parameters of study identified, the proper sensor and experimental setup should be designed and installed. In this regard, wireless communication allows the locating of sensors without an action that interferes with the normal use of the building. The possibility of installation during construction as a work control system, and retaining the SHM system afterwards for performance monitoring, is an additional advantage of wireless systems [27]. An initial system evaluation

should be conducted to ensure that all sensors work correctly in the specific scenario.

- *Research and parameter control.* Protocols for data recording, transmission and storage should be defined. In short-term systems, data would be analyzed afterwards in a final report, with all the structural models necessary for a future intervention design. In long-term monitoring, threshold values for each parameter are useful as a control in the event that an emergency intervention is necessary. In this case, automated analysis tools would be helpful for optimizing the system. Permanent monitoring objectives should also define the data acquisition frequency, which necessitates a compromise between data volume and analysis requirements.
- *Maintenance and decommissioning.* Long-term monitoring systems will require periodic testing to assess SHM robustness. Wireless devices may also need additional care, depending on the power source selected (e.g. battery replacement). In this regard, recent research has focused on the development of power harvesters to eliminate the need for battery power supply [24, 25]. Finally, the last stage involving the removal of the SHM system should be carefully designed to avoid worker injury and damage to SHM components or the structure itself.

The SHM industry is challenged with implementing wireless networks in order to improve traditional monitoring strategies. A complete guideline was recently published [49], showing the different actors involved in a wireless SHM system and a general overview of the technical problems associated with each stage of monitoring.

Pathologies and Non-destructive Testing (NDT) Techniques

A pathology study is useful for assessing the state of the structure that will be the object of the monitoring system. A complete pathology report comprises not only geometric measurements and representation of the damage distribution, but also a set of physical, chemical and mechanical tests. Microstructural characterization and chemical composition testing to characterize the mineralogy of materials are commonly used in this field to detect the cause of various pathologies. However, they are beyond the scope of the current chapter, in which NDT techniques are presented only insofar as they will benefit from automation and technological development, as will be illustrated in various real-world applications.

First, a brief description of the most frequent types of physico-mechanical damage is presented. Heritage structures are usually constructed in masonry; therefore, the main pathologies that can be distinguished at this stage are cracks and physical alterations of materials (masonry or mortar). In steel structures, the equivalent pathology is corrosion, while reinforced concrete elements involve a

Fig. 1 The most frequently observed pathologies in built heritage: **a** discoloration, **b** spalling, **c** mortar loss, **d** cracking

combination of both scenarios. The main types of deterioration can be summarized as follows:

- *Discoloration of masonry materials*, Fig. 1a which can be directly linked to changes originated by chemical effects. Buildings located in heavily polluted environments (e.g. industrial or road traffic contamination) present color changes more frequently and more visibly. Some differences are dependent on the level of exposure, for example, to the prevailing winds or moist wind. In addition, differences in the characteristics of each material (e.g. origin, composition or porosity) can be detected after exposure. This type of damage can include humidity-related pathologies, such as efflorescence (which appears due to salt precipitation) or growth of microorganisms (due to moisture). Although the cause of these pathologies is not the same, for the purpose of this chapter, the detection techniques will be identical.

- *Spalling in the surface of stone-like materials*, Fig. 1b. The loss of material can occur due to salt content in the internal water. Damage will be more severe with greater wind exposure or groundwater filtration. In addition, materials with higher porosity are prone to this type of damage.
- *Mortar loss in masonry joints or plaster*, Fig. 1c. Similar to spalling, this pathology is an important factor in the stability of slender structures such as industrial chimneys or bell towers, especially when they are built in masonry bricks, because it can generate loose brick elements. If an external plaster layer is present, its debonding and possible spalling is also a frequent pathology related to hygrothermal effects.
- *Cracks*, Fig. 1d. Unreinforced masonry has almost no tensile strength. Hence, crack initiation and propagation must be controlled to guarantee structural stability. The causes can differ widely—for example, ground movements or seismic events. Crushing cracks in columns can appear due to high compressive stress, which can be a problem if water content is high, as stone strength diminishes in saturated conditions [50]. Because structural performance can be severely compromised if cracks remain active, crack monitoring is an important basic parameter in long-term heritage SHM systems.

Table 1 includes a summary of the most common NDT techniques used in heritage characterization, from visual methods to physical techniques such as thermography or radar. A brief description is given below, highlighting those techniques that are currently being improved through the use of automation and new technologies.

Visual inspection can provide a general idea of the structural conditions and the parameters of interest for monitoring. Traditionally, human inspections have been limited to areas of access or have required specific machinery or equipment for this purpose (e.g. cranes). Today, the development of UAV allows this task to be performed more safely and at lower cost. Figure 2 includes examples of different visual inspections. In Fig. 2a, b the use of cranes was necessary to reach the areas for inspection, with consequent temporary disruption of road traffic in some cases. The UAV shown in Fig. 2c was used to complement the inspection of the masonry chimney in Fig. 2a, d. This robot-aided inspection can even be performed inside buildings, as shown in Fig. 2e.

Image processing or DIC and UAV are being implemented for crack detection, especially in concrete bridges and tall engineering facilities (chimneys or wind

Table 1 NDT techniques for geometric and physical evaluation of heritage buildings

Technique	Applications	References
Human visual inspection	Crack detection, superficial pathologies	[27]
Automated visual inspection	Crack detection, superficial pathologies (digital image correlation)	[30, 36]
Survey	Geometry, drift, deflections	[39, 59]
3D scan	3D geometric modeling, superficial pathologies	[33, 35]
Thermography	Layer and delamination detection, joint and masonry detection, water filtration or condensation	[14, 32]
Ground-penetrating radar (GPR)	Layer definition, void detection	[5, 41]
Synthetic-aperture radar (SAR)	Displacements	[53, 58]
Ultrasound	Mechanical characterization, damage detection	[8, 40]
Vibration-based	Characterization of mechanical properties, damage detection	[9, 21, 38]

turbines). For heritage preservation, crack and spalling detection of a masonry bell tower can also be achieved by combining these techniques [36].

Traditional survey techniques can be useful for geometric definition of buildings or control during structural retrofitting [39]. Structural drift due to subsidence or any type of displacement can be easily assessed, and even their evolution over time can be monitored [59]. However, photogrammetric restitution and 3D scanning are gaining importance for this purpose because more detailed models can be obtained. Moreover, complete cracking and surface defect mapping can be drawn with the aforementioned image-processing algorithms [35].

Both active and passive thermographic techniques are useful for detecting defects in several types of structures, not only masonry. Thermography has already been used commercially as an accessory to UAV for structural monitoring. Typical applications in heritage include the detection of voids (e.g. mortar loss in joints), humidity filtration or moisture content [55], and plaster delamination or cracks [32]. This technique is a good complement to the aforementioned 3D scanning for detecting defects difficult to see with the naked eye [14].

Two radar-based techniques have been included: ground-penetrating radar (GPR) and synthetic-aperture radar (SAR). The former requires direct contact with the element to be measured, and provides in-depth information about material composition [5]. The latter can control displacements across a wide area, and has been used to monitor ground subsidence in public works [53] and heritage [58]. In addition, SAR has recently been incorporated into UAV to widen its range of application [3].

Fig. 2 Human visual inspection: **a** masonry chimney and **b** bell tower, both protected structures in Agost (Alicante). **c** UAV for automated inspection used in **d** masonry chimney and **e** interior inspection of Saint Peter's Church, both in Agost

Structural Analysis of Heritage Structures

This chapter does not pretend to provide an exhaustive assessment of the current state of numerical modeling in architectural heritage. A good review of the evolution of structural analysis of historic constructions can be found in the work of Roca et al. [47].

In SHM design, the objective of structural analysis is twofold: First, numerical modeling is often necessary for the proper design of monitoring systems. The optimal sensor location [20], expected measurement ranges, threshold values and data interpretation can be assessed using a previous model. Second, this same model can be calibrated with experimental data from the monitoring system and will be used in the diagnosis [22] and retrofitting proposal.

Table 2 includes a summary of Spanish heritage structures (mainly towers and churches) for which numerical models were used to assess their stability or as a tool during their restoration. These examples have been selected to show different levels of complexity, from numerical models with only an elastic linear analysis [50], to nonlinear analyses including damage criteria for predicting cracking patterns in masonry structures [18, 46]. The selection of the type of analysis will determine the necessary resources, the reliability of the results and the possible conclusions that can be drawn afterwards.

Structural Health Monitoring in Spain

The aforementioned MHS [Monitoring Heritage System] project of the Santa María la Real Foundation is one of the best examples of how heritage management can benefit from technological development and innovation. The MHS philosophy is premised on a three-legged system: preliminary study, SHM system design and performance monitoring [28]. The preliminary study may comprise a visual inspection and eventually numerical modeling. The objectives of the SHM will then be defined and the parameters of interest determined, after which the most appropriate measuring technique and sensor performance may be selected. Finally, once the system is implemented and running, the use of automated analysis software will aid the decision-making process. These whole-package monitoring solutions are becoming increasingly important for long-term monitoring. In addition to the MHS algorithm example [12], other cloud-based systems can be found in the Spanish market (e.g. www.kbuilding.es). A continuous knowledge transfer between industry and research groups at universities should also be promoted in this field (e.g. www.cal-sens.com).

Table 3 includes examples of SHM systems in heritage structures in Spain reported during the twenty-first century. This list is not exhaustive, but merely affords a view of the different types of monitoring. In some cases, such as MHS, which has been successfully applied in more than 20 heritage sites in northwestern

Table 2 Examples of numerical models of Spanish architectural heritage structures

Site	Structural and analysis details	References
Cathedral of Valencia	Gothic cathedral from 12th century, numerical models only of the cimborio (octagonal lantern) and dome from 15th century. Geometric 3D model by laser scanner and numerical model with solid elements. Linear and nonlinear (isotropic damage) analysis for self-weight hypothesis	[4]
San Martín tower (Toledo)	Mudéjar brick-masonry tower from 13th century. Presented vertical cracks, local buckling of interior elements and tilt. 3D numerical model with solid elements for masonry, and beam and shell elements for reinforced concrete retrofitting. Two linear analyses (vertical and tilted tower) with self-weight and loads transmitted by the annexed church. Reanalysis introducing cracking effect as disconnected elements	[19]
Cathedral of Tortosa (Barcelona)	Fourteenth-century Gothic cathedral, study of a lateral oval dome (18th century) constructed in two brick layers, which failed in 2011. 3D geometry by photogrammetric survey, numerical model with solid elements and nonlinear damage analysis	[26]
Cuatrovitas tower (Sevilla)	Twelfth-century Almohad minaret, 14.8-m-high brick-masonry structure. 3D numerical model with solid elements. Two static analyses: linear and nonlinear (plasticity criteria and cracking effect). Modal analysis for dynamic evaluation, nonlinear time history analyses for dynamic response of two models (3D solid elements and simplified beam elements)	[42]
Árchez tower (Málaga)	Andalusian minaret from 14th century, 16.85-m-high brick-masonry tower. 3D model with solid elements, linear and nonlinear static analyses (plasticity and cracking), modal analysis and seismic analyses. Pushover analyses to evaluate an earlier structural retrofitting	[43]
Cathedral of Mallorca	Fourteenth-century Gothic cathedral. Models consider viscoelastic behavior of masonry, tension-compression damage model, and smeared vs. localized damage. 3D numerical model with solid elements for stage construction analysis and long-term deformations. 2D numerical model for seismic analysis with various damage criteria	[46]
Church of Apostle Santiago (Jerez de la Frontera)	Fifteenth-century church, which suffered several column collapses, and currently showed vertical cracking and damage due to dampness. 3D model with shell and beam elements for linear analysis and gravity loads, and evaluation of retrofitting solution	[50]

(continued)

Table 2 (continued)

Site	Structural and analysis details	References
Basilica of Pilar (Zaragoza)	Original 9th century church, central dome from 19th century and towers in the perimeter from 20th century. Cracking in arches, deterioration of towers, fissures and humidity affecting valuable frescoes. Global 3D model, including soil interaction, for the entire temple, and local model to study the dome's behavior. Different construction stages and retrofitting were considered. Linear and nonlinear (two different criteria) analyses were performed with gravity and thermal loads. Global model considered damage only in tension with a smear crack approach. Local model of the dome also considered compression damage by crushing	[51]
Royal Monastery (Sta. Mª de Poblet)	Twelfth-century monastery with a Romanesque church, which exhibits extensive deformations and material damage due to past fires and corrosion of embedded steel elements. A complete 3D geometric model with the deformed shape was obtained by terrestrial laser scanning. A detailed 3D model of a single bay (considering its deformed shape) was constructed with solid elements. Nonlinear analyses based on damage continuum mechanics under different loading hypotheses (self-weight, ground settlement and seismic pushover analyses)	[52]
Homenaje tower (Granada)	Twelfth-century tower inside the Alhambra complex, 24.8-m-high brick-masonry and rammed earth structure. Graphical limit analysis based on thrust lines. 3D numerical model with solid elements for linear (gravity loads) and nonlinear pushover (concrete damage plasticity) analyses. Simplified 2D model with shell elements for nonlinear pushover analysis	[60]

Spain [13], only the most significant example has been included—in this case, the Cathedral of Palencia. The first system classification relates to the SHM objectives and the forecasted time span (i.e. long-term vs. short-term monitoring). The reasons for long-term measurements include:

- *Health control after retrofitting*, especially for parameters related to the damage actually repaired. Castillo et al. [10] report the complete process (i.e. diagnosis, repair, and SHM installation and performance) involved in the singular reinforced concrete grandstands of the La Zarzuela racecourse in Madrid. In this case, severe corrosion damage had to be repaired in the steel rebars of the roof. During this intervention, electrochemical sensors were embedded in key points in order to control the corrosion levels of the repaired structure.

Table 3 Examples of structural health monitoring of heritage elements in Spain in the twenty-first century

Site	Material	Sensors and techniques	SHM system	References
Fillaboa bridge (Salvaterra do Mino)	Masonry	DIC (digital camera, survey, GPR, FEM)	Single monitoring diagnosis	[5]
Sagrada Familia (Barcelona)	Masonry + RC	Optical fiber, LVDT, Temperature, survey	Preventive during tunnel construction	[7, 23]
La Zarzuela racecourse (Madrid)	RC	Corrosion, water, temperature	Permanent after retrofitting	[10]
Casa Milà (Barcelona)	Masonry + steel	LVDT, survey	Preventive during tunnel construction	[23]
Seminario Mayor (Comillas)	Masonry	Inclinometer, LVDT, temperature, RH, wind.	Pre-retrofitting 1.5-year wireless	[27]
Cernadela bridge (Mondariz)	Masonry	DIC + GPR + FEM	Structural assessment with automated crack detection	[30]
Cathedral of Santiago de Compostela	Masonry	Photogrammetry	Surface pathology mapping	[33]
S Juan Bautista church (Talamanca del Jarama)	Masonry	Temperature, RH	2-week temporary wireless	[34]
Santo Domingo de Silos church (Pinto)	Masonry	Temperature, RH	3-week temporary wireless	[34]
Santa Marina church (Córdoba)	Masonry	LiDAR + DIC	Single surface pathology mapping	[35]
S. Juan de los Caballeros church (Cádiz)	Masonry	Accelerometer (OMA), 3D scan	Preliminary study for structural retrofitting	[38]
Teatro Romano (Cádiz)	Masonry	Survey + accelerometer (OMA)	Control during restoration	[39]
La Giralda (Sevilla)	Masonry + steel	Acceleration, temperature, wind, inclinometer, strain gages, RH, corrosion	Permanent (2-year report)	[54]

(continued)

Table 3 (continued)

Site	Material	Sensors and techniques	SHM system	References
Roman arch bridge (Lugo)	Masonry	GPR	Single-day monitoring for structural characterization	[56]
Cathedral of Palencia	Masonry	MHS (LVDT + temperature + RH + accelerometer)	Permanent wireless	[17]
Sta. Justa and Rufina church (Orihuela)	Masonry	SAR + accelerometer	Structural assessment and long-term subsidence control	[22, 58]
Sta. Mª la Blanca church (Agoncillo)	Masonry	Survey	2-year displacement data	[59]

Another example of monitoring for preventive maintenance can be found in La Giralda, Sevilla [54], in which an SHM system was installed after the restoration of the vane atop the tower, a singular sculpture known as "the Giraldillo," and the steel structure that connects it to the tower. The SHM comprises several types of sensors, and will be used to determine when the next structural maintenance should occur. Because the system must function outdoors in harsh climate conditions, a five-year revision is scheduled in order to replace the sensors (due to life-cycle exhaustion).

- *Structural control during retrofitting or external works.* This preventive monitoring features continuous control throughout the intervention that may affect the monitored structure. For example, during the construction of an underground railway tunnel in Barcelona, various singular heritage buildings were monitored to ensure that subsidence did not affect the Sagrada Familia [7] or Casa Milà [23].
  The work to uncover the roman theatre in Cádiz required permanent control of the residential buildings currently over the theatre, while structural reinforcement and digging operations were cast [39].
- *Preliminary study for structural diagnosis.* Lombillo et al. [27] developed a wireless monitoring system for a church in Comillas. The objective of this preliminary monitoring phase was twofold: designing and installing a wireless sensor network with remote access, and health monitoring prior to the structural repairs to evaluate the detected damage. The possibility for performance during retrofitting work is another advantage of wireless systems. Preliminary monitoring also serves as calibration for the post-intervention SHM.
- *Time evolution based on periodic measurements.* Another type of long-term monitoring involves the use of intermittent measurements—repeating the same procedure once every several months or years—and comparing geometric differences between measurements. Satellite images obtained by SAR can be useful for detecting structural displacement due to ground movements. This technique was applied by Tomás et al. [58] to evaluate the effect of subsidence in Santa Justa and Rufina Church in Orihuela. Traditional survey methods can also be applied for this geometric control purpose [59].
- *Preventive conservation.* The MHS project is based on the concept of preventive maintenance. The MHS algorithm [12] calibrates specific decay curves for each monitored structure in order to determine the optimal intervention duration as a compromise between structural health and economic investment.

Short-term monitoring can be understood as measurements registered in just one day or over a period of a few days. Incompatibility with the normal use of the structure makes it impossible to apply these techniques over longer periods. Therefore, they focus on obtaining a particular characterization of the structural condition. GPR has been used to detect different layers based on differences in the material's behavior. It has been applied in several historic masonry bridges in northwestern Spain as the only characterization NDT [56]. In addition, multidisciplinary diagnosis combines GPR with other techniques, such as GPR plus thermography for moisture content in masonry [55], and GPR plus image correlation obtained with digital cameras for a structural 3D model [5] or crack detection [30].

Martínez-Garrido and Fort [34] also performed short-term monitoring (two or three weeks) in several heritage constructions. The purpose in these cases was to develop a wireless monitoring system applicable in different situation. Therefore, they tested the system under real conditions to determine the best protocols, guaranteeing the robustness of the measurements.

A second classification can be made based on the origin of the variables to control—mechanical, ambient or chemical, for example.

- *Ambient conditions*. Most of the damage that affects this type of heritage construction (i.e. masonry structures) is directly related to humidity or temperature [34]. Even the preservation of other works of art, such as frescoes [51], makes the control of these two parameters essential in order to avoid undesirable effects. Thermohygrometer measurements are also necessary to guarantee the precision of other sensors (e.g. strain gauges, fiber optics), whose values have to be corrected depending on the service ambient conditions.
- *Mechanical characteristics of materials*. For structural analysis, the differences between layers of diverse materials should be determined, including layer depth, density and mechanical strength. NDT techniques such as GPR [5], ultrasonic devices [40] or thermography [55] are usually used for that purpose. Typically, this sampling is performed only once during the preliminary studies.
- *Dynamic properties monitoring*. Dynamic behavior analysis can also be a powerful tool for structural assessment of heritage buildings, as was noted in some of the numerical modeling examples. This SHM can be performed as single measures [22] or in long-term monitoring for continuous damage detection.
- *Chemical control*. The presence of certain gases (due to excessive pollution) that can affect heritage buildings or works of art is the object of some monitoring actions [17]. Electrochemical measures are necessary to control rebar corrosion in reinforced concrete elements [10].

Heritage Monitoring Examples: Static Versus Dynamic Monitoring

Two monitoring approaches are discussed below. First, an SHM system was used to control the movements in a sixteenth-century church with severe cracking in different parts of the structure. Second, dynamic monitoring was used to evaluate the possible effects of extreme loading conditions involving blast-induced vibrations on a historic fountain. These two examples are currently being developed by the research group of testing, simulation and modeling of structures (GRESMES) of the University of Alicante.

Church of Saint Andrew Apostle, Jaén

The church of Saint Andrew in Villanueva del Arzobispo (Jaén) was built in the sixteenth century. The structure exhibits several cracks in the area of the transept and the dome. Figure 3a shows structural cracks in one of the arches; in addition, horizontal cracking can be observed in the area of contact with the masonry diaphragm. The possible settlement of the arches and dome also affects the nave next to the transept, and cracks appear in the arch and window, Fig. 3b. In order to study the evolution of these movements, an SHM system was implemented. The instrumentation comprised a total of eight sensors (six linear variable displacement transducers (LVDT) for crack opening-closing measurement, and two thermocouples for temperature control). Data were able to be accessed remotely to obtain real-time images of the structural displacements.

Figure 4 includes images of sensor installation. Figure 4b shows an LVDT that was located in the vertical crack of the keystone of one arch, which supports the dome. Because all monitored cracks were still active after a number of days of monitoring, emergency measures were taken to ensure structural stability. Hence, the scaffolding system shown in Fig. 4c was cast to support the dome's weight in the damaged arches while the structural reinforcement was designed and cast.

An SHM system was also used to evaluate the performance of the scaffolding solution. Figure 5 shows the time evolution of two displacement sensors located in the arch above the high altar, one in the lower part of the keystone and another in the joint between the arch and the upper diaphragm. Temperature measurements obtained in a thermocouple next to the LVDTs are also included in order to exclude thermal effects in the structural displacements. Results for the first five months of

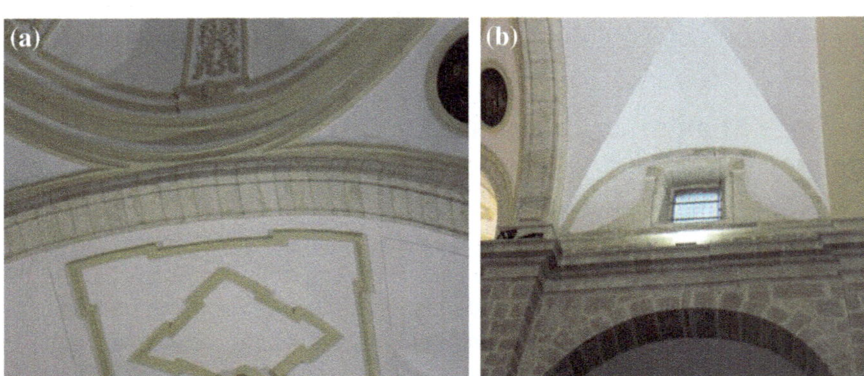

Fig. 3 Main structural cracks: **a** arch and dome junction; **b** nave next to the transept and dome

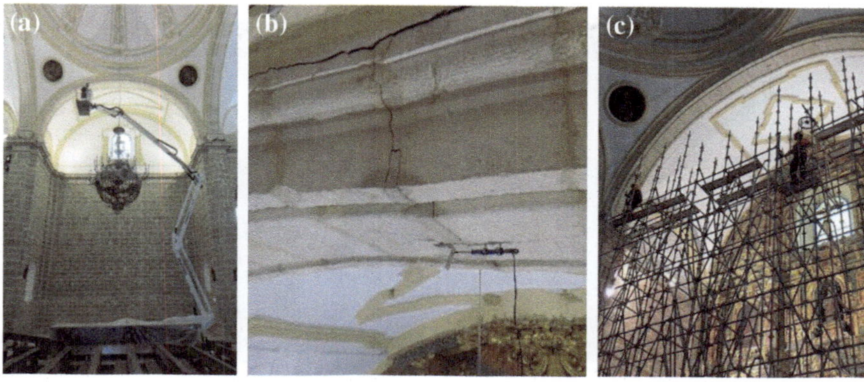

Fig. 4 **a** Sensor installation. **b** LVDT sensor attached to the lower part of the arch over the high altar. **c** Casting of the scaffolding system to unload one of the arches supporting the dome

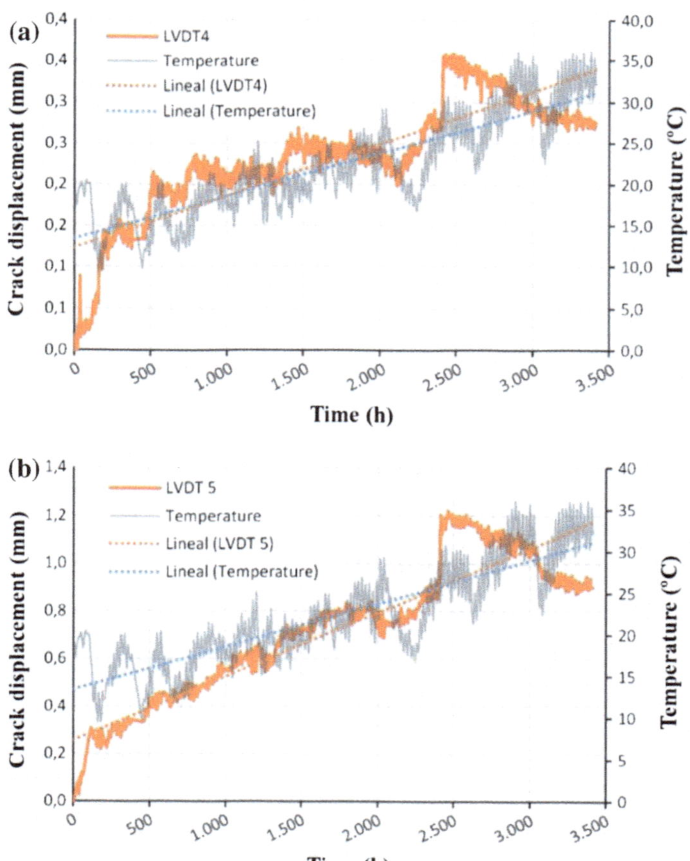

Fig. 5 Temperature and displacement measurements of LVDT located in the arch over the high altar, **a** in the lower part of the keystone and **b** in the joint between the arch and the upper diaphragm

monitoring are included. The effect of the scaffolding can be observed in the changes registered after 2000 h. At around 2500 h, a readjustment of the scaffolding was necessary after the detection of a sudden displacement increase.

Fig. 6 **a** General view of the fountain and the protective screens; **b** sensor installation; **c** triaxial setup; **d** relative position of the fountain and explosives; **e** detail of the fireworks devices; **f** view of the fountain during the spectacle

Subsequently, the widening trend in the crack opening has been reversed. This type of monitoring will also be useful during foundation reinforcement work and structural reloading when the scaffolding is dismantled.

Historic Fountain, Alicante

The fountain in the center of the Plaza de los Luceros in Alicante, Fig. 6a, is one of the city's most important monuments. This modernist structure, built in 1930, is composed of a central body and four horse sculptures. It was repaired several times in the early 2000s, and a retrofitting using composite materials is currently being evaluated. One of these interventions consisted in the total dismantlement and

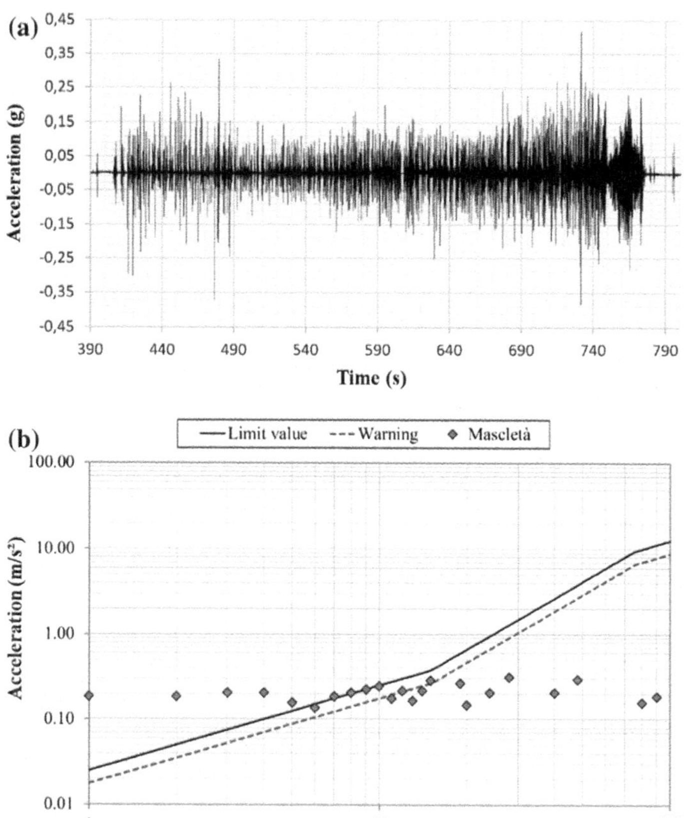

Fig. 7 **a** Vertical acceleration time history registered by a sensor attached to the fountain structure. **b** Maximum accelerations and UNE 22381:1993 limits versus frequency for data registered in a sensor attached to the ground near the structure

subsequent reconstruction of the entire monument, due to the work on an underground railway station under the plaza.

The most common types of damage are superficial cracking and spalling as a result of nearby road traffic, water from the fountain and corrosion of the embedded steel elements of the fountain installation. However, another possible cause should be taken into consideration: Each year during the city festivities, several fireworks displays (called mascletàs) are held in the plaza, near the fountain. These mascletàs, and their location, have led to a public debate on their possible effects on the fountain's health. As a temporary measure, each horse sculpture has been protected with composite protective panels during these events, Fig. 6a.

In this case, dynamic monitoring has been selected to evaluate the effect of the blasts on the fountain elements. A monitoring system comprising eight accelerometers attached to the fountain and on the ground was installed to measure vibration levels due to the fireworks. The sensor installation can be observed in Fig. 6b; an example of a triaxial setup, attached to the ground near the fountain, is included in Fig. 6c. The relative location of the explosive charges with respect to the monument and a detail of them are include in Fig. 6d and e, respectively. An example of the pyrotechnic arrangement for this performance is shown in Fig. 6f.

Accelerations on all sensors were measured at 500 Hz throughout the blast spectacle, lasting approximately 6–7 min. Figure 7a includes the accelerations recorded at the fountain, with peak acceleration of around 0.4 g. After frequency analysis, these values will be compared with different recommendations found in standards UNE 22381:1993 [1] or DIN 4150-3:2015 [16] and in scientific reports. The analysis shown in Fig. 7b represents the maximum vertical accelerations registered in a sensor attached to the ground, which are higher than the limits included in UNE 22381:1993 [1] for low frequencies, in a range of 1–10 Hz. Similar conclusions can be derived for the sensors directly attached to the monument. Therefore, these blasts may produce excessive damage to the fountain, and preventive measures should be taken.

Conclusions

This chapter has presented various systems developed by Spanish research groups and companies in the field of SHM, applied to historic constructions and civil engineering facilities. In recent years, techniques such as wireless or fiber optic sensors, continuous monitoring and trigger alerts have been successfully used for preventive maintenance and serviceability limit states. In some cases, as in the aforementioned spin-off Spanish companies, this research led to a transfer of knowledge between research groups and industry. The future of this industry will require automated processes to aid experts and engineers in the rapid analysis of large volumes of information to identify the correct course of action. Furthermore, new monitoring techniques can lead to significant advances in SHM applications—

for example, UAV equipped with sensor technologies such as hyperspectral cameras, radar or thermography.

Acknowledgements The authors would like to thank the University of Alicante for its financial support, VIBGROB-212-(GRESMES). F. J. Baeza also wants to acknowledge the Spanish Government, Ministerio de Educación, Cultura y Deporte, for the financial support, Programme José Castillejo, grant CAS15/00223.

References

1. AENOR UNE 22381:1993 (1993) Control of vibrations made by blastings
2. AENOR UNE 178104:2015 (2015) Smart cities. Infrastructures. Comprehensive systems for a Smart City management
3. Aguasca A, Acevo-Herrera R, Broquetas A, Mallorqui JJ, Fabregas X (2013) ARBRES: Light-Weight CW/FM SAR sensors for small UAVs. Sensor 13:3204–3216
4. Alonso A, Martínez A, Llopis V, Moreno J (2013) Construction and structural analysis of the dome of the Cathedral of Valencia. In: VIII international conference on fracture mechanics of concrete and concrete structures FraMcos-8, Toledo, Spain, 10-14 Mar 2013
5. Arias P, Armesto J, Di-Capua D, González-Drigo R, Lorenzo H, Pérez-Gracia V (2007) Digital photogrammetry, GPR and computational analysis of structural damages in a mediaeval bridge. Eng Fail Anal 14:1444–1457
6. Barbat A, Pujades LG, Lantada N (2006) Performance of buildings under earthquakes in Barcelona, Spain. Comp-Aided Civ Infrastruct Eng 21:573–593
7. Bounatian L, Sánchez F, Poy I (2011) The monitoring of Barcelona's Sagrada Familia church with fiber optic technology: control of the construction of a nearby tunnel. In: 5th International Conference on Structural Health Monitoring of Intelligent Infrastructure SHMII-5. Cancún, Mexico, pp 11–15, Dec 2011
8. Brotóns V, Tomás R, Ivorra S, Grediaga A (2014) Relationship between static and dynamic elastic modulus of calcarenite heated at different temperatures: the San Julián's stone. B Eng Geol Environ 73:791–799
9. Brownjohn JMW, De Stefano A, Xu YL, Wenzel H, Aktan AE (2011) Vibration-based monitoring of civil infrastructure: challenges and successes. J Civ Struct Health Monit 1:79–95
10. Castillo A, Andrade C, Martínez I et al (2011) Assessment and monitoring of durability of shell structures in Zarzuela Racecourse Madrid. Inf Constr 63(524):33–41
11. Ceci AM, Contento A, Fanale L, Galeota D, Gattulli V, Lepidi M, Potenza F (2010) Structural performance of the historic and modern buildings of the University of L'Aquila during the seismic events of April 2009. Eng Struct 32:1899–1924
12. Chiriac M, Basulto D, Prieto JC (2014) Development and implementation of the MHS algorithm for the preventive conservation of heritage monuments. In: Rogerio-Candelera MA (ed) Science, Technology and Cultural Heritage. CRC Press, pp 417–422
13. Chiriac M, Basulto D, López E et al (2013) The MHS system as an active tool for the preventive conservation of Cultural Heritage. In: Rogerio-Candelera MA, Lazzari M, Cano E (eds) Science and technology for the conservation of cultural heritage. CRC Press, pp 383–386
14. Costanzo A, Minasi M, Casula G et al (2015) Combined use of terrestrial laser scanning and IR thermography applied to a historical building. Sensor 15:194–213. https://doi.org/10.3390/s150100194
15. Deloitte Consulting (2015) Estudio y guía metodológica sobre ciudades inteligentes. ONTSI

16. Deutsches Institut für Normung (2015) DIN 4150–3: Vibration in building—Part 3: Effects on structures
17. Fundación Santa María la Real (2016) Monitoring Heritage System Project. http://www.mhsproject.com. Accessed 14 Jul 2016
18. González R, Caballé F, Domengé J, Vendrel M, Giráldez P, Roca P, González JL (2008) Construction process, damage and structural analysis. Two case studies. In: D'Ayala D, Fodde E (eds) Structural analysis of historic construction. CRC Press, London, pp 643–651
19. Hernández S, Romera LE, Díaz J (2007) Twenty years of experience in numerical models of historical constructions: a temporal perspective. In: Brebbia CA (ed) WIT Transactions on the Built Environment, vol 95, WIT Press, pp 385–397. https://doi.org/10.2495/STR070361
20. Ivorra S. Baeza FJ, Bru D, Varona FB (2015) Seismic behavior of a masonry chimney with severe cracking condition: preliminary study. Key Eng Mater 628:117–122
21. Ivorra S, Foti D, Bru D, Baeza FJ (2015) Dynamic behavior of a pedestrian bridge in Alicante, Spain. J Perform Constr Facil 29(5):04014132 p 10. https://doi.org/10.1061/(ASCE)CF.1943-5509.0000556
22. Ivorra S, Pallarés FJ, Adam JM, Tomás R (2010) An evaluation of the incidence of soil subsidence on the dynamic behaviour of a Gothic bell tower. Eng Struct 32:2318–2325
23. Jäger W, Burkert T, Boekhoff B, Bakeer T (2011) The monitoring of world heritage sites during construction works in their vicinity: the case of Casa Milà and of the Church of Sagrada Familia in Barcelona, Spain. In: Brebbia CA and Binda L (eds) WIT transactions on the built environment, vol 118, pp 357–374
24. Jia Y, Yan J, Feng T et al (2015) A vibration powered wireless mote on the Forth Road Bridge. J Phys Conf Ser 660:012094 5p. https://doi.org/10.1088/1742-6596/660/1/012094
25. Jia Y, Yan J, Soga K, Seshia AA (2014) A parametrically excited vibration energy harvester. J Intell Mater Syst Struct 25:278–289
26. Lluís i, Ginovart J, Costa A, Fortuny G (2013) Assessment and restoration of a masonry dome in the cathedral of Tortosa enclosure. In: Brebbia CA (ed) WIT transactions on the built environment, vol 131, WIT Press, pp 391–401. https://doi.org/10.2495/STR130331
27. Lombillo I, Blanco H, Pereda J et al (2016) Structural health monitoring of a damaged church: design of an integrated platform of electronic instrumentation, data acquisition and client/server software. Struct Control Health Monit 23:69–81
28. López E, Chiriac M, Basulto D et al (2012) Desarrollo e innovación en los sistemas de gestión del patrimonio. MSH (Monitoring Heritage System) sistema de monitorización del patrimonio como herramienta de gestión integral. In: Congreso internacional AR&PA. Valladolid, Spain, 24–27, May 2012
29. Lozano-Galant JA, Payá-Zaforteza I, Xu D et al (2012) Forward algorithm for the construction control of cable-stayed bridges built on temporary supports. Eng Struct 40:119–130
30. Lubowiecka I, Arias P, Riveiro B, Solla M (2011) Multidisciplinary approach to the assessment of historic structures based on the case of a masonry bridge in Galicia (Spain). Comput Struct 89:1615–1627
31. Macías-Bernal JM, Calama-Rodríguez JM, Chávez MJ (2014) Prediction model of the useful life of a heritage building from fuzzy logic. Inf Constr 66:e006
32. Maierhofer C, Roellig M (2009) Active thermography for the characterization of surfaces and interfaces of historic masonry structures. In: 7th International symposium on non-destructive testing in civil engineering. Nantes, France, 30 June–3 July 2009
33. Martínez S, Ortiz J, Gil ML, Rego MT (2013) Recording complex structures using close range photogrammetry: the Cathedral of Santiago de Compostela. Photogramm Rec 28(144):375–395. https://doi.org/10.1111/phor.12040
34. Martinez-Garrido I, Fort R (2016) Experimental assessment of a wireless communications platform for the built and natural heritage. Measurement 82:188–201
35. Meroño JE, Perea AJ, Aguilera MJ, Laguna AM (2015) Recognition of materials and damage on historical buildings using digital image classification. S Afr J Sci 111:Art.#2014-0001 9p

36. Morgenthal G, Hallermann N (2014) Quality assessment of unmanned aerial vehicle (UAV) based visual inspection of structures. Adv Struct Eng 17(3):289–302
37. Oses N, Dornaika F, Moujahid A (2014) Image-based delineation and classification of built heritage masonry. Remote Sens 6:1863–1889. https://doi.org/10.3390/rs6031863
38. Pachón P, Compán V, Rodríguez E et al (2015) Definition and characterization of a historical building by using digital photogrammetry and operational modal analysis. San Juan de los Caballeros Church (Cádiz, Spain). In: Proceedings of 3rd international conference on mechanical models in structural engineering, Sevilla, Spain, 24–26 June 2015
39. Pachón P, Rodríguez-Mayorga E, Cobo A, Yanes E (2014) Control of the structural intervention in the area of the Roman Theatre of Cadiz. Inf Constr 66 (EXTRA 1):m003
40. Palaia L, Tormo S, López V, Mofort J (2011) NDT assessment of timber structures: a case study—Villa Ivonne, Meliana. In: Brebbia CA, Binda L (eds) WIT transactions on the built environment, vol 118, WIT Press, pp 529–540
41. Pérez-Gracia V, Caselles O, Clapés J et al (2009) Radar exploration applied to historical buildings: a case study of the Marques de Lló palace, in Barcelona (Spain). Eng Fail Anal 16:1039–1050
42. Pineda P, Sáez A (2009) Assessment of ancient masonry slender towers under seismic loading: dynamic characterization of the Cuatrovitas tower. In: Brebbia CA (ed) WIT transactions on the built environment, vol 109, WIT Press, pp 615–629. https://doi.org/10.2495/STR090541
43. Pineda P, Robador MD, Gil-Martí MA (2011) Seismic damage propagation prediction in ancient masonry structures: an application in the non-linear range via numerical models. Open Constr Build Technol J 5(Suppl 1-M4):71–79
44. Potenza F, Federici F, Lepidi M et al (2015) Long-term structural monitoring of the damaged Basilica S. Maria di Collemaggio through a low-cost wireless sensor network. J Civil Struct Health Monit 5:655–676. https://doi.org/10.1007/s13349-015-0146-3
45. Prieto AJ, Macías-Bernal JM, Chávez MJ, Alejandre FJ (2016) Expert system for predicting buildings service life under ISO 31000 standard. Application in architectural heritage. J Cult Herit 18:209–218
46. Roca P, Cervera M, Pelà L et al (2013) Continuum FE models for the analysis of Mallorca Cathedral. Eng Struct 46:653–670
47. Roca P, Cervera M, Gariup G, Pelà L (2010) Structural analysis of masonry historical constructions. Classical and advanced approaches. Arch Comput Method Eng 17:299–325
48. Roca P, González JL, Aguerri F, Aguerri JI (2003) Monitoring of long-term damage in Gothic Cathedrals. In: Brebbia CA (ed) WIT transactions on the built environment, vol 66, WIT Press, pp 109–119
49. Rodenas-Herráiz D, Soga K, Fidler P, de Battista N (2016) Wireless sensor networks for civil infrastructure monitoring: a best practice guide. ICE Publishing
50. Rodríguez-Mayorga E, Yanes-Bustamante E, Sáez-Pérez A (2015) Analysis and diagnosis of the Church of Santiago in Jerez de la Frontera (Spain). Inf Constr 67:e127
51. Romera LE, Hernández S, Reinosa JM (2005) A comprehensive structural study of the Basilica of Pilar in Zaragoza (Spain). In: Brebbia CA (ed) WIT transactions on the built environment, vol 83, WIT Press, pp 103–113
52. Saloustros S, Pelà L, Roca P, Portal J (2015) Numerical analysis of structural damage in the church of the Poblet Monastery. Eng Fail Anal 48:41–61
53. Serrano-Juan A, Vázquez-Suñè E, Monserrat O et al (2016) Gb-SAR interferometry displacement measurements during dewatering in construction works. Case of La Sagrera railway station in Barcelona, Spain. Eng Geol 205:104–115
54. Solís M, Romero A, Galvín P (2010) Monitoring the mechanical behavior of the weathervane sculpture mounted atop Seville Cathedral's Giralda tower. Struct Control Health Monit 9 (1):41–57
55. Solla M, Lagüela S, Riveiro B, Lorenzo H (2012) Non-destructive testing for the analysis of moisture in the masonry arch bridge of Lubians (Spain). Struct Health Monit 22:334–339

56. Solla M, Lorenzo H, Rial FI, Novo A (2011) GPR evaluation of the Roman masonry arch bridge of Lugo (Spain). NDT&E Int 44:8–12
57. Stajano F, Hoult N, Wassell I et al (2010) Smart bridges, smart tunnels: transforming wireless sensor networks from research prototypes into robust engineering infrastructure. Ad Hoc Netw 8(8):872–888
58. Tomás R, García-Barba J, Cano M et al (2012) Subsidence damage assessment of a Gothic church using differential interferometry and field data. Struct Health Monit 11(6):751–762
59. Valle JM, Rodríguez A, Pérez P (2008) Evaluation of the conventional surveying equipment applied to deformation analysis of heritage buildings. A case study: the bell tower of Santa María la Blanca church in Agoncillo (La Rioja, Spain). In: 13th FIG international symposium on deformation measurements and analysis, Lisbon, Portugal, 12–15 May 2008
60. Villegas D, Cámara M, Compán V (2014) Assessment of procedures for the structural analysis of Homenaje tower in the Alhambra in Granada (Spain). Inf Constr 66(EXTRA 1): m017

Data-Driven Approach to Structural Health Monitoring Using Statistical Learning Algorithms

Debarshi Sen and Satish Nagarajaiah

Abstract Condition assessment and prediction of existing infrastructure systems is a major objective in civil engineering. Structural health monitoring (SHM) achieves this goal by continuous data acquisition from an array of sensors deployed on the structure of interest. SHM constructs ubiquitous damage features from the acquired data, ensuring maximum sensitivity to the onset of damage and robustness to noise and variability in environmental and operational conditions. Traditionally, SHM has used a model-based approach, wherein a high-fidelity model of a structure is constructed and studied in detail to aid engineers in detecting the onset of damage based on deviations from an undamaged model of the system. However, given the complexity of structures and the inability to perfectly model all aspects of a system, a data-driven approach becomes an attractive alternative. In data-driven approaches, a surrogate model, constructed using acquired data from a system, is substituted for a real model. Although such models do not necessarily capture all the physics of a system, they are efficient for damage detection purposes. Statistical learning algorithms aid in the construction of such surrogate models, and their use has now been extensively documented in the literature. This chapter provides a brief review of applications of statistical learning algorithms, both supervised and unsupervised, in SHM for real-time condition assessment of civil infrastructure systems.

Introduction

Structural health monitoring (SHM) is an essential component in both the maintenance of existing civil infrastructure systems and the planning of new systems. It entails condition assessment by data acquisition from an array of sensors deployed on the system of interest [1]. Condition assessment is typically followed by

D. Sen (✉) · S. Nagarajaiah
Rice University, 6100 Main Street, Houston TX 77005, Texas, USA
e-mail: debarshi.sen@rice.edu

S. Nagarajaiah
e-mail: satish.nagarajaiah@rice.edu

E. Ottaviano et al. (eds.), *Mechatronics for Cultural Heritage and Civil Engineering*, Intelligent Systems, Control and Automation: Science and Engineering 92,
https://doi.org/10.1007/978-3-319-68646-2_13

damage prediction that involves estimation of the remaining useful life [2]. The focus of this chapter is on SHM only. Condition assessment typically employs damage detection that requires quantitative and qualitative appraisal of changes in system properties with respect to a known undamaged state of the system. The comparison can be performed using either an undamaged high-fidelity model of the system or a pseudo-model constructed solely with acquired data from the system. This gives rise to two classes of SHM techniques: model-based [3] and data-driven approaches [4]. Barthorpe [5] compares these approaches and discusses the advantages and limitations of both.

A model-based approach involves the construction of a high-fidelity model of the system, typically a finite element (FE) model. Data acquired from an undamaged state of the system may be used to verify or update such a model. One then performs damage detection by comparing responses from the model and the actual system, and observing discrepancies. The major drawback of such an approach is that evaluation of responses from these models is generally computationally prohibitive. For example, when simulating guided ultrasonic waves for damage detection, a finite element model with a high number of elements is necessary for reliable simulation, which increases the computational effort dramatically [6].

In a data-driven approach, however, a surrogate model constructed from acquired data replaces the high-fidelity model. This derives from the idea that capturing the physics involved in a system becomes more challenging as the complexity of the system increases. The aim of a surrogate model, on the other hand, is not necessarily to replicate system behavior, but only to achieve efficient damage detection by accentuating and quantifying the change in a system due to the onset of damage. Moreover, because of the vast improvements in sensor technology, the volume of acquired data is on the rise as well [7]. In light of this, the surrogate model is typically a statistical model equipped for any of the first three levels of SHM (Rytter [8] defines the four levels of SHM).

Statistical learning algorithms are equipped for construction of such models necessary for data-driven approaches, with the help of data acquired from sensors deployed on systems. Typically, these algorithms are classified as either supervised or unsupervised [9]. Supervised statistical learning algorithms require labels for the data used for surrogate model construction. Unsupervised algorithms, however, do not require data labels, and are used primarily for pattern recognition in the data. In addition, unsupervised algorithms can be used in conjunction with supervised learning algorithms.

This chapter is a brief review of the applications of statistical learning algorithms in data-driven SHM. We discuss the utility of using these algorithms for the purpose of damage detection. We also present a few examples of applications of these algorithms. However, algorithmic details of specific learning algorithms have been avoided.

Statistical Learning

In this section, we briefly discuss statistical learning and illustrate the applicability of such algorithms for damage detection. As Bousquet et al. [10] state, *The main goal of statistical learning theory is to provide a framework for studying the problem of inference, that is of gaining knowledge, making predictions, making decisions or constructing models from a set of data.* This resonates with the concepts of SHM and damage detection. As discussed earlier, for data-driven approaches to SHM, we use acquired data for model construction, and decisions are made based on predictions from these models.

Data are generally one of two types: one with labels and the other unlabeled. Data labels are typically used to signify the origin of the data from a specific source or source condition. When a statistical learning algorithm uses labeled data for model construction and label predictions for future data, it is called a supervised learning approach. The alternative is an unsupervised learning scheme, where data labels are either unavailable or simply not used. These algorithms, although they do not make direct predictions about data labels, are capable of revealing inherent structures and patterns in the data that may be crucial. We will later show how this class of learning algorithms can be efficiently implemented for damage detection.

Supervised learning can also be divided into two types, namely regression and classification. The key distinction between the two approaches is in the output variables obtained, which are continuous for the former and discrete for the latter. All supervised learning algorithms require two stages. The first is training. This involves the use of part of the data, referred to as the *training data*, for model construction. Predictions are made using the model for the *test data*. These data are used as an input to the constructed model for obtaining the desired predictions or outputs. For the selection of algorithmic parameters, a process known as *cross validation* is performed to optimize the parameters to the given data, using the training data set. For further details, the reader is referred to the books by Hastie et al. [9] and Witten et al. [11].

Based on the above discussion, it is clear that statistical learning algorithms are best suited for data-driven SHM. A primer on the applicability of the various approaches discussed above can be found in Worden and Manson [12]. The following sections will provide a few examples where statistical learning algorithms have been applied for SHM. While these examples are by no means exhaustive, we try to cover a range of applications of a variety of learning algorithms.

Applications of Statistical Learning Algorithms

Supervised Learning

Classification Techniques

In this section, we discuss a few applications of supervised learning algorithms. As explained earlier, supervised learning algorithms can be either regression or classification problems. We will discuss some examples of classifiers first, followed by illustrations of applications of various regression algorithms in damage detection.

The Naïve-Bayes (NB) classifier is one of the most elemental base learners available in a data scientist's arsenal. Once trained, it classifies data based on conditional probability density functions constructed from the training data. The idea is to define class probabilities. If X is the data set, Y is the class variable, and there are K possible classes, then the classification rule for an NB classifier (probability that variable $Y = k$), based on the Bayes theorem, is defined as:

$$P(Y = k|X) = \frac{P(X|Y = k)P(Y = k)}{\sum_{i=1}^{K} P(X_i|Y = k)P(Y = k)} \tag{1}$$

Addin et al. [13] used an NB classifier for damage detection in composite materials using Lamb waves. The authors used underlying distinction in Lamb wave signals generated by different kinds of damage. Muralidharan and Sugumaran [14] also used an NB classifier for fault detection in centrifugal pumps; however, they demonstrated that a Bayes net classifier produced higher accuracy for their experiments. An extension of the NB classifier is linear discriminant analysis (LDA) with Fisher's discriminant. The key idea is to evaluate optimum discriminant functions that maximize the difference between various classes of data. For this, LDA assumes a Gaussian distribution for the random variable $X|Y = k$, with the same covariance for all classes. Farrar et al. [15] used LDA on acceleration data acquired from a concrete column for damage detection. Guadenzi et al. [16] also used LDA for low- velocity impact damage detection in laminated composite plates.

Support vector machines (SVM) are another classifier that constructs linear or nonlinear functions that separate various classes of data in the sample space. It is based on maximizing the distance of data points, belonging to distinct classes, to these separating hyperplanes. SVM solves the following optimization problem in a sample space:

$$\begin{aligned} \min_{\beta_0,\beta} \quad & \|\beta\|_2^2 \\ \text{subject to} \quad & y_i\left(x_i^T\beta + \beta_0\right) \geq 1 \quad \forall i = 1, \dots, n. \end{aligned} \tag{2}$$

where x_i is a data point in the sample space, β is the slope of a normal to optimal separating hyperplane, β_0 is an intercept, and y_i is a class variable that can take values of either +1 or −1. This variable essentially aids in the binary classification of the data.

Typically, SVMs are designed for solving two-class classification problems. However, they can be extended to solve multi-class problems as well. He and Yan [17] used an SVM-based damage detection scheme for a spherical lattice dome, where feature extraction was performed using wavelet transform of the ambient vibration data of the structure. Shimada et al. [18] employed SVM for damage detection in power distribution poles. Bulut et al. [19] proposed an SHM system for the Humboldt Bay Bridge using SVM, in which data acquired from a numerical impact test on the bridge were used for damage detection. Chattopadhyay et al. [20] used a nonlinear version of SVM for damage detection in composite laminates. A nonlinear version implies that the optimal separating hyperplane is a nonlinear function. Bornn et al. [21] employed SVM in conjunction with an autoregressive (AR) model for damage detection as well as localization by signal reconstruction and residual estimation. They reconstructed the signal using an AR model and used SVM for damage classification. For nonlinear SVM, they used a Gaussian kernel. Worden and Lane [22] demonstrated the general efficacy of SVM as a statistical learning algorithm for classification problems in engineering.

Trees are another type of popular base learner for classification. They are top-down recursive partitions of the feature or data space via binary splits. Although they are poor predictors, they are adept at making interpretations from data sets that eventually aid in decision making. Kilundu et al. [23] employed decision trees to develop an early detection system for bearing damage, in which they used features extracted from vibration data of bearings to train the trees. Vitola et al. [24] also demonstrated the use of trees for damage detection in aluminum plates using high-frequency waves, and showed the efficacy of *bagging* and *boosting* trees. Bagging, which stands for bootstrap aggregation, as the name suggests, involves bootstrapping (re-sampling) of training data and application of multiple trees to the enlarged data set. The key idea is to reduce the variance in estimates based on the law of large numbers. Boosting, on the other hand, does not involve data duplication. It is a weighted congregation of results from weak base learners trained using a randomly allocated portion of the training data.

Regression Techniques

Regression algorithms is also used for the purpose of damage detection. As discussed earlier, regression algorithms are generally used to build models consisting of continuous variables using only available data. The output of such regression models is hence not discrete. The most classical form of regression is linear regression or least squares, which attempts to solve the following:

$$\mathbf{Y}_{tr} = \mathbf{X}_{tr}\beta \tag{3}$$

where $\mathbf{Y}_{tr} \in \mathbb{R}^{n\times 1}$ is the vector of training labels, $\mathbf{X}_{tr} \in \mathbb{R}^{n\times p}$ is the training data matrix, and $\beta \in \mathbb{R}^{p\times 1}$ are the coefficients to be estimated by the following optimization problem:

$$\min_{\beta} \quad \|\mathbf{Y}_{tr} - \mathbf{X}_{tr}\beta\|_2^2 \tag{4}$$

where $\|.\|_2$ is the ℓ_2-norm. The optimal β obtained from the above equation is:

$$\hat{\beta} = (\mathbf{X}_{tr}^T\mathbf{X}_{tr})^{-1}\mathbf{X}_{tr}^T\mathbf{Y}_{tr} \tag{5}$$

The $\hat{\beta}$ obtained above is then used for making label predictions for a test data set:

$$\mathbf{Y}_{tst} = \mathbf{X}_{tst}\hat{\beta} \tag{6}$$

where the subscript *tst* stands for test data sets. Pan [25] and Shahidi et al. [26] used linear regression for damage detection in structures. They formulated the modal expansion of a dynamic response as a linear regression problem, and used the deviation in slope between the damaged and undamaged models as a damage feature. They demonstrated the use of both single- and multi-variable linear regression.

However, least squares regression suffers drawbacks related to collinearity (results in a singular $\mathbf{X}_{tr}^T\mathbf{X}_{tr}$), high dimensionality of data, and computational issues associated with matrix inversion. To overcome these problems, regularization terms are added to the optimization problem. Two popular regression techniques that can overcome these issues can be found in the literature: Tikhonov regularization or ridge regression, and Sparse Regression.

Ridge regression is based on constraining the elements of vector β evaluated using least squares. The updated optimization problem is as follows:

$$\begin{aligned} \min_{\beta} \quad & \|\mathbf{Y}_{tr} - \mathbf{X}_{tr}\beta\|_2^2 \\ \text{subject to} \quad & \|\beta\|_2^2 \leq t \end{aligned} \tag{7}$$

The more popular Lagrange equivalent form of ridge regression is:

$$\min_{\beta} \quad \|\mathbf{Y}_{tr} - \mathbf{X}_{tr}\beta\|_2^2 + \lambda\|\beta\|_2^2, \quad \lambda \geq 0 \tag{8}$$

where λ is the regularization parameter. The solution to the ridge regression optimization problem is:

$$\hat{\beta}_{ridge} = \left(\mathbf{X}_{tr}^T\mathbf{X}_{tr} + 2\lambda\mathbb{I}\right)^{-1}\mathbf{X}_{tr}^T\mathbf{Y}_{tr} \tag{9}$$

Sparse regression is another approach for overcoming the drawbacks of least squares regression. A vector of length n is k-sparse, if and only if $k \ll n$. If the need of a problem statement is a sparse β, the above optimization problem can be further modified by replacing the ℓ_2-norm by an ℓ_1-norm. Although the ℓ_1-norm is a weak definition of sparsity relative to the ℓ_0-norm, it allows for the optimization problem to remain convex, guaranteeing a global optimum. The use of ℓ_1-norm instead of ℓ_0-norm for sparse regression is referred to as a convex relaxation. The optimization problem associated with sparse regression is:

$$\min_{\beta} \quad \|\mathbf{Y}_{tr} - \mathbf{X}_{tr}\beta\|_2^2 + \lambda\|\beta\|_1, \quad \lambda \geq 0 \tag{10}$$

where λ again is the regularization parameter. The choice of the parameter λ for both ridge and sparse regression is made by performing cross-validation studies on the training data set.

Zhang and Xu [27] compared both regularized algorithms for application to vibration-based damage detection and showed that sparse regularization performed better. Yang and Nagarajaiah [28] used sparse regression to solve the damage detection problem in a sparse representation framework. The key idea was to construct a dictionary of features for a variety of possible damage scenarios. A sparse vector was then used to point to a specific element (a specific damage scenario) of the dictionary which best defined a given test signal. Yang and Nagarajaiah [29] also use sparse regression in a novel sparse component analysis technique that performs blind identification using limited sensor data.

Unsupervised Learning

In this section, we discuss the various applications of unsupervised learning in damage detection. As mentioned earlier, unsupervised learning does not use data labels; rather, they exhibit inherent patterns and structures useful for damage detection in the data. The most common unsupervised algorithms are principal component analysis (PCA), independent component analysis (ICA), and clustering algorithms. Clustering algorithms can range from *k*-means clustering to hierarchical clustering.

PCA is well equipped for performing two tasks, namely, data compression and data visualization. The notion of principal components is mathematically defined as the eigenvectors of the covariance matrix of a data set. Physically, they represent the directions of maximum variances. The data, when projected onto these eigenvectors, help to observe patterns in the data that aid in damage detection. For a data matrix $\mathbf{X} \in \mathbb{R}^{n \times p}$, the covariance matrix is defined as $\mathbf{X}^T\mathbf{X}$. It can be shown that the eigenvectors of the covariance matrix are equivalent to the right singular vectors $\mathbf{V}$ of $\mathbf{X}$, defined as $\mathbf{X} = \mathbf{UDV}^T$. The projection of the data in the eigenvector space is then defined by $\hat{\mathbf{X}} = \mathbf{XV} = \mathbf{UD}$.

As discussed earlier, PCA is used in most damage detection studies for reducing the dimensions of the data. This helps increase the numerical efficiency of most algorithms. The idea is to reconstruct the data matrix using eigenvectors with significant singular values only. Zang and Imregun [30] used PCA for frequency response function (FRF) data reduction for application to an artificial neural network (ANN)-based damage detection scheme. Yan et al. [31, 32] employed PCA for damage detection in bridges under varying environmental conditions. They used a portion of the undamaged data for obtaining the principal component directions, which were subsequently used for estimating changes between damaged and undamaged data from projections along these directions. Tibaduiza et al. [33] included a brief review of applications of PCA for SHM.

Similar to PCA, ICA too is a matrix factorization algorithm. The algorithm was first developed for solving the blind source separation (BSS) problem. It is defined as the retrieval of a set of k independent signals of length n, $\mathbf{S} \in \mathbb{R}^{k \times n}$ and a corresponding mixing matrix $\mathbf{A} \in \mathbb{R}^{k \times k}$ from a set of recorded signals $\mathbf{X} \in \mathbb{R}^{k \times n}$ that are assumed to be a mixture of the independent signals. Mathematically, this can be expressed as:

$$\mathbf{X} = \mathbf{A}\mathbf{S} \tag{11}$$

The key distinction between PCA and ICA is that PCA evaluates projection directions for maximum variance, whereas ICA evaluates most statistically independent vectors that can construct the given data set. Tibaduiza et al. [34] compared the performance of PCA with ICA for damage detection by active sensing in plates using high-frequency waves. They reported that ICA did not necessarily hold an advantage over PCA for their application. Similar to that with PCA, Zang et al. [35] employed ICA for data reduction for use in FRF damage classification using an ANN. Yang and Nagarajaiah [36] used ICA in conjunction with wavelet transform for blind damage identification.

Another class of unsupervised learning algorithms is the clustering algorithms. Although they cannot perform data reduction like PCA and ICA, they are effective in recognizing different patterns in data that might be useful for damage detection. Da Silva et al. [37] compared the performance of two fuzzy clustering algorithms for damage classification from vibration-based data. Park et al. [38] used a k-means clustering algorithm for damage detection using electromechanical impedance. The clustering algorithm was used on principal component projections of the impedance data acquired from an experimental beam. Santos et al. [39] employed a mean shift clustering algorithm for damage detection on the data from the Z-24 bridge. They also compared the performance of the proposed algorithm to that of k-means clustering, fuzzy c-means clustering, and Gaussian mixture model (GMM)-based techniques. Nair and Kiremidjian [40] demonstrated the use of GMMs for damage detection using vibration data from a benchmark building. Sen and Nagarajaiah [41] used hierarchical clustering for developing a semi-supervised learning approach to damage detection in steel pipes using guided ultrasonic waves. The objective was to detect the presence of damage using a minimum number of piezoelectric actuators and sensors.

Figure 1 shows a typical result of the application of the semi-supervised learning algorithm. The data are acquired from a single actuator and sensor pair. A narrowband pulse ensures minimal dispersion of Lamb waves propagating in the steel pipes. The central frequency of the pulse used for actuation is appropriately selected to min-

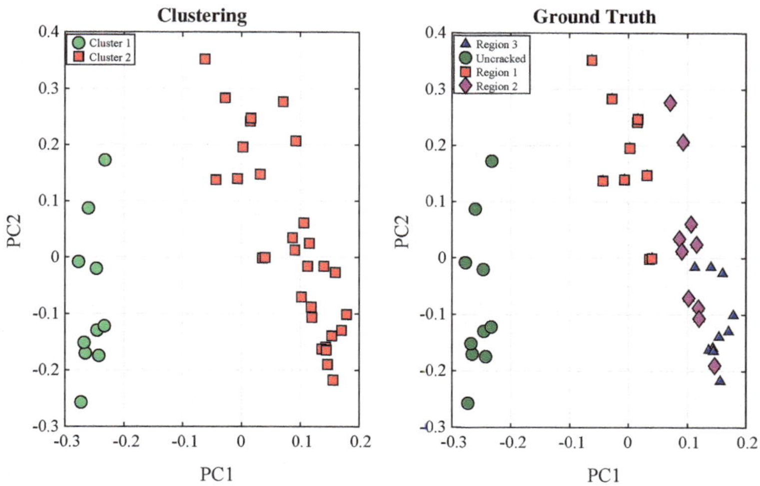

Fig. 1 Results from Sen and Nagarajaiah [41]. This figure shows the efficacy of hierarchical clustering in detecting the presence of damage in a steel pipe

imize the number of propagating wave modes. In both figures, data are projected onto a two-dimensional plane whose basis vectors are the first two principal components. Figure 1b shows the actual labels of each data point, the damaged and the undamaged data acquired from a steel pipe. The damaged data points are of three types, namely, regions 1, 2, and 3. These regions specify the location of the damage in relation to the location of the actuator and sensor. Figure 1a shows the clusters obtained from hierarchical clustering. Clearly, clusters 1 and 2 consist of all the undamaged data and damaged data, respectively, thus demonstrating the efficacy of the proposed approach.

Conclusions

In this chapter, we have reviewed some of the most popular techniques from statistical learning that have been used for damage detection, as well as SHM in general. We discussed the application of both supervised and unsupervised learning algorithms in this field. With the advent of a data deluge in the field of SHM, a move from model-based algorithms to data-driven approaches is necessary. The use of statistical learning algorithms not only makes for robust and efficient SHM systems, but also leads to a reduction in computational effort relative to model-based methods, which typically involve high-fidelity modeling of systems. This lays the foundation for developing online SHM systems aimed at performing damage detection in real time, with minimal human interference.

References

1. Farrar C, Worden K (2007) An introduction to structural health monitoring. Philos Trans Roy. Soc. A 365:303–315
2. Farrar CR, Lieven NAJ (2007) Damage prognosis: the future of structural health monitoring. Philos Trans Roy. Soc. A 365:632–632
3. Park KC, Reich GW (1999) Model-based health monitoring of structural systems: progress, potential and challenges. In: Proceedings of the 2nd international workshop on structural health monitoring, pp. 82–95
4. Farrar CR, Worden K (2012) Structural health monitoring: a machine learning perspective. Wiley
5. Barthorpe RJ (2010) On model-and data-based approaches to structural health monitoring. PhD thesis, University of Sheffield
6. Gopalakrishnan S (2009) Modeling aspects in finite elements. In: Encyclopedia of structural health monitoring. Wiley
7. Nagarajaiah S, Yang Y (2017) Modeling and harnessing sparse and low-rank data structure: a new paradigm for structural dynamics, identification, damage detection and health monitoring. Struct Control Health Monit 24(1). http://doi.org/10.1002/stc.1851
8. Rytter A (1993) Vibration based inspection of civil engineering structures. PhD thesis, Aolborg University
9. Hastie T, Tibshirani R, Friedman J (2009) The Elements of Statistical Learning: Data Mining, Inference and Prediction. Springer
10. Bousquet O, Boucheron S, Lugosi G (2004) Introduction to statistical learning theory. Lect Notes Artif Intel 3176:169–207
11. James G, Witten D, Hastie T, Tibshirani R (2013) An introduction to statistical learning. Springer
12. Worden K, Manson G (2006) The application of machine learning to structural health monitoring. Proc R Soc A 365:515–537
13. Addin O, Sapuan SM, Mahdi E, Othman M (2007) A naïve-bayes classifier for damage detection in engineering materials. Mater Des 28:2379–2386
14. Muralidharan V, Sugumaran V (2012) A comparative study of naïve bayes classifier and bayes net classifier for fault diagnosis of monoblock centrifugal pump using wavelet analysis. Appl Soft Comp 12:2023–2029
15. Farrar CR, Nix DA, Duffey TA, Cornwell PJ, Pardoen GC (1999) Damage identification with linear discriminant operators. In: Proceedings of the international modal analysis conference. 599–607
16. Gaudenzi P, Nardi D, Chiapetta I, Atek S, Lampani L, Sarasini F, Tirillo J, Valente T (2015) Impact damage detection in composite laminate plates using an integrated piezoelectric sensor and actuator couple combined with wavelet based features extraction approach. In: Proceedings of the 7th ECCOMAS thematic conference on smart structures and materials
17. He H-X, Yan W-M (2007) Structural damage detection with wavelet support vector machine: introduction and applications. Struct Control Health Monit 14:162–176
18. Shimada M, Mita A, Feng MQ (2006) Damage detection of structures using support vector machines under various boundary conditions. In: Proceedings of SPIE, 6174(61742K-1)
19. Bulut A, Singh AK, Shin P, Fountain T, Jasso H, Yan L, Elgamal A (2004) Real-time nondestructive structural health monitoring using support vector machines and wavelets. In: Technical report, University of California, Santa Barbara
20. Chattopadhyay A, Das S, Coelho CK (2007) Damage diagnosis using a kernel-based method. Insight-Non-Destr Test Cond Monit 49:451–458
21. Bornn L, Farrar CR, Park G, Farinholt K (2009) Structural health monitoring with autoregressive support vector machines. J Vibr Acoust 139:021004
22. Worden K, Lane AJ (2001) Damage identification using support vector machines. Smart Mater Struct 10:540–547

23. Kilundu B, Dehombreux P, Letot C, Chiementin X (2009) Early detection of bearing damage by means of decision trees. J Autom Mob Robot Intel Syst 3(3):70–74
24. Vitola J, Tibaduiza M, adn Anaya D, Pozo F (2016) Structural damage detection and classification based on machine learning algorithms. In: 2016 Proceedings of the 8th european workshop on structural health monitoring EWSHM
25. Pan Y (2012) Linear regression based damage detection algorithm using data from a densely clustered sensing system. Master's thesis, Lehigh University
26. Shahidi SG, Nigro MB, Pakzad SN, Pan Y (2015) Structural damage detection and localisation using multivariate regression models and two-sample control statistics. Struct Infrastruct Eng 11(10):1277–1293
27. Zhang CD, Xu YL (2016) Comparative studies on damage identification with tikhonov regularization and sparse regularization. Struct Control Health Monit 23:560–579
28. Yang Y, Nagarajaiah S (2014) Structural damage identification via a combination of blind feature extraction and sparse representation classification. Mech Syst Sig Process 45(1):1–23
29. Yang Y, Nagarajaiah S (2013) Output-only modal identification with limited sensors using sparse component analysis. J Sound Vibration 332:4741–4765. http://doi.org/10.1016/j.jsv.2013.04.004
30. Zang C, Imregun M (2001) Structural damage detection using artificial newral networks and measured frf data reduced via principal component projection. J Sound Vib 242(5):813–827
31. Yan A-M, Kerschen G, De Boe P, Golinval J-C (2005) Structural damage diagnosis under varying environmental conditions-part I: a linear analysis. Mech Syst Sig Process 19:847–864
32. Yan A-M, Kerschen G, De Boe P, Golinval J-C (2005) Structural damage diagnosis under varying environmental conditions-part II. Mech Syst Sig Process 19:865–880
33. Tibaduiza DA, Mujica LE, Rodellar J (2013) Damage classification in structural health monitoring using principal component analysis and self-organizing maps. Struct Control Health Monit 20:1303–1316
34. Tibaduiza DA, Mujica LE, Anaya M, Rodellar J, Güemes A (2012) Principal component analysis vs independent component analysis for damage detection. In: Proceedings of the 6th European workshop on structural health monitoring
35. Zang C, Friswell MI, Imregun M (2004) Structural damage detection using independent component analysis. Struct Health Monit 3(1):69–83
36. Yang Y, Nagarajaiah S (2014), Blind identification of damage in time-varying systems using independent component analysis with wavelet transform. Mech Syst Signal Process 47:3–20. http://doi.org/10.1016/j.ymssp.2012.08.029
37. da Silva S, Júnior MD, Junior VL, Brennan MJ (2008) Structural damage detection by fuzzy clustering. Mech Syst Sigl Process 22:1636–1649
38. Park S, Lee J-J, Yun C-B, Inman DJ (2008) Electro-mechanical impedance-based wireless structural health monitoring using pca-data compression and k-means clustering algorithms. J Intel Mater Syst Struct 19:509–520
39. Santos A, Silva M, Santos R, Figueiredo E, Sales C, Costa JCWA (2016) Output-only structural health monitoring based in mean shift clustering for vibration-based damage detection. In: Proceedings of the 8th European workshop on structural health monitoring
40. Nair KK, Kiremidjian AS (2007) Time series based structural damage detection algorithm using gaussian mixtures modeling. J Dyn Syst Measur Control 129:285–293
41. Sen D, Nagarajaiah S (2016) Damage detection in steel pipes using a semi-supervised statistical learning algorithm. In: ASNT annual conference. Long Beach, CA

Vibration-Based Monitoring of Civil Structures with Subspace-Based Damage Detection

Michael Döhler, Falk Hille and Laurent Mevel

Abstract Automatic vibration-based structural health monitoring has been recognized as a useful alternative or addition to visual inspections or local non-destructive testing performed manually. It is, in particular, suitable for mechanical and aeronautical structures as well as on civil structures, including cultural heritage sites. The main challenge is to provide a robust damage diagnosis from the recorded vibration measurements, for which statistical signal processing methods are required. In this chapter, a damage detection method is presented that compares vibration measurements from the current system to a reference state in a hypothesis test, where data-related uncertainties are taken into account. The computation of the test statistic on new measurements is straightforward and does not require a separate modal identification. The performance of the method is firstly shown on a steel frame structure in a laboratory experiment. Secondly, the application on real measurements on S101 Bridge is shown during a progressive damage test, where damage was successfully detected for different damage scenarios.

Keywords Subspace methods · Damage detection · Statistical tests · Vibrations Structural health monitoring

M. Döhler (✉) · L. Mevel
Inria, I4S/IFSTTAR, COSYS, SII, Campus Universitaire de Beaulieu,
35042 Rennes, France
e-mail: michael.doehler@inria.fr

L. Mevel
e-mail: laurent.mevel@inria.fr

F. Hille
BAM Federal Institute of Materials Research and Testing, Unter den Eichen 87,
12205 Berlin, Germany
e-mail: falk.hille@bam.de

E. Ottaviano et al. (eds.), *Mechatronics for Cultural Heritage and Civil Engineering*,
Intelligent Systems, Control and Automation: Science and Engineering 92,
https://doi.org/10.1007/978-3-319-68646-2_14

Introduction

Structural health monitoring (SHM) has the goal of determining the system health, i.e. the occurrence and characterization of damages, based on the measurement data recorded on the monitored structure [16]. With statistical signal processing techniques, one can find damages automatically within a monitored structure, before they grow to dangerous extents. Automatic inspection is of enormous value for structures with high failure consequences or poor accessibility, such as bridges, structures of cultural heritage, offshore constructions, large engineering structures, power plants, etc. With these means, maintenance operations can be optimized and costly downtime kept to a minimum, while the structural safety is increased.

Vibration-based SHM has shown great potential for automated damage monitoring [8]. Methods for damage assessment are categorized into four groups of increasing difficulty and complexity: (1) detection, (2) localization, (3) quantification, and (4) prognosis of remaining life [21]. Many methods have been proposed in the literature (see e.g. reviews [9, 15]), with however few methods taking into account the statistical nature of the vibration measurements, which is essential for robustness under realistic (noisy) operational conditions in practice. Among the methods for damage assessment, the detection methods are currently the most developed and some of them have matured to an industrial application level [22].

Vibration-based damage detection is performed on the rationale that damage affects the stiffness of the monitored structure and, thus, changes its dynamic properties. The analysis of the dynamic properties can be performed entirely based on measurements, without the need of a finite element model. Many of the currently used vibration-based methods for early damage detection perform modal system identification and compare the obtained modal parameter of the actual state with those of the undamaged state, e.g. by using control charts [19, 20]. In this context, especially the natural frequencies are used for a comparison, as they can be reliably identified. However, the automated estimation of modal parameters from a system identification method and matching them from measurements of different states of the structure for their comparison might require an extensive preprocessing step. Other methods avoid the system identification step in the possibly damaged state. For instance, such methods include non-parametric change detection based on novelty detection [24, 25], whiteness tests on Kalman filter innovations [7] or other data-driven features that are sensitive to changes in the modal parameters.

The subspace-based damage detection approach [4, 6, 11–13], which will be considered in this chapter, also falls into the latter category. In this method, a data-driven model is obtained in the reference state and compared to measurements from an unknown, possibly damaged state. This comparison is performed using a subspace-based residual function and a χ^2-test built on it for a hypothesis test, without actually estimating the modal parameters in the unknown, possibly damaged states. In this way, the entire structural response is taken into account. This approach provides a complete statistical framework for analyzing a damage residual for both reference and possibly damaged structures. The asymptotic probability distribution of the considered damage residual is given and a statistical hypothesis test to detect damage in the structure has been proposed.

In this chapter, the subspace-based damage detection method is presented. The performance of the method is firstly shown on a steel frame structure in a laboratory experiment. Secondly, the application on real measurements on S101 Bridge is shown during a progressive damage test, where damage was successfully detected for different damage scenarios.

Subspace-Based Damage Detection

In this section, the subspace-based damage detection algorithm is introduced. With this global automated damage detection method, vibration measurements from the current system are compared to a reference state in a subspace-based residual vector. In a hypothesis test, the uncertainties of the residual are taken into account and the respective test statistic is compared to a threshold in order to decide if the structure is damaged or not.

Dynamic Structural System Model

The vibration behaviour of the monitored structure is assumed to be described by a linear time-invariant dynamical system

$$\mathcal{M}\ddot{z}(t) + \mathcal{C}\dot{z}(t) + \mathcal{K}z(t) = v(t), \qquad (1)$$

where t denotes continuous time; $\mathcal{M}, \mathcal{C}, \mathcal{K} \in \mathbb{R}^{m \times m}$ are mass, damping, and stiffness matrices, respectively; the state vector $z(t) \in \mathbb{R}^m$ is the displacement vector of the m degrees of freedom of the structure; and $v(t)$ is the external force. Since the excitation is usually ambient and unmeasured for long-term monitoring, the external force is modelled as white noise.

Observed at r sensor positions (e.g. by displacement, velocity or acceleration sensors) at discrete time instants $t = k\tau$ (with sampling rate $1/\tau$), system (1) can be transformed into a discrete-time state space system model [18]

$$\begin{cases} x_{k+1} = Ax_k + v_k \\ y_k = Cx_k + w_k \end{cases} \qquad (2)$$

with the state vector $x_k = \left[z(k\tau)^T \; \dot{z}(k\tau)^T\right]^T \in \mathbb{R}^n$, the measured outputs $y_k \in \mathbb{R}^r$, the state transition matrix

$$A = \exp\left(\begin{bmatrix} 0 & I \\ -\mathscr{M}^{-1}\mathscr{K} & -\mathscr{M}^{-1}\mathscr{C} \end{bmatrix} \tau\right) \in \mathbb{R}^{n \times n}$$

and the observation matrix

$$C = \begin{bmatrix} L_d - L_a \mathscr{M}^{-1}\mathscr{K} & L_v - L_a \mathscr{M}^{-1}\mathscr{C} \end{bmatrix} \in \mathbb{R}^{r \times n},$$

where $n = 2m$ is the model order and $L_d, L_v, L_c \in \{0, 1\}^{r \times m}$ are selection matrices indicating the positions of displacement, velocity or acceleration sensors at the degrees of freedom of the structure, respectively. The state noise v_k and output noise w_k are unmeasured and assumed to be centered and square integrable.

Damages in the monitored system correspond to changes in the matrices $\mathscr{M}$, $\mathscr{C}$ or $\mathscr{K}$ in (1), e.g. loss of mass or loss of stiffness, and affect the system matrices (A, C) in (2) and the modal parameters (natural frequencies, damping ratios, mode shapes). Hence, damages in model (1) can be equivalently detected as changes in the system matrices in (2). Based on measurements y_k of the monitored system from a reference and from a possibly damaged state, a residual vector is defined based on subspace properties, as outlined in the following section.

Description of the Method

In [4, 6] a residual vector was proposed whose mean is zero in the reference state, and non-zero in the damaged state. It is computed from the measurements y_k without actually identifying the modal parameters in the unknown, possibly damaged state. The considered residual is associated with a covariance-driven output-only subspace identification algorithm as follows.

Define the block Hankel matrix $\mathscr{H}_{p+1,q}$ of the output covariances $R_i = \mathbf{E}(y_k y_{k-i}^T)$ as

$$\mathscr{H}_{p+1,q} \stackrel{\text{def}}{=} \begin{bmatrix} R_1 & R_2 & \dots & R_q \\ R_2 & R_3 & \dots & R_{q+1} \\ \vdots & \vdots & \ddots & \vdots \\ R_{p+1} & R_{p+2} & \dots & R_{p+q} \end{bmatrix} \stackrel{\text{def}}{=} \text{Hank}(R_i),$$

where the parameters p and q are chosen such that $\min\{pr, qr\} > n$ with usually $p + 1 = q$. It possesses the well-known factorization property

$$\mathscr{H}_{p+1,q} = \mathscr{O}_{p+1} \mathscr{C}_q \tag{3}$$

into the matrices of observability and controllability

$$\mathcal{O}_{p+1} = \begin{bmatrix} C \\ CA \\ \vdots \\ CA^p \end{bmatrix}, \quad \mathcal{C}_q = \begin{bmatrix} G & AG & \dots & A^{q-1}G \end{bmatrix}, \tag{4}$$

where $G = \mathbf{E}(x_{k+1}y_k^T)$ is the cross-covariance between the states and the outputs. From a singular value decomposition of $\mathcal{H}_{p+1,q}$, the left null space matrix S can be obtained with the characteristic property

$$S^T \mathcal{H}_{p+1,q} = 0 \tag{5}$$

in the reference state of the structure. If the system is damaged, the matrices (A, C) and thus $\mathcal{H}_{p+1,q}$ change, thus the product $S^T \mathcal{H}_{p+1,q}$ is not zero anymore in the damaged state.

Based on these properties, a residual vector is defined as follows. From a set of output measurements y_k, $k = 1, \dots, N$, estimates of the output covariances and the Hankel matrix are computed as

$$\widehat{R}_i = \frac{1}{N} \sum_{k=1}^{N} y_k y_{k-i}^T, \quad \widehat{\mathcal{H}}_{p+1,q} = \text{Hank}(\widehat{R}_i),$$

and the residual vector is defined as

$$\zeta = \sqrt{N} \, \text{vec}(S^T \widehat{\mathcal{H}}_{p+1,q}), \tag{6}$$

where vec$(\cdot)$ denotes the column stacking vectorization operator. Due to property (5), the mean of ζ is zero in the reference state of the structure, and non-zero in the damaged state. The residual vector ζ is asymptotically Gaussian distributed [4]. Let Σ be the asymptotic residual covariance matrix. Then, the corresponding hypothesis test for a decision between $\mathbf{H}_0$: the system is in the reference state, and $\mathbf{H}_1$: the system is in a damaged state, leads to the test statistic

$$s = \zeta^T \Sigma^{-1} \zeta, \tag{7}$$

which is asymptotically χ^2 distributed with a non-centrality parameter in the damaged state. To decide if the monitored structure is damaged or not, the test statistic s is compared to a threshold t, and damage is announced if $s > t$. The threshold t is typically chosen such that the probability of false alarms (type I error) is below a certain level. An example of the probability distributions of the test statistic s in the reference and damaged states is given in Fig. 1. Computational details for obtaining estimates of the null space matrix S, covariance matrix Σ and the threshold t in the reference state of the monitored structure, which are required for the damage detection test, are given in the Appendix.

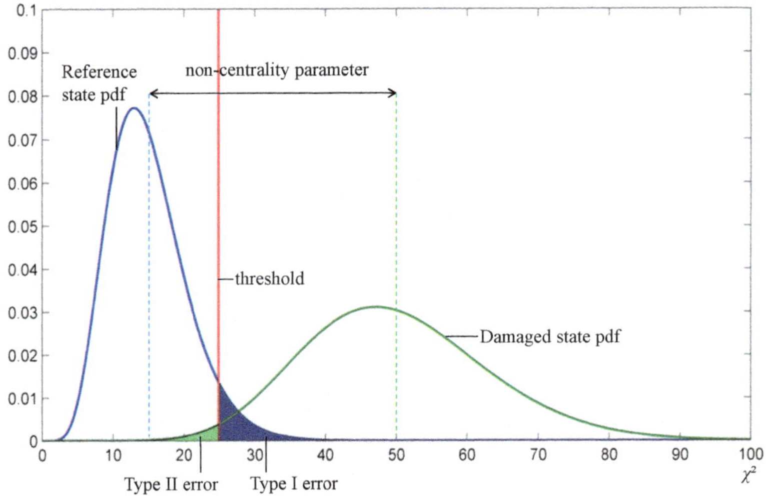

Fig. 1 Scheme of the probability density functions (pdf) of the χ^2-distributed damage detection test statistic s in the reference and in a damaged state

Note that the multiplication with $\sqrt{N}$ in the definition of the residual vector (6) ensures its asymptotic Gaussianity through the central limit theorem, and thus the definition of a χ^2-distributed test statistic in (7). Moreover, this statistical framework allows the detection of very small damages if the number of measurements N is large enough. A detailed description of the statistical framework and more theoretical insight of this method is given in [4, 12–14], including parametric test variants.

Robustness to Environmental Nuisances

In the previous section, the subspace-based damage detection test is given in its most basic version. The test is set up entirely on measurement data without the need of a numerical model, and no modal analysis is required at all. It takes into account the natural variability that is related to the measurements, which is due to unknown ambient excitation and measurement noise. While it is convenient for many applications, additional robustness to natural environmental changes like excitation or temperature variations can be easily accounted for.

Changes in the ambient excitation properties do not affect the dynamic properties of the structure, but may have an influence on the test statistic (7) that is directly computed on the measurements. Robustness to such changes is achieved by a slight modification of the residual vector (6), where the Hankel matrix estimate $\widehat{\mathscr{H}}_{p+1,q}$ is replaced by the matrix U_1 of its principal left singular vectors as in (9), leading to the robust residual

$$\tilde{\zeta} = \sqrt{N}\,\mathrm{vec}(S^T U_1).$$

Since the principal left singular vectors of $\mathscr{H}_{p+1,q}$ also satisfy the characteristic property (5) of the reference state, the robust residual $\tilde{\zeta}$ similarly detects changes in the dynamic system properties, with the advantage that U_1 is an orthonormal matrix and thus independent from scaling due to excitation changes. The respective test statistic with the appropriate covariance matrix $\tilde{\Sigma}$ is, analoguously,

$$\tilde{s} = \tilde{\zeta}^T \tilde{\Sigma}^{-1} \tilde{\zeta}. \tag{8}$$

Further theoretical and implementation details are given in [12, 13].

Temperature variations can be taken into account in a non-parametric or in a parametric way. In the non-parametric way, the left null space matrix S and the residual covariance matrix Σ are computed on several datasets that are recorded under different temperature scenarios [1]. In this way, the variation of the residual due to temperature changes is eliminated in the test statistic. In the parametric way, a model describes the influence of the temperature on the monitored structure. Then, variations in the temperature parameter can be rejected in the test statistic as nuisance or the left null space matrix S can be adjusted with the temperature [2, 5].

Application to a Steel Frame

In this section, the damage detection method is applied to a scaled two-dimensional section of a jacket-type support structure of an offshore wind turbine in a laboratory experiment. In practice, these support structures are used in deep sea waters of 30 m and more. They generally include a significant number of highly stressed joints and, therefore, possess a high vulnerability to fatigue damage. In addition, most of the structural components are situated under sea level which makes damage investigation costly. This leads to the application of robust SHM techniques.

Laboratory Test Configuration

The developed model structure out of steel pipe components represents a scaled two-dimensional section of a jacket-type support structure of an offshore wind turbine as they are common for wind park installations. The scale of the general dimensions is approximately 1:10. The lab model is designed with one and a half diagonal bracings between two legs. The upper completion is formed by an I-sectional girder. At the lower end the legs end in steel foot plates. With exception of the damage sections all structural parts of the model are welded. The foot plates are screwed via a transition piece to the lab floor. In the third direction the structure is held on the head girder by an auxiliary construction, whose two fork-like support bars allow for the horizontal

Fig. 2 Laboratory test stand at BAM

movement of the head girder. Figure 2 shows the laboratory test structure in the proof testing facilities at BAM.

The damage to be modeled is supposed to represent fatigue cracks according to the characteristic cyclic stressing which can even escalate by resonance effects as well as due to the notches in the material from the welding process. The descriptiveness of the damage in location and quantification for comparison of different test series within the study as well as reversibility of the inserted damage for the ability to repeat test series were considered for the damage modeling.

For the difficulty in designing a reversible damage in the joint area of a gusset, the damage areas of the laboratory structure were constructed with a shift into the steel tube domain. A crack-like damage can be induced artificially by the loosening of bolted flange connection at one K-type gusset of the model structure. Four of such flange connections are designed, each at the four ends of the K-type gusset. So there were two damage locations at the leg and two at the bracing diagonal. In Fig. 2 the four flanges at the end of the red drawn component on the left side of the model can be seen at each end of the red K-type gusset.

During the test series, dynamic excitations and cyclic loading of the laboratory structure were deployed. With different excitations the dependency of the detection of simulated damage on the frequency and energy input of the excitation source was examined. The excitation for the tests was provided by an electrodynamic shaker. A broadband random acceleration signal with a frequency content between 10 Hz and 1000 Hz was produced by a shaker control unit and was induced after amplification via the shaker at the top of the structure. The shaker was equipped with a 5 kg excitation mass and the excitation direction was horizontal and approximately 30 ° rotated out of the in-plane-direction. Eight piezoelectric accelerometers were applied at the structure for the vibration measurements. The sensors were connected with the structure by permanent magnets and measured in the direction perpendicular to the surface they were applied to. The signals were amplified, recorded and stored with a 20-channel DIFA measurement unit.

Artificial Damage Detection Tests

In a first test stage, the general reaction of the test as well as its robustness to changing ambient excitation levels were investigated. The damage location was chosen at the lower brace of the flange. In the reference state all bolts were screwed tight. For inducing damage, successively one, two, three, five or seven adjacent bolts were unscrewed, each representing a fatigue crack of increasing length. Thereby, the loosening of three bolts is comparable with a reduction of the moment of inertia by 3%, whereas the unscrewing of seven bolts means a loss of 30% of the bending stiffness.

To validate the damage detection algorithm under changing excitation, three excitation (power) levels were used for each damage state. Besides the full scale level, a reduction of 5 dB and a reduction of 10 dB with respect to the full scale level were performed and are denoted as "Excitation 1", "Excitation 2" and "Excitation 3" in the tests, respectively. A reduction of 5 dB accounts for a power ratio of $\approx$0.31 and an amplitude ratio of $\approx$0.56 and a reduction of 10 dB for a power ratio of 0.1 and an amplitude ratio of $\approx$0.31. The full-scale excitation power level was chosen preliminary in dependence of the shaker performance. For each damage level and each excitation level, four signals of 16.4 s length was measured.

In Fig. 3a, the test statistic s from (7) is computed on the datasets with excitation 1. The test values increase with the damage size, where the damage cases of five and seven loosened bolts are well above the threshold that was established in the reference state. The robustness to unknown changing excitation properties is tested with the robust test variant (8). Results for the three different excitation levels are shown in Fig. 3b. There are small fluctuations of the test values in the reference state and in the states with small damage, but these fluctuations are not significant with respect to the established threshold. For the damages of three or more loosened bolts, these variations seem to be independent from the excitation level, which validates the robustness of the test to changing excitation. The relatively small damage of only three loosened bolts can be detected, and the further damages of five and seven loosened bolts are clearly above the threshold. Further details and results of this experiment are given in [10].

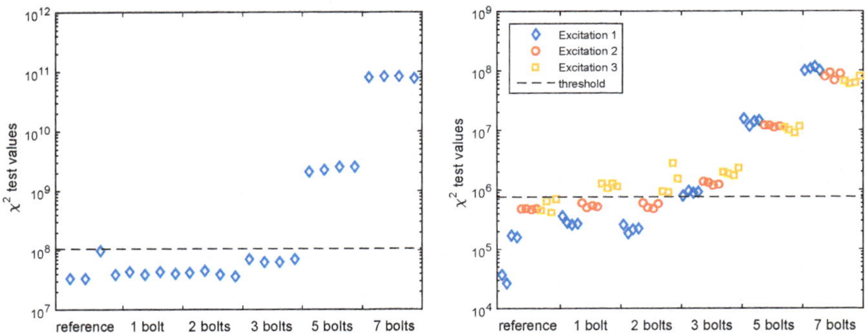

(a) Test (7) under constant excitation levels. (b) Robust test variant (8) under different excitation levels.

Fig. 3 Damage detection tests on steel frame structure for different damage extents (log-scale)

Fatigue Damage Detection Tests

In the next step, a fatigue test was accomplished to test the applicability of the damage detection method to real damage. The test procedure consisted of two alternating tasks: the insertion of the fatigue relevant load cycles and the measurement of dynamic response series for determination of the damage detection test value. A load series with a constant alternating horizontal force of ±50 kN was applied at the top girder of the structure with a frequency of 3 Hz. After completion of each load series, a damage test was conducted, where a broadband random excitation signal with a frequency content between 10 and 1000 Hz was produced by the shaker. The vibration response was measured with nine piezoelectric accelerometers at the structure. A total of ten load series (with varying number of load cycles) and, therefore, ten damage states were applied. The fatigue damage did not occur at the expected location of one gusset, but at the welded connection of the legs to the footplate. After approximately 152,000 load cycles, the loading was stopped, since the cracks had a considerable length of approximately $a = 100$ mm each (Fig. 4).

The result of the subspace-based damage detection is shown in Fig. 5. The bars represent the mean of 10 to 30 tests executed in one damage state and the error bars represent the related standard deviation. The diagram shows a continuous increase of the χ^2-test value, except for the final load series, where a significant drop of the value is noticeable. The reason for this evolution is, that a first fatigue crack has evolved at one leg-footplate connection during load series 3 to 9. Due to stress redistribution in the last loading series, a second fatigue crack developed at the other leg-footplate connection of the structure. During progression of the first crack, the modal system changed gradually with the effect of a continuous increase of the damage test value, while the second crack resulted in a distinct switch back of the modal system and thus led to a decrease of the test value. A detailed description of the fatigue test and discussion of its result is given in [17]. Note that the fatigue damage was successfully

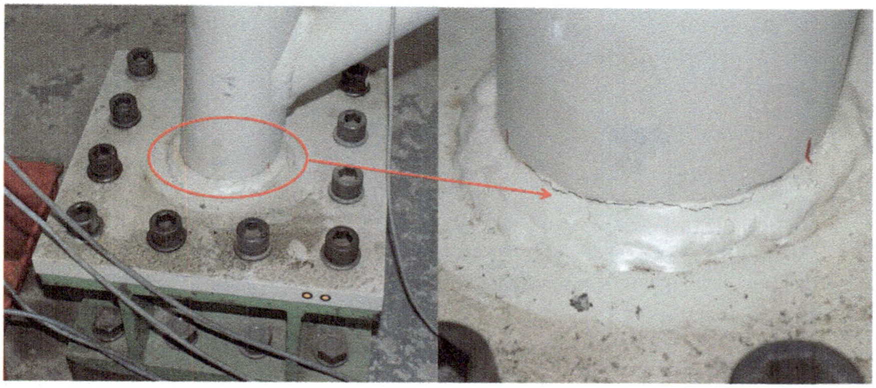

Fig. 4 First fatigue crack at one jacket foot

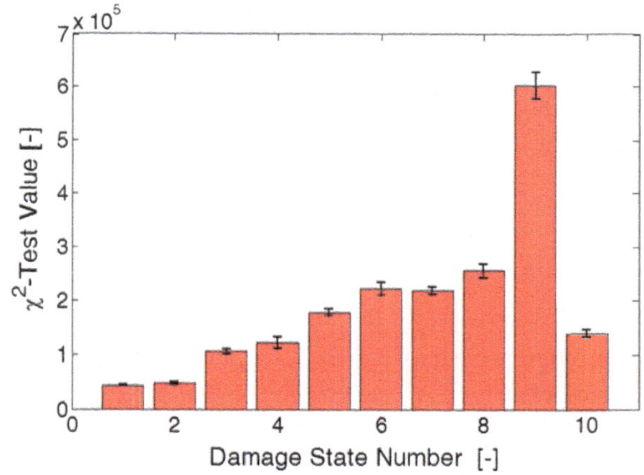

Fig. 5 Mean and standard deviation of damage detection test values for the 10 damage states

detected with the presented technique. The relationship between the test value and the actual damage extent is non-trivial and requires the usage of further damage quantification techniques [14].

Application to S101 Bridge in Progressive Damage Test

Within the European research project, "Integrated European Industrial Risk Reduction System (IRIS)" a prestressed concrete bridge was artificially damaged [23]. The intention was to provide a complete set of monitoring data during a defined loss of structural integrity for testing and evaluation of various SHM methods and applications. Therefore, the static and dynamic behavior of the structure was measured permanently during the three-day damaging process. The progressive damage campaign was planned and organized by the Austrian company VCE. The characteristics of the structure under observation, the measurement campaign and the introduced damages are presented briefly.

The S101 Bridge

The S101 Bridge (Fig. 6) was a prestressed concrete bridge from the early 1960 s and, therefore, a characteristic representative of the partly invalid highway infrastructure asset in Europe. Despite a general lack of experience in the time of their design, prestressed concrete was a very popular construction type in those days. Retrospectively, some of the major design assumptions proved to be erroneous and after short periods of operation significant and characteristic damage patterns occurred at the structure. The load bearing capacity and especially the durability of the bridges remained mostly limited despite costly retrofitting activities. In addition, an increase of heavy load vehicle traffic has been ongoing since the start of operation. Thus, prestressed bridge structures in central Europe have been exposed to degradation processes initiated by poor design, while the operational loading and the associated dynamic stressing increases steadily in the same time.

Damage Description

The progressive damage test took place from 10–13 December 2008. During the test the highway beneath the bridge was open in one direction. The second direction was closed for traffic because of construction works, which in addition took place near the bridge.

Two major damage scenarios were artificially induced. First, a significant damage of one of the four columns was inserted by cutting through the column on its lower

Fig. 6 Bridge S101 during damage test [23]

Fig. 7 Progressive damage of column and tendons on bridge S101 [23]

end. With this action a change in the global structural system was implemented. After a second cut a 5 cm thick slice of the column was removed and the column was lowered for altogether 3 cm. Afterwards the column was uplifted again to its original position and secured there by steel plates. Then, prestressing tendons of one of the beams were cut successively for a second damage scenario. All in all three and a quarter of a wire bundle were cut through. Between each intersection pauses of several hours were kept to let the structural system change into a new state of equilibrium. Pictures of both scenarios of the progressive damaging are shown in Fig. 7. In Table 1 all damage actions are sorted in chronological order.

Measurement Description

For the vibration measurements a BRIMOS® measurement system with a permanent sensor grid was used. The grid consisted of 15 sensor locations on the bridge deck (see Fig. 8), in each location three sensors for measurements in the bridge deck's vertical, longitudinal and transversal direction. Altogether, 45 acceleration sensors were applied.

Table 1 Notation of consecutive damage actions

Damage action		Damage action	
A	Begin of cutting through column	G	Uplifting column
B	End of second cut through column	H	Exposing cables and cutting of 1st cable
C	Lowering of column – 1st step (10 mm)	I	Cutting through 2nd cable
D	Lowering of column – 2nd step (20 mm)	J	Cutting through 3rd cable
E	Lowering of column – 3rd step (27 mm)	K	Partly cutting of 4th cable
F	Inserting steel plates		

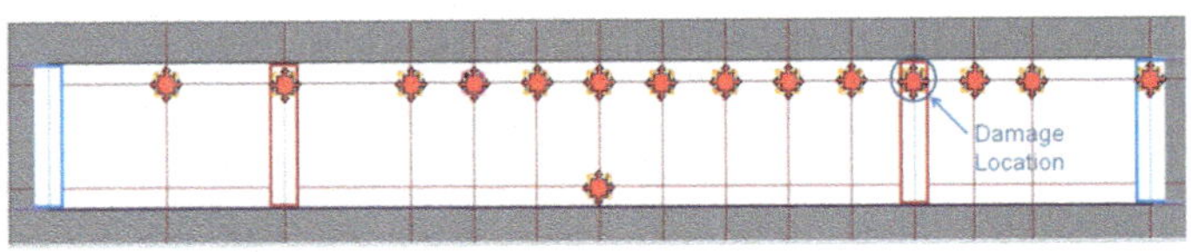

Fig. 8 Location of the sensors on the bridge deck [23]

The data were recorded permanently with a sampling frequency of 500 Hz. During the three days measurement campaign, 714 data files each containing 45 channels with 165 000 data points were produced, corresponding to 5.5 min of measurements each.

Data Analysis

The vibration measurements were recorded throughout the whole three day long damage test, including the nights. The measurement campaign started approximately 12 h before the damage of the column was introduced to monitor the undamaged state for an adequate time period.

The reference state of the undamaged structure was set up by computing and averaging the reference matrices from several datasets. With this approach [1], a possible disturbance by single excitation events or different environmental conditions is minimized. In the test stage, the test statistics s in (7) is computed for every data set, which in real time means an indicator of damage for every 5.5 min. Note that no modal parameter estimation or tracking is necessary for this damage detection approach.

Results of Damage Detection

Figure 9 shows a bar plot of values as damage indicators of all consecutive tests within the three days campaign, where all 45 sensors ($r = r_0 = 45$) were used. For the computation of the residual covariance matrix 100 datasets of the undamaged state were used. The abscissa of the plot describes the chronological sequence of the damage activities as noted in Table 1 as well as the 6:00 am and 6:00 pm points of time for orientation.

The reaction of the damage detection test at the main damage events can be easily observed. From the measurements in the reference state (before damage event A), a threshold of the damage indicator can be set easily, such that all the subsequent scenarios after damage event B would be correctly classified as damaged. This qualifies the proposed method for damage detection. In the following, the obtained results are analyzed in more detail. It can be seen that the damage indication is interfered by noise in the ambient excitation of the bridge. The *s* values periodically swell up in the morning and ebb away in the evening. It is assumed that the traffic going underneath the bridge and/or the construction work nearby are the source of the noise. Since the measurements in the reference state were only taken during 12 h in the night, an extension of the measurement time period of the undamaged state to a whole day circle might reduce these disturbances significantly when taken into account in the residual covariance matrix. Influences by solar radiation and/or temperature alternation can be excluded, since during the three-day campaign misty winter weather with only moderate temperature changes just below freezing was dominant.

Figure 10 shows the damage indicator during the several steps of the first damage scenario, the cutting and settling of one of the four bridge columns. With exception of the time periods of the direct mechanical destruction processes, the displayed sequence of damage indicators has a consecutive course. As can be seen in Fig. 10,

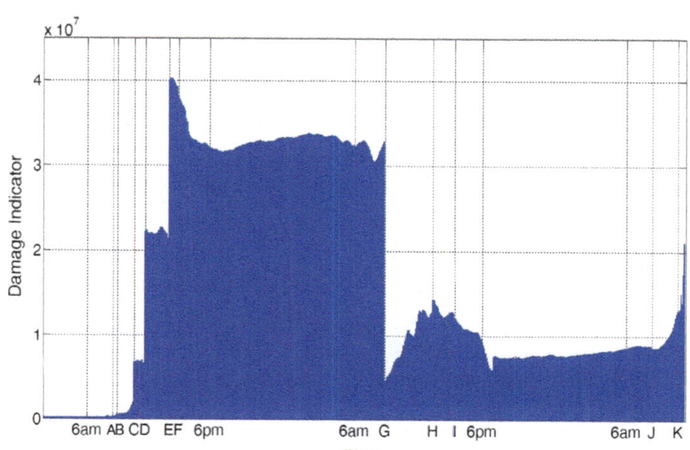

Fig. 9 Course of the damage indicator over the three-day damage test

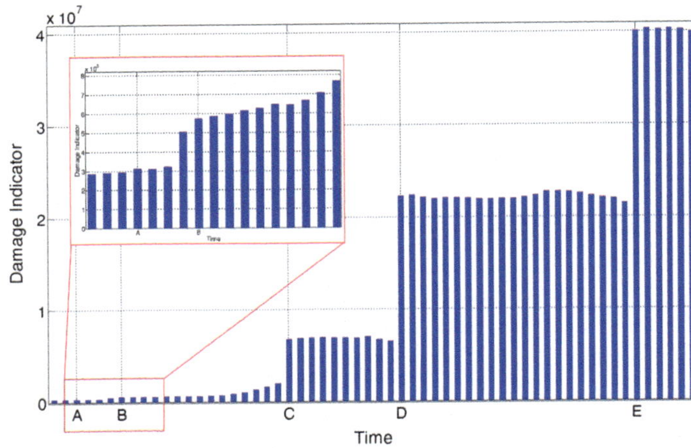

Fig. 10 Damage indicator for cutting and settling of one bridge column, detail of damage indicator for cutting through one bridge column

the three steps of the column settlement action are very distinctive in their influence on the computed damage indicator. Obviously, the dynamic system has changed to quite some extent, and the elastic settlement of altogether 27 mm can be clearly detected. Although not that strongly visible, the cutting of the column (A+B) also caused an increase of the indicator of approximately 75 %. However it has to be mentioned that the absolute effect is superimposed by noise effects. The inlet in Fig. 10 shows a detail of that time period.

The column remained in the settled condition for approximately one day and was then uplifted again in its former position (event G). The effect of the uplifting is again clearly visible in Fig. 9 by a drop of the damage indicator. However, the indicator did not drop completely to its original value, which is certainly due to the fact that the lowering of one column has led to cracking within the concrete structure to some extent and hence to a change of the dynamic signature of the system.

As one can see in Fig. 9, the cutting of the prestressing tendons did not lead to a significant change in the damage indicator after the single cutting steps. Nevertheless, a distinctive increase of the indicator could be observed at the end of the measurement. Figure 11 shows the last time period in detail. It is possible that a change of the bridge's structural system took place with a time delay after cutting partly the fourth tendon.

For a prestressed concrete structure, the loss of prestressing is a major damage that comes along with a significant loss of its load bearing capacity. One reason that the cutting of the tendons does not affect the proposed damage indicator might be that the overall prestressing is designed for combinations of dead and traffic loads. Since the bridge deck of S101 has a quite slender cross section, the dead load is not that high in comparison with the maximum design traffic loading. An additional dead load reduction comes from the removal of the asphalt surface before the dam-

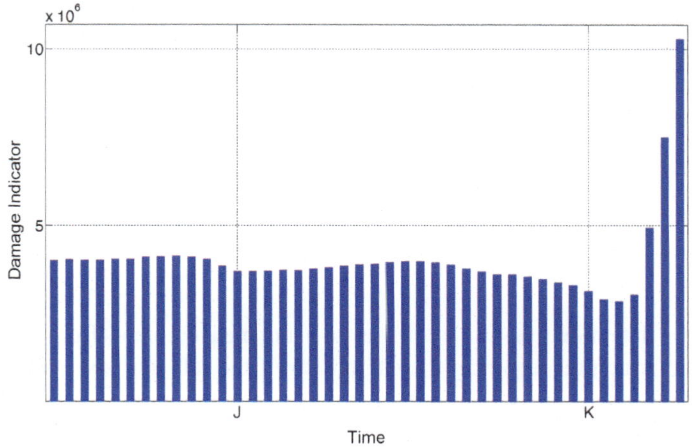

Fig. 11 Detail of damage indicator at the end of the tendon cutting process

age test. Furthermore it has to be recalled that the prestressing was designed with adequate safety margins. Also, the cutting results only in local loss of prestressing because of the bound between tendons and concrete. For these specified reasons it is assumed that the cutting of the tendons during the damage test did not lead to a significant change of the global structural system, since the loading of the structure was not high enough to activate the damage right after its insertion. The increase of the indicator at the end of the test series might be the result of a delayed rearrangement of the structural system by a reduction of stresses under generation of cracks in the concrete of the bridge deck. Though, an evidence of that assumption, for instance by an increase of the measured bridge deck deflection, could not be found.

Conclusions

In this chapter, a method for vibration-based damage detection has been presented that combines statistical robustness with simplicity of application. A damage test statistic is computed entirely from measurement data of a (healthy) reference system and the current system, without the need of a finite element model. The method is designed for the detection of small damages before they grow to dangerous extents, while taking the intrinsic uncertainties of vibration measurements due to unknown excitation and noise into account. The damage detection method has been validated successfully on a laboratory test case and on a full scale damage test on a real highway bridge. Thanks to its properties, it is well suited for being embedded in automated structural health monitoring systems. Extensions of the damage detection method to damage localization and quantification in the same framework [3, 14] are possible in connection with a finite element model, being a step towards the development of robust damage assessment systems that will combine measurement data with physical models in the future.

Acknowledgements We thank Dr. Helmut Wenzel, VCE, and the FP7 IRIS project for providing the data from S101 Bridge.

Appendix

The damage detection test statistic s is set up using measurements from the reference state of the structure, requiring the estimation of the left null space matrix S, the residual covariance matrix Σ and setting up a threshold t.

First, the Hankel matrix $\widehat{\mathscr{H}}^{0}_{p+1,q}$ is computed from the reference measurements. Then, the left null space matrix S can be estimated from its SVD

$$\widehat{\mathscr{H}}^{0}_{p+1,q} = \begin{bmatrix} U_1 & U_0 \end{bmatrix} \begin{bmatrix} \Delta_1 & 0 \\ 0 & \Delta_0 \end{bmatrix} \begin{bmatrix} V_1 \\ V_0 \end{bmatrix} \tag{9}$$

as $\widehat{S} = U_0$, where the SVD is truncated at the desired model order n with $\Delta_1 \in \mathbb{R}^{n \times n}$ and $\Delta_0 \approx 0$.

Second, the residual covariance matrix Σ is obtained by separating the available reference dataset into n_b data blocks of length N_b, such that the total data length yields $N = n_b N_b$. On each block, the Hankel matrix is computed as

$$\widehat{R}_i^{(j)} = \frac{1}{N_b} \sum_{k=1+(j-1)N_b}^{jN_b} y_k y_{k-i}^T, \quad \widehat{\mathscr{H}}^{(j)}_{p+1,q} = \mathrm{Hank}(\widehat{R}_i^{(j)})$$

Then, $\widehat{\mathscr{H}}^{0}_{p+1,q} = \frac{1}{n_b} \sum_{j=1}^{n_b} \widehat{\mathscr{H}}^{(j)}_{p+1,q}$ and the covariance estimate of the residual follows from the covariance of the sample mean as

$$\widehat{\Sigma} = \frac{N_b}{n_b - 1} \sum_{j=1}^{n_b} \mathrm{vec}\left(S^T \widehat{\mathscr{H}}^{(j)}_{p+1,q} - S^T \widehat{\mathscr{H}}^{0}_{p+1,q}\right) \mathrm{vec}\left(S^T \widehat{\mathscr{H}}^{(j)}_{p+1,q} - S^T \widehat{\mathscr{H}}^{0}_{p+1,q}\right)^T.$$

Finally, compute the test statistic s for several training datasets of length N from the reference state, and determine the threshold t from the test values for a desired type I error.

References

1. Balmès E, Basseville M, Bourquin F, Mevel L, Nasser H, Treyssède F (2008) Merging sensor data from multiple temperature scenarios for vibration-based monitoring of civil structures. Struct Health Monitor 7(2):129–142
2. Balmès E, Basseville M, Mevel L, Nasser H (2009) Handling the temperature effect in vibration-based monitoring of civil structures: a combined subspace-based and nuisance rejection approach. Control Eng Pract 17(1):80–87
3. Balmès E, Basseville M, Mevel L, Nasser H, Zhou W (2008) Statistical model-based damage localization: a combined subspace-based and substructuring approach. Struct Control Health Monitor 15(6):857–875
4. Basseville M, Abdelghani M, Benveniste A (2000) Subspace-based fault detection algorithms for vibration monitoring. Automatica 36(1):101–109
5. Basseville M, Bourquin F, Mevel L, Nasser H, Treyssède F (2010) Handling the temperature effect in vibration monitoring: two subspace-based analytical approaches. J Eng Mech 136(3):367–378
6. Basseville M, Mevel L, Goursat M (2004) Statistical model-based damage detection and localization: subspace-based residuals and damage-to-noise sensitivity ratios. J Sound Vibration 275(3):769–794
7. Bernal D (2013) Kalman filter damage detection in the presence of changing process and measurement noise. Mech Syst Signal Process 39(1–2):361–371
8. Brownjohn J, De Stefano A, Xu Y, Wenzel H, Aktan A (2011) Vibration-based monitoring of civil infrastructure: challenges and successes. J Civil Struct Health Monitor 1(3):79–95
9. Carden E, Fanning P (2004) Vibration based condition monitoring: a review. Struct Health Monitor 3(4):355–377
10. Döhler M, Hille, F (2014) Subspace-based damage detection on steel frame structure under changing excitation. In: Proceedings of 32nd International Modal Analysis Conference. Orlando, FL, USA
11. Döhler M, Hille F, Mevel L, Rücker W (2014) Structural health monitoring with statistical methods during progressive damage test of S101 Bridge. Eng Struct 69:183–193
12. Döhler M, Mevel L (2013) Subspace-based fault detection robust to changes in the noise covariances. Automatica 49(9):2734–2743
13. Döhler M, Mevel L, Hille F (2014) Subspace-based damage detection under changes in the ambient excitation statistics. Mech Syst Signal Process 45(1):207–224
14. Döhler M, Mevel L, Zhang Q (2016) Fault detection, isolation and quantification from Gaussian residuals with application to structural damage diagnosis. Ann Rev Control 42:244–256
15. Fan W, Qiao P (2011) Vibration-based damage identification methods: a review and comparative study. Struct Health Monitor 10(1):83–111
16. Farrar C, Worden K (2007) An introduction to structural health monitoring. Philoso Trans Royal Soc A Math Phys Eng Sci 365(1851):303–315
17. Hille F, Petryna Y, Rücker W (2014) Subspace-based detection of fatigue damage on a steel frame laboratory structure for offshore applications. In: Proceedings of the 9th International Conference on Structural Dynamics, EURODYN 2014. Porto, Portugal, July 2014
18. Juang JN (1994) Applied system identification. Prentice Hall, Englewood Cliffs, NJ, USA
19. Kullaa J (2003) Damage detection of the Z24 Bridge using control charts. Mech Syst Signal Process 17(1):163–170
20. Ramos L, Marques L, Lourenço P, De Roeck G, Campos-Costa A, Roque J (2010) Monitoring historical masonry structures with operational modal analysis: two case studies. Mech Syst Signal Process 24(5):1291–1305
21. Rytter A (1993) Vibrational based inspection of civil engineering structures. Ph.D. thesis, Aalborg University, Denmark
22. Structural Vibration Solutions A/S: ARTeMIS modal pro–damage detection plugin (2015). www.svibs.com

23. VCE (2009) Progressive damage test S101 Flyover Reibersdorf/draft. Tech. Rep. 08/2308, VCE
24. Worden K, Manson G, Fieller N (2000) Damage detection using outlier analysis. JSound Vibr 229(3):647–667
25. Yan A, De Boe P, Golinval J (2004) Structural damage diagnosis by Kalman model based on stochastic subspace identification. Struct Health Monitor 3(2):103–119

Numerical and Experimental Investigations of Reinforced Masonry Structures Across Multiple Scales

Eleni N. Chatzi, Savvas P. Triantafyllou and Clemente Fuggini

Abstract This review chapter outlines the outcomes of a combined experimental-numerical investigation on the retrofitting of masonry structures by means of polymeric textile reinforcement. Masonry systems comprise a significant portion of cultural heritage structures, particularly within European borders. Several of these systems are faced with progressive ageing effects and are exposed to extreme events, as for instance intense seismicity levels for structures in the center of Italy. As a result, the attention of the engineering community and infrastructure operators has turned to the development, testing, and eventual implementation of effective strengthening and protection solutions. This work overviews such a candidate, identified as a full-coverage reinforcement in the form of a polymeric multi-axial textile. This investigation is motivated by the EU-funded projects Polytect and Polymast, in the context of which this protection solution was developed. This chapter is primarily concerned with the adequate simulation and verification of the retrofitted system, in ways that are computationally affordable yet robust in terms of simulation accuracy. To this end, finite element-based mesoscopic and multiscale representations are overviewed and discussed within the context of characterization, identification and performance assessment.

E. N. Chatzi (✉)
Institute of Structural Engineering, ETH zürich, Stefano-Franscini-Platz 5,
8093 Zurich, Switzerland
e-mail: chatzi@ibk.baug.ethz.ch

S. P. Triantafyllou
Centre for Structural Engineering and Informatics, The University of Nottingham,
University Park, NG7 2RD, UK
e-mail: Savvas.Triantafyllou@nottingham.ac.uk

C. Fuggini
Industrial Innovation Division, DAppolonia S.p.A., Via San Nazaro 19, 16145
Genova, Italy
e-mail: clemente.fuggini@dappolonia.it

E. Ottaviano et al. (eds.), *Mechatronics for Cultural Heritage and Civil Engineering*,
Intelligent Systems, Control and Automation: Science and Engineering 92,
https://doi.org/10.1007/978-3-319-68646-2_15

Introduction

In recent years awareness has been raised among both engineers and authorities on the importance of shielding infrastructure against extreme events, such as earthquakes, particularly with respect to those items that form part of our cultural heritage. Recent pronounced catastrophes associated with the earthquakes in Tohoku (2011), Christchurch (2011), Modena (2012) and Central Italy (2016), formed a wake up call regarding the potential impact of such events in terms of material damage and, more importantly, human loss. Societies have since turned towards the notion of resilience, which encourages the search for new materials and methodologies able to strengthen and protect structures against natural hazards, including seismicity.

Masonry is a composite material comprising distinct units of various natural or industrial materials e.g., stone, brick, concrete etc. [1, 2]. The constituents usually demonstrate a brittle and, in general, anisotropic behavior in a micro-level. The latter eventually propagates to the macro-level [3–6], and is further enhanced by inherent "weak" planes along the head and bed joints, leading to in- and out-of-plane failure mechanisms, which are often activated by seismic loading. The prevention of structural collapse due to seismic events, and the enhancement of structural integrity, are still open points of discussion within the scientific community as a number of different techniques may be considered suitable, depending on the characteristics of the input load.

Masonry structures comprise an essential feature of global infrastructure heritage. Many of these structures, however, have not necessarily been designed against seismic loads, but rather with the primary goal of withstanding gravity loads [7]. Nonetheless, underestimation of the effects of seismicity may be detrimental, particularly when associated with increased in-plane and out-of-plane forces, which may ultimately lead to failure [8]. To accurately simulate the behavior of masonry, a thorough accounting of the damage and failure mechanisms need be put in place. To this end, effective nonlinear modeling tools are essential, albeit often laborious to establish, owing to the inherent complexity in either the local (constituents) or global (system) level.

For alleviating collapse and failure incidents, retrofitting techniques have been proposed that are particularly suited for the special class of masonry structures. Standard methods of intervention include strengthening through external pre-stressing [9], externally bonded strips or overlays [10], and near-surface mounted reinforcement [11]. During recent years however, textile composites have surfaced as an effective means for the retrofitting and strengthening of reinforced concrete and masonry structures [12]. As opposed to other fiber-based materials such as fiber reinforced polymers (FRPs) that are implemented onto masonry as a set of individual strips, textile materials are produced as fabric meshes of fibre rovings in at least two directions. As such, textile composites provide wide area coverage, can prevent falling debris, help distribute loads, are low-cost, easy to apply, provide a high strength to weight ratio, are resistant to electro-chemical corrosion, are non-invasive, are fatigue resistant, are non-magnetic, and have the potential to reduce seismic retrofitting costs.

Such composite materials in the form of sheets or wrappers have been been implemented for the effective strengthening of masonry structures [13, 14].

Textile solutions are often implemented in strips of fiber-reinforced composites (carbon or glass fibers), which are attached via adhesive (resin) onto localized areas of the wall (CNR-DT 200/2004) [15]. A shortcoming to this approach lies in the potential local increase of stiffness leading to undesired stress concentrations and an unfavorable redistribution of dynamic loads. A remedy to this approach, in the form of a full coverage solution, has been tested and validated within the context of the EU funded project Polytect-Polyfunctional Technical Textiles for Protection against Natural Hazard [16–18]. The purpose of implementing a full coverage solution as in the case of composite textiles is two fold. First of all, the system acts as a mechanical adhesive bridging cracks that may exist on the body of the masonry. Although each individual fiber demonstrates brittle material behavior with an elastic branch until failure, the overall behavior of the textile is ductile as failure propagates in a successive manner during cyclic loading. As a result, textiles significantly increase the ductility and shear strength of retrofitted masonry, while additionally achieving an improved load distribution. Finally, an added advantage to this solution is delivered via its embedded monitoring technology enabled via fiber optical strain sensors. Monitoring of these solutions is further suggested by the European Guidelines (CNR-DT 200/2004), as a consequence of the reduced knowledge base available and rather limited experience gathered regarding their long-term behavior.

The use of testing and the subsequent extraction of structural feedback in the form of measurements allows for enhancing current understanding on such complex systems. When investigated within a laboratory setting, a number of mixed numerical-experimental techniques have been proposed for simulation and verification of polymer reinforcing materials [19–21]. On the other hand, and as was originally conceived in the Polytect project, structural feedback may also be obtained from structures in-operation via use of appropriate sensors. In both cases, it is only via coupling of the recorded information with an appropriate structural model that the info stemming from raw measurements may be transcribed into meaningful insight regarding the condition and behavior of the system.

This is commonly achieved with an inverse problem setting, where response response observations serve as the starting point of an iterative procedure for identifying the underlying structural properties of the investigated system. However, when discussing retrofitting solutions for the purpose of seismic protection, it becomes evident that availability of measurements from dynamic loads or seismic events is of particular importance. The latter naturally necessitates a nonlinear dynamic analysis, which nonetheless comes at increased computational costs with obvious drawbacks when cast in an inverse problem setting. It is, therefore, imperative to develop and employ model structures that are adept in counterpoising the desired level of precision with the associated computational cost.

This paper provides a review of methods available for the modeling and characterization of retrofitted masonry systems. The Polytect project is exploited as a reference case-study for demonstrating the steps that are necessary for the monitoring, simulation and verification of protection solutions for cultural heritage structures. In a first step, we demonstrate the use of finite-element based models of diverse fidelity, and their inverse calibration via heuristic approaches, with a genetic algorithm serving as the optimization tool. The procedure proves successful, albeit severely draining in terms of both computational power and time. In a second stage, this review paper demonstrates how a multiscale-analysis scheme can enhance the computational efficiency of numerical simulation tools, while retaining their level of refinement. The latter results in a dramatic acceleration of the problem solution, which can be exploited in the context of demanding tasks such as identification, uncertainty quantification and reliability assessment.

State-of-the-Art in Simulation

Masonry Modeling

The modeling of materials that are multi-phase in nature is non-trivial. Masonry is a composite material whose diverse constituents exhibit an anisotropic and generally brittle microscopic behavior. The anisotropy at the micro level, together with the inherent weak directions along joints, results in highly anisotropic macroscopic behavior [3]. This anisotropy is also influenced by the spatial distribution of the joints, as well as the mechanical properties of the mortar. An immediate consequence of this anisotropy is significant variability in the bending strength of a masonry wall when in-plane or out-of plane bending is considered [22]. Furthermore, fabrication and construction processes induce notable variability in mechanical properties [23].

It is therefore apparent that, depending on the problem at hand and the particular goal of the simulation study, appropriate modeling tools and corresponding assumptions should be adopted. There are three main modeling approaches for the modeling of masonry response: micro-models, macro-models and multiscale models (Fig. 1). Macro-modeling is a phenomenological approach where the response of a structural element, e.g. a wall, is examined rather than the response of its constituents. In macro-modeling, masonry is treated as a continuum with smeared material properties. To account for different properties along the main axes of the element, the material is regarded as an anisotropic composite, and a relationship is established between generalized stress and strain measures. In the case of masonry, an accurate macro-model must reproduce an orthotropic material with different tensile and compressive strengths along separate material axes. To this end, several macro-elements have been proposed [24–26]. Relatively recently, Chen et al. [27] developed a macro-element for the in-plane nonlinear analysis of unreinforced masonry piers.

In contrast, refined masonry models (micro-models) include a distinct representation of the separate constituents, i.e. units (bricks), mortar and the brick/mortar interface, with fusion of continuum and discontinuous elements. The units (bricks) are often represented by continuum elements, while discontinuous elements approximate the behavior of joints and interfaces. Each joint, consisting of mortar and the

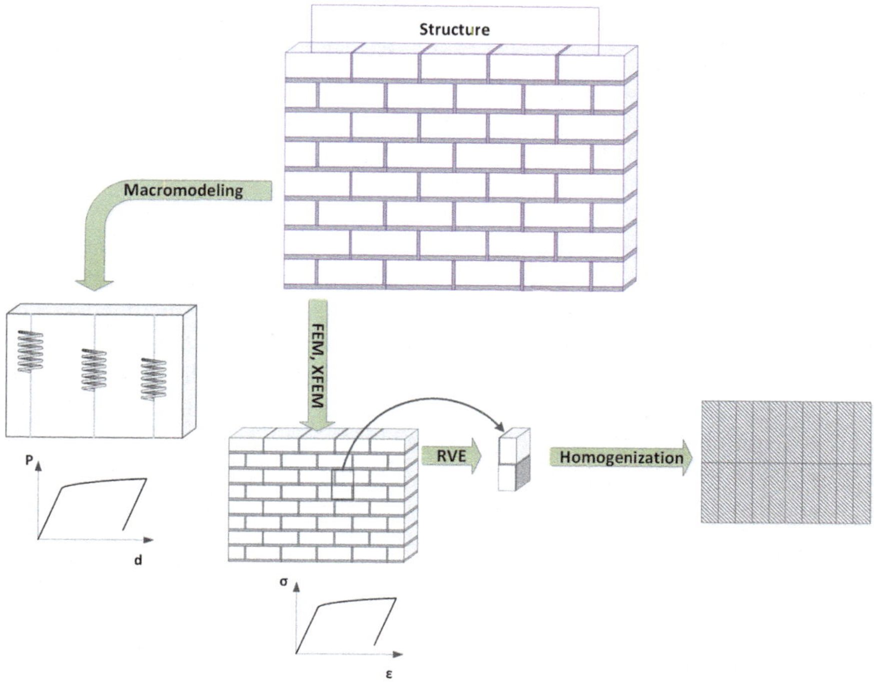

Fig. 1 Different modeling approaches for the structural analysis of composite structures

two unit-mortar interfaces, is lumped into an "average" interface, while the units are expanded in order to keep the geometry unchanged.

Due to their obvious computational advantages as opposed to more refined approaches, macro-modeling procedures are practically the only computational tool implemented at the design office. However, the associated macro-modeling parameters need to be properly identified to provide acceptable results. This calibration is usually based on actual experimental data that is hard to secure and typically comes at a high cost. In alleviating these issues, the multiscale method offers a compromise between the macro and micro approaches, and is in this work presented as a viable alternative.

Multiscale Modeling

Multiscale modeling is a rigorous mathematical approach for the scaling of complex numerical problems, significantly reducing the required computational cost [28]. Multiscale and homogenization schemes [29] are commonly adopted in the modeling of materials whose macroscopic behavior is influenced by, and dependent upon, changes in the micro-structure, e.g. granular [30] or cellular and honeycomb materials [31, 32]. In terms of constitutive modeling, multiscale methods have proven their versatility and accuracy where complex plasticity behaviors are observed [33]. The major benefit of multiscale methods lies in their ability to significantly enhance the computational performance of conventional computational mechanics schemes, such as the finite element method. This feature justifies their popularity across a broad spectrum of implementation domains, from molecular mechanics to computational mechanics [34–37].

Considerable work has been performed in this field with respect to the modeling of composites (see e.g. [38–40]). Massart et al. [41, 42] developed a mesoscale constitutive model for masonry that accounts for anisotropic plasticity effects and damage to the constituents by implementing a generalized plane-stress state assumption. However, multiscale schemes relying on homogenization place the rather limiting assumption of periodicity of not only the micro-structure, but additionally to what concerns damage propagation throughout the macro-scale. A further assumption is placed concerning the full separation of the micro and macro scales considered. These limiting assumptions are eliminated in the multiscale finite element method (MsFEM) introduced by Efendiev and Hou [43]. In [44], an enhanced multiscale finite element (EMsFEM) scheme is presented that extends the applicability of the classical MsFEM into the realm of nonlinear mechanics. The work presented here builds upon this latter method, and extends it for implementation in problems of nonlinear plasticity.

Smooth Hysteretic Modeling

Like cementitious materials, masonry is only adept in handling compression, with limited capabilities in the tensile or shear fronts. It further exhibits orthotropic behavior in directions that lie parallel and orthogonal to the mortar joints. In developing a model able to account for the resulting nonlinear behavior, a multiplicity of effects must be taken into account, including cracking, crushing, tension softening, compression softening and shear transfer across crack slips. Naturally, the modeling task becomes more complex when moving from monotonic to cyclic or dynamic loads, since nonlinearity is then linked to energy dissipation and hysteresis, with manifestation of stiffness degradation and strength deterioration effects. In tackling this issue from a macroscopic perspective, Karapitta et al. [45] have employed a smeared-crack approach for the modeling of unreinforced masonry walls, while Lourenço [25] introduced a model relying on plasticity theory for the modeling of masonry walls loaded in-plane.

From a microscopic perspective, on the other hand, a hysteretic finite element formulation may be adopted, relying on the concept of smooth hysteretic models, such as the Bouc-Wen model [46] or Preisach-type models [47]. As illustrated in Fig. 2,

Fig. 2 The Bouc Wen model: additive decomposition of the elastic and hysteretic component

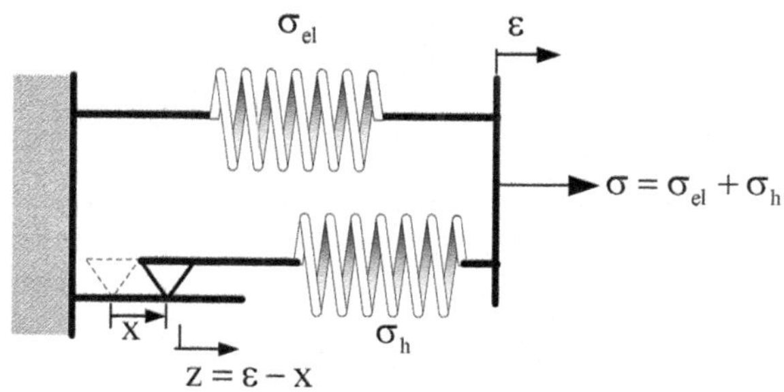

the Bouc-Wen model in particular tackles the representation of hysteresis via superposition of (1) an elastic and (2) a hysteretic component, corresponding to an elastic stress component σ_{el} and a hysteretic stress component σ_h. Furthermore, the strain corresponding to the hysteretic spring is established in terms of total strain, ϵ, and slip on the slider, x. Incorporation of hysteretic finite elements enables a more robust and computationally efficient formulation, able to represent the full hysteretic response cycle, which is particularly well suited for nonlinear dynamic analysis. Furthermore, smooth hysteretic models are capable of compactly simulating damage-induced phenomena, including stiffness degradation, strength deterioration and pinching [48, 49]. Due to their robust mathematical background, implementation of smooth hysteretic laws has been proven to yield computationally efficient formulations in different disciplines, ranging from solid physics and ferromagnetism [50, 51] to stochastic engineered systems [52], and to hysteretic finite element schemes [53], with the latter demonstrated in what follows.

The Hysteretic Multiscale Finite Element Method (HMsFEM)

In coupling the capabilities offered by the state-of-the-art formulations overviewed previously, Triantafyllou and Chatzi [54] have proposed the Hysteretic Multiscale FEM (HMsFEM) formulation, which merges the computational multiscale scheme with the implementation of the hysteretic finite elements in the micro-scale.

The Enhanced Multiscale Formulation (EMsFEM)

A computational multiscale approach is utilized in this review paper, namely the Enhanced Multiscale Finite Element method (EMsFEM) introduced in [44]. The method exploits patterns of periodicity present at the micro-scale level for grouping sets of micro-elements into clusters, herein referred to as Representative Volume

Elements or RVEs (macro-elements). A visualization is offered in Fig. 3, where the coarse RVEs are indicated in Subfigures 3c and 3d. Discretization on the basis of the RVEs defines a mesh that is significantly coarser than the respective micro-scale mesh. The benefit of the EMsFEM approach lies in the redefinition of the space where governing equations are solved, shifting the analysis from the fine mesh of Fig. 3a to the coarser mesh of Fig. 3e. To this end, it is useful to distinguish between the micro- and macro-scale via use of an appropriate notation; $\mathbf{u_m}$ will be used to denote the displacements at the micro-nodes, while $\mathbf{u_M}$ stands for the macro-displacement field. As in the standard finite element context, it is possible to interpolate the micro-displacement vector at the nodes by using an interpolation scheme as in [55]. For a hex-element this would imply:

$$\mathbf{u_m} = [N]_m d^i_m \quad where \quad d^i_m = \underbrace{\left\{ u_{m(1)} \; v_{m(1)} \; \cdots \; v_{m(8)} \right\}^T}_{1x24} \tag{1}$$

is the vector of nodal displacements of the ith micro-element, and $[N]_m$ is the interpolation matrix of the hex-element.

The RVEs also comprise hex-elements whose nodal displacements may be aggregated in the following macro-displacement vector:

$$d^i_M = \underbrace{\left\{ u_{M(1)} \; v_{M(1)} \; \cdots \; v_{M(8)} \right\}^T}_{1x24} \tag{2}$$

where (i) designates the ith macro-node of the coarse (RVE) mesh. In what follows, m will be used to designate components at the micro-scale, while M will be used to designate components at the macro-scale.

For transitioning from the micro- (1) to the macro-scale (2), a mapping is constructed relying on suitable basis functions. In matrix form this is established as:

$$\{d\}_m = [N]_m \{d\}_M \tag{3}$$

where $\{d\}_m$ is the $(3n_{micro} \times 1)$ vector of the micro-mesh nodal displacements, $[N]_m$ is the micro-basis shape function matrix evaluated at the nodes of the micro-mesh (x_j, y_j, z_j), while $\{d\}_M$ is the vector of the macro-node displacements. Per the standard definition of shape function, each column of $[N]_m$ represents a deformed configuration of the RVE, with a value of unit at the corresponding macro-degree of freedom, and null values at the remaining macro-degrees of freedom.

The micro-basis functions occur via solution of the following boundary value problem:

$$\begin{aligned} [K]_{\mathrm{RVE}} \{d\}_m &= \{\emptyset\} \\ \{d\}_S &= \{\bar{d}\} \end{aligned} \tag{4}$$

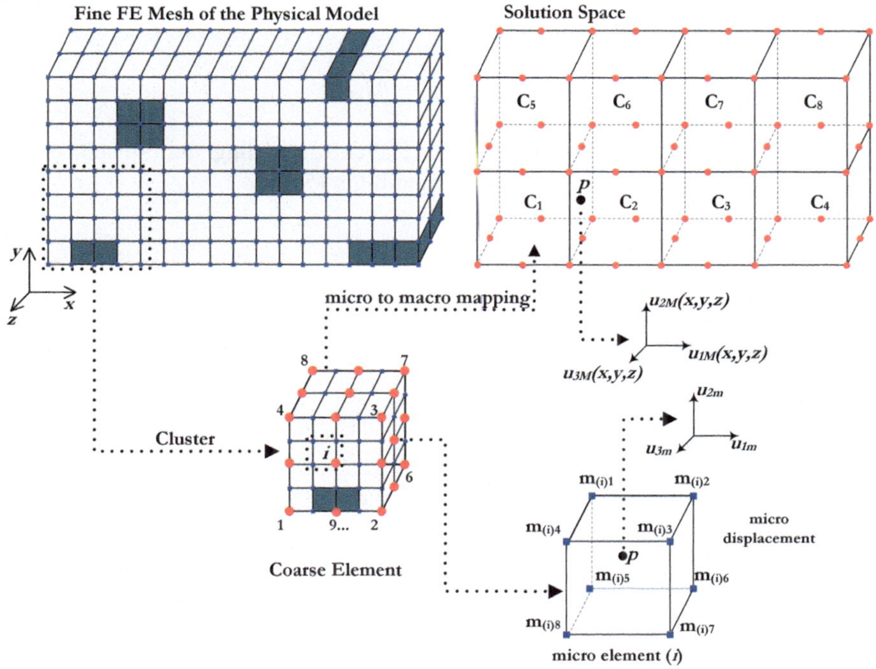

Fig. 3 The MsFE modeling scheme

where $[K]_{RVE}$ designates the RVE stiffness matrix, $\{d\}_S$ is the vector of degrees of freedom along the boundary S of the RVE, and $\{\bar{d}\}$ is a predefined displacement vector. Different options are available for specifying the boundary conditions, a feature critical to the performance of the method, including the linear, periodic and oscillatory boundary conditions, as elaborated upon in [43, 44].

The Hysteretic Multiscale Formulation (HMsFEM)

This Section offers a brief overview of the HMsFEM scheme, however the interested reader is referred to [56, 57] for further details on this formulation. The hysteretic feature of the formulation relies on the additive decomposition of the total strain into a reversible elastic and an irreversible inelastic component [58] (Fig. 2):

$$\{\dot{\varepsilon}\}_{m(i)} = \left\{\dot{\varepsilon}^{el}\right\}_{m(i)} + \left\{\dot{\varepsilon}^{pl}\right\}_{m(i)} \tag{5}$$

where $\{\varepsilon\}_{m(i)}$ is the total strain tensor, $\left\{\varepsilon^{el}\right\}_{m(i)}$ is the tensor of the elastic, reversible, strain and $\left\{\varepsilon^{pl}\right\}_{m(i)}$ denotes the tensor of plastic strains, while $m\,(i)$ indexes the ith micro-element within the RVE. The (.) symbol denotes a time-derivative.

The isoparametric interpolation scheme is here considered for the displacement field

$$\{d\}_{m(i)} = [N]\{u\}_{m(i)} \tag{6}$$

where $[N]_{m(i)}$ is the shape function matrix. The corresponding strain-displacement relationship is inferred on the basis of compatibility [55] as:

$$\{\varepsilon\}_{m(i)} = [B]\{u\}_{m(i)} \tag{7}$$

where $[B]$ is the strain-displacement matrix.

Moreover, the interpolation scheme defined for plastic deformations assumes the form:

$$\left\{\dot{\varepsilon}^{pl}\right\}_{m(i)} = [N]_e \left\{\dot{\varepsilon}^{pl}_{cq}\right\}_{m(i)} \tag{8}$$

where $\left\{\varepsilon^{pl}_{cq}\right\}_{m(i)}$ denotes the strain vector evaluated at appropriately defined points (collocation points) and $[N]_e$ is the respective interpolation matrix.

By plugging Eq. (5) into the Principle of Virtual Work [59], and via use of the former interpolation schemes, it is straightforward to derive the elastic $\left[k^{el}\right]_{m(i)}$ and hysteretic $\left[k^{h}\right]_{m(i)}$ stiffness matrices:

$$\begin{aligned} \left[k^{el}\right]_{m(i)} &= \int_{V_e} [B]^T [D]_{m(i)} [B]\, dV_e \\ \left[k^{h}\right]_{m(i)} &= \int_{V_e} [B]^T [D]_{m(i)} [N]_e\, dV_e \end{aligned} \tag{9}$$

The governing equation of the problem may then be written as:

$$\left[k^{el}\right]_{m(i)} \left\{\dot{d}\right\}_{m(i)} - \left[k^{h}\right]_{m(i)} \left\{\dot{\varepsilon}^{pl}_{cq}\right\}_{m(i)} = \frac{1}{v_\eta} \left\{\dot{f}\right\}_{m(i)} \tag{10}$$

where v_η is a degradation parameter, which increases with plastic deformation (reference value equals 1 for no yielding, see [60] for further details). It need be mentioned that both $\left[k^{el}\right]_{m(i)}$ and $\left[k^{h}\right]_{m(i)}$ are constant matrices. The evolution of plastic deformations, and thereby nonlinearity, is evaluated at the Gauss points of the corresponding micro-scale element, and is defined as:

$$\left\{\dot{\varepsilon}^{pl}\right\}_{m(i)} = \mathscr{F}\left(\left\{\varepsilon^{el}\right\}_{m(i)}, \left\{\dot{\varepsilon}^{el}\right\}_{m(i)}, \{\sigma\}_{m(i)}\right) \tag{11}$$

where $\mathscr{F}$ is a hysteretic operator [50, 61, 62]. A multi-axial Bouc-Wen type smooth plasticity model [63] is herein adopted as the aforementioned operator. The benefit of employing such a rule is that the nonlinear behavior may be accounted for in a compact representation, fully governed by a finite set of parameters; the parameters

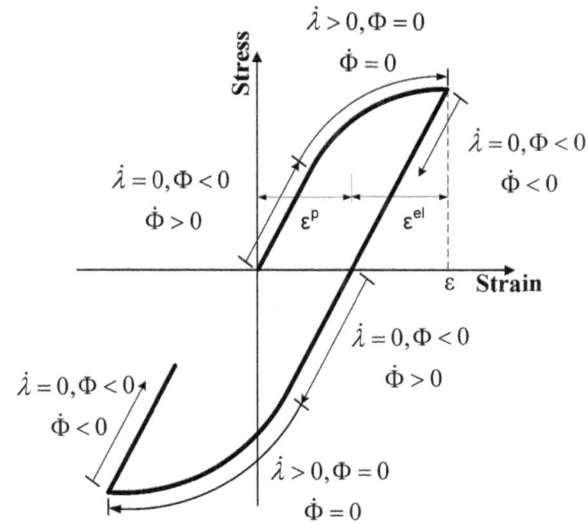

Fig. 4 The classical plasticity loop accounted for by the Bouc Wen model

Table 1 Smooth hysteretic model parameters

Effect	Parameter
Hysteretic shape	β, γ, N
Deterioration	c_η (stiffness), c_s (strength)
Pinching	$\zeta_1^0, \psi_0, \delta_\psi, \mu, p, q$

of the hysteretic model. Figure 4 exemplifies how a typical hysteretic loop may be accounted for via adoption of an appropriate yield flow rule Φ and a kinematic hardening rule, further regulated by the plastic multiplier $\dot{\lambda}$, as described in the work of [64]. For further details, the interested reader is referred to [60].

The smooth hysteretic model additionally offers the potential of including damage induced phenomena such as damage degradation, strength deterioration and pinching, as elaborated upon in [48, 65]. Table 1 summarizes the defining parameters of the smooth hysteretic model, with details on the corresponding formulations found in [54].

Next, by following the reasoning adopted under the EMsFEM scheme, one can demonstrate [54] that the RVE equilibrium equation now assumes the following form:

$$[K]^M_{RVE(j)}\{d\}_M = \{f\}_M - \{f_h\}_M \tag{12}$$

where $[K]^M_{RVE(j)}$ is the RVE stiffness matrix derived as:

$$[K]^M_{RVE(j)} = \sum_1^i \left[k^{el}\right]^M_{m(i)} \tag{13}$$

while $\{f\}_M$ is the nodal force vector, comprising mapped contributions from the micro-nodal force components:

$$\{f\}_{m(i)}^{M} = \frac{1}{v_\eta} [N]_{m(i)}^{T} \{f\}_{m(i)} \tag{14}$$

and $\{f_h\}_M$ is the plastic force vector evaluated as:

$$\{f_h\}_M = \sum_{i=1}^{m_{el}} \left[k^h\right]_{m(i)}^{M} \left\{\varepsilon_{cq}^{pl}\right\}_{m(i)} \tag{15}$$

Equation (12) occurs through the principle of energy equivalence between the deformation energy of the RVE and the corresponding micro-mesh [54]. The major benefit of this approach versus standard multiscale schemes is that the recalculation of the micro-basis functions, which would be required every time the structure shifts from the elastic to plastic regimes, is no longer necessary. Instead, the micro-basis functions are now calculated only once in the beginning of the analysis, while the employed smooth hysteretic law accounts for the evolution of nonlinearity.

The direct stiffness method is then implemented for casting the governing equation of the dynamic problem in the macro-level:

$$[M]\left\{\ddot{U}\right\}_M + [C]\left\{\dot{U}\right\}_M + [K]\{U\}_M = \{P\}_M \tag{16}$$

The nodal load vector $\{P\}_M$ for the RVE mesh $\left(ndof_M \times 1\right)$ in Eq. (16) occurs as:

$$\{P\}_M = \{F\}_M + \left\{F_h\right\}_M \tag{17}$$

where $\{F\}_M$ is the $\left(ndof_M \times 1\right)$ external loads vector and $\left\{F_h\right\}_M$ is the $\left(ndof_M \times 1\right)$ hysteretic load vector at the global level. $[M]$, $[C]$ and $[K]$ designate the $\left(ndof_M \times ndof_M\right)$ mass, viscous damping and elastic stiffness matrices, which are evaluated at the RVE mesh (macro-scale).

The solution of the global equations of motion is therefore carried out at the macro-scale. This in turn implies that the resulting macro-displacements $\{U\}_M$ should be downscaled (mapped to the micro-scale) in order to derive the corresponding local strains. The computational aspects of the HMsFEM overviewed herein are offered in detail in [54].

Case Study—Polymeric Textiles for Seismic Retrofitting of Masonry

Within the framework of the Seismic Engineering Research Infrastructures for European Synergies (Series) initiative (https://www.series.upatras.gr, Polymast 2011) and as part the EU co-funded projects Polytect [66] and Polymast [17], textile reinforcement solutions have been developed. These were tested under full-scale seismic tests carried out in the European Center for Training and Research in Earthquake Engineering (Eucentre). As part of the conducted experimental campaign, a previously damaged unreinforced two-storey stone building served as a reference case-study for assessment of a "seismic wallpaper" reinforcing solution (Fig. 5). The purpose was to establish a benchmark for the retrofitting and repair of masonry-type structures, which have been damaged due to seismic events. The building's dimensions along the length, width and height were 5.80 m × 4.40 m × 5.80 m, respectively. It features concrete foundation, a wooden roof and a wood first-storey slab.

While still in the unreinforced (URB) stage, the structure was tested under dynamic loads on a shake-table until damage. Preliminary repairs were performed by filling cracks with epoxy resin and stiffening the wood slab. The damage-repaired building (DRB) was again non-destructively tested and subsequently reinforced via application of a full-cover solution, namely a composite seismic wallpaper. The composite wallpaper (polymeric textile), which is illustrated in Fig. 6, features the following components:

1. Multiaxial, warp-knitted, alkali-resistant (AR)-glass and polypropylene (PP) fibers.
2. Nanoparticle-enhanced coatings for the polymeric textile; Nanoparticle-enhanced mortar for ensuring the bond to the protected structure.
3. Fiber optic strain sensors embedded into the textile, with dynamic sampling capability up to 1 kHz.

Fig. 5 3D View of the unreinforced (left) and reinforced (right) stone building, Reproduced from [17, 66]

Fig. 6 Composition of the quadri-axial polymeric textile employed as a full-coverage seismic protection, reproduced from [17, 66]

An epoxy mortar compound ensured an almost perfect bond between the textile and the masonry structure; an assumption which was later experimentally validated. This further allows for simplification of the adopted numerical analysis alleviating the necessity for assumptions on interface/contact laws. The properties of the primary constituents involved here, i.e., masonry, and the polymeric textile as provided by the manufacturer, are listed in Table 2. It should be noted that as is typically the case for polymer materials, the nominal specifications do not necessarily comply with the in-situ properties. A detailed outline of the further material properties of the stone building is provided in [18].

Upon retrofitting with the polymeric textile, the building was assessed under seismic input on the Eucentre shake-table at increasing peak accelerations of 0.1, 0.3, 0.4, 0.5 and 0.6 g. The third test at 0.4 g was, in fact, the first to induce damage and nonlinear response behavior. The resulting response was recorded via a set of accelerometers mounted in several positions on the structure.

Table 2 Mechanical properties of the case-study materials

Property	Masonry	Textile
Density	2579 kg/m^3	2000 kg/m^3
Young modulus	$E = 2550$ MPa	$E_x = 40000$ MPa, $E_{y,z} = 32000$ MPa
Poisson coefficient	$\nu = 0.4$	$\nu_{xy} = 0.14$, $\nu_{xz} = \nu_{yz} = 0.2$
Shear modulus	840 MPa	4500 MPa
Maximum compression stress σ_c	3.28 MPa	
Maximum tensile stress σ_t	0.137 MPa	400 MPa
Maximum shear stress σ_s		10 MPa

Numerical—Experimental Characterization of the Composite Structure

The data recorded, by means of the described campaign, offer a multiplicity of information over various stages of the building's condition, namely prior to damage (URB), after damage-minor repair (DRB), and upon retrofitting with the polymeric textile (REB).

Step 1: Identification of Elastic Properties—OMA

In identifying the properties of the building in the elastic range for various condition stages, operational modal analysis (OMA) may be employed [67]. OMA requires sets of ambient (broadband) vibration records on the basis of which modal properties, such as natural frequencies, mode shapes and damping ratios may be inferred. These were, in this case, extracted via four three-axis geophones, located on the parapet of the building's windows. Half-hour records were processed, acquired at a sampling rate of 265 Hz per construction phase (URB, DRB, REB). In addition to the ambient records, forced hammer (impact) tests were further carried out, for corroborating the OMA results. The records were processed via use of well-established modal identification techniques, namely the frequency domain de-composition (FDD) and the natural excitation technique combined with the eigensystem realization algorithm (Next-ERA) [68–70]. The identified modal frequencies are summarized in Table 3 for the first three modes of the building.

Step 2: Preliminary Numerical Analysis

A preliminary numerical Finite-Element-based model was set up in ANSYS [71]. The material properties were assigned according to the manufacturers specifications (Table 2 and [18]), while the modeling of damage and repairs was based on the close

Table 3 Experimental modal frequencies under various construction phases

Mode #	URB frequency [Hz]	DRB frequency [Hz]	REB frequency [Hz]
1	11.55	8.51	9.50
2	12.07	9.31	10.30
3	16.27	15.69	18.25

Table 4 Numerical modal frequencies under various construction phases—no model calibration

Mode #	URB frequency [Hz]	DRB frequency [Hz]	REB frequency [Hz]
1	12.62	10.93	13.77
2	13.51	12.28	14.73
3	20.22	19.25	22.35

observation of the structure between construction phases. This included the recording of major crack patterns (size and orientation). Both masonry and the textile are, in this preliminary simulation attempt, modeled as anisotropic yet homogeneous materials. The first three modes identified via this rough numerical model are summarized in Table 4, where it is evident that the uncalibrated numerical model significantly overestimates the building's eigenfrequencies (Table 3).

While results do not appear significantly divergent for the first two modes of the unreinforced phase (URB), the error increases for the damaged/minor repairs phase (DRB), and it becomes quite significant for the retrofitted phase (REB). The discrepancy may be attributed to numerous uncertainties relating to lack of prior information, the simplifying assumptions in the modeling of the masonry and polymer constitutes, as well as the assumptions relating to the damage and intervention modeling (crack filling and diaphragm stiffening).

For ameliorating the estimation results, it is clear that a more refined approach need be enforced. At the same time, the nonlinear range of the response, available as part of the conducted shake-table tests, has not so far been exploited. In utilizing these results, a time history analysis is required, albeit implying significant computational toll. To this end, it is of the essence to utilize computational models, which may offer sufficient flexibility for the calibration (updating) process, while alleviating unreasonable computational toll owing to excessive refinement. Such a procedure is described next.

Step 3: Inverse Problem Formulation via Heuristic Optimization

In compromising the above-mentioned conflicting objectives, i.e., sophistication and model adaptability against computational cost, four different FE setups were put together corresponding to essentially four types of RVEs, as illustrated in Fig. 7. The employed setups, which are materialized in ANSYS, correspond to varying modeling refinement levels, and therefore allow for diverse adaptability with respect to the actual case-study. More specifically,

Setup 1 (FE1)
In the first configuration masonry is discretized by means of an eight-noded solid element (SOLID65), which is capable of cracking/crushing and sliding (shear) effects across the crack face. The multiaxial textile is modeled as a homogeneous anisotropic material, with different properties along different wall faces, since walls placed perpendicular to the direction of the seismic load were dressed with more than one layers of textile. The SOLID46 layered element is adopted for the textile model, allowing for the combination of a layer of mortar (matrix) and a layer of the assumed anisotropic textile material. A total of 32 material parameters serve as inputs to be calibrated for this configuration.

Setup 2 (FE2)
In the second configuration masonry is once again modeled via SOLID65 elements. The textile is in this case approximated by a meso-scale representation, with the layered SOLID46 finite element now updated to include a total of five layers. The first layer corresponds to the mortar compound matrix, while the remaining four correspond to each one of the glass and polypropylene fiber layers, accounted for with their corresponding orientation angles. The fibers are represented via dedicated anisotropic materials. The different behavior along perpendicular wall directions is now accounted for via independent thickness parameters. A total of 31 material parameters serve as inputs to be calibrated for this configuration.

Setup 3 (FE3)
In the third configuration masonry is now discretized using the SOLID45 3D structural solid element, which allows for plasticity, creep, swelling, stress stiffening, large deflection, and large strain effects. It further accommodates anisotropic material properties, a feature which is not offered by the SOLID65 option used in the previous setups. The textile is modeled via the anisotropic macroscopic representation adopted in the 1st setup (FE1). A total of 39 material parameters serve as inputs to be calibrated for this configuration.

Setup 4 (FE4)
In the fourth configuration masonry is again modeled via the SOLID45 element, while the multi-axial reinforcing textile follows the meso-scale representation adopted in setup 2 (FE2). The total number of input parameters is now 40.

FE1 corresponds to the setup of the preliminary model, with the added option of calibrating the involved material properties. FE2 breaks the textile material further down to its constituents (meso-scale representation), permitting dedicated failure criteria per fiber class and orientation. This breakdown has been motivated from the mode of failure occurring during the shake-table tests, which lied along specific fibers and orientations. FE3 allows for refinement on the masonry front by introducing anisotropic properties for the masonry element. Finally, FE4 is the more refined

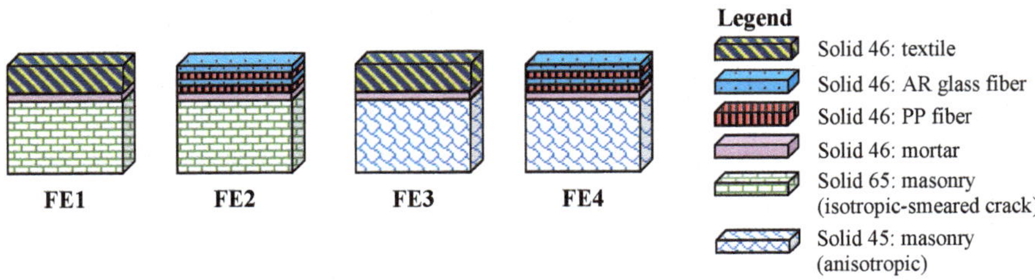

Fig. 7 Adopted FE setups

modeling option, combining an anisotropic material for masonry and a meso-scale model for the textile.

For updating the parameters of the candidate numerical models, a multistage evolutionary approach was implemented in [18]. A genetic algorithm (GA) is adopted as the heuristic optimization tool, albeit this being interchangeable with other alternatives, such as particle swarm optimization (PSO) or the covariance matrix adaptation evolutionary strategy algorithm (CMA-ES). A heuristic is deemed as appropriate for the inverse formulation discussed herein, due to lack of a straightforward functional relationship between the optimization function, i.e., the discrepancy between experiments and simulation, and the input parameters, i.e., the material properties. Given availability of a diverse set of reference (experimental) data, involving both operational modal analysis & nonlinear dynamic tests across different construction phases of the case-study building, a multi-stage optimization procedure is set up. The unreinforced test data (URB) is discarded, given that damage essentially reinitializes the structural properties in the DRB case. The optimization process comprises the following stages:

Stage 1:
In this stage the elastic properties of the repaired masonry building (DRB) model are updated in order to yield an improved estimation, ω_s^{DRB}, of the experimentally identified frequencies, ω_e^{DRB}.
Stage 2:
In this stage, the experimentally identified frequencies of the retrofitted structure (REB) serve for updating the elastic properties of the textile on the basis of the matching between experimental, ω_e^{REB}, and simulated, ω_s^{REB}, eigenfrequencies.
Stage 3:
While the former two analysis stages pertain to identification of elastic material properties, this stage performs a time history analysis. This allows for further configuration of the nonlinear material properties on the basis of the matching between experimental, a_e^i, and simulated, a_s^i, acceleration time histories in eight measured locations.

The final objective function (residual), r to be minimized via the GA occurs via superposition of the target approximations for each of the three aforementioned stages, and may be written as:

$$r = \frac{\|\omega_s^{DRB} - \omega_e^{DRB}\|}{\|\omega_e^{DRB}\|} + \frac{\|\omega_s^{REB} - \omega_e^{REB}\|}{\|\omega_e^{REB}\|} + \sum_{i=1}^{8} \frac{\|a_e^i - a_s^i\|}{\|a_e^i\|} \tag{18}$$

For all simulations ANSYS serves as the modeling platform, iteratively called from a genetic algorithm module coded in a Fortran environment [72].

Step 4: Updated Model Results

The convergence of the natural frequency approximation, during the GA-based optimization is presented in Fig. 8 for each of the four candidate models, and contrasted to the experimentally identified values. The upper row of plots correspond to the damaged/repair (DRB) case, whilst the bottom row correspond to the modal estimates for the retrofitted (REB) case. The plots reveal that the updated modal estimates better approximate the experimental results, when contrasted against the preliminary (uncalibrated) model estimates (Tables 3 and 5). Setups FE3 and FE4 return a closer approximation of the reference (experimental) values. This may be attributed to the higher flexibility of these setups; FE3 accounts for the anisotropic behavior of masonry, while FE4 achieves a more refined representation of the polymeric textile.

The third residual term, or optimization criterion, pertains to the approximation of the dynamic time history measurements, which further allows for calibration of the nonlinear material properties. Figure 9 illustrates results from FE4, which performs the best, from two characteristic building nodes (A & F), lying along the perimeter of the first and second floor of the structure respectively. Although not perfect, the simulation adequately approximates the observed dynamics. In further refining the achieved results, the material damping parameters ought to enter the optimization process, while the monitoring capability offered by the textile should be exploited. It is worth noting that the textile features embedded with fiber optic strain sensors are able to provide information on the micro-scale level.

However, the refinement of the simulation via introduction of further parameters into the optimization problem, or further refinement of the scale of the representation will eventually render the optimization procedure infeasible. The current modal analysis run-time lies in the order of ca. 10 s against an order of ca. 15 min for the nonlinear dynamic analysis part (with the cost increasing for more complex FE setups). It is thus necessary to come up with efficient methods for accelerating the simulation procedure, particularly regarding the dynamic nonlinear analysis component.

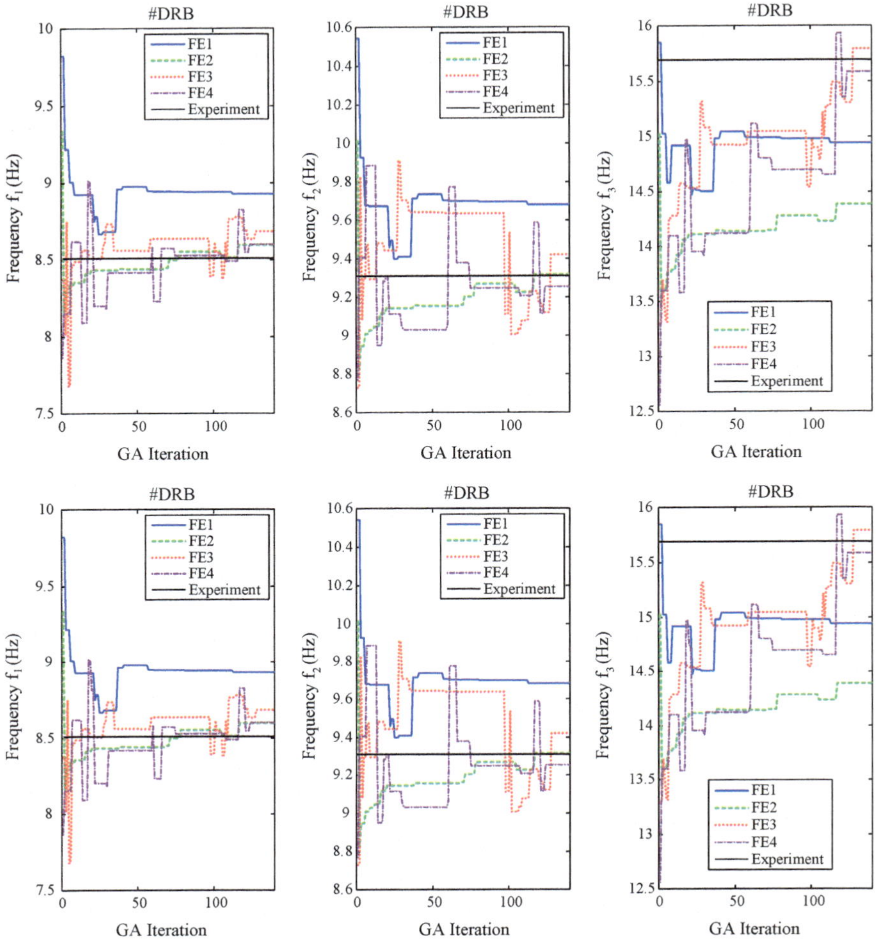

Fig. 8 Convergence plot for the candidate models (FE Setups) in terms of approximation of the elastic (eigenfrequency) properties of the damage/repaired (DRB-upper plots) and retrofitted (REB-lower plots) building

Multiscale Analysis of Textile-Retrofitted Masonry Wall

As discussed in the earlier section, despite limiting the candidate models to meso-scale representations, a heavy computational toll occurs, particularly with respect to the time history nonlinear analysis. The latter necessitates about 15 min for 10 s of a single transient analysis, carried out on an average performance double core PC, versus only a few seconds required by a modal analysis. Nonetheless, adoption of such an analysis is necessary when dynamic testing results are available, and identification of nonlinear properties is sought. In tackling this, Triantafyllou and Chatzi, introduced the HMsFEM discussed and overviewed in section

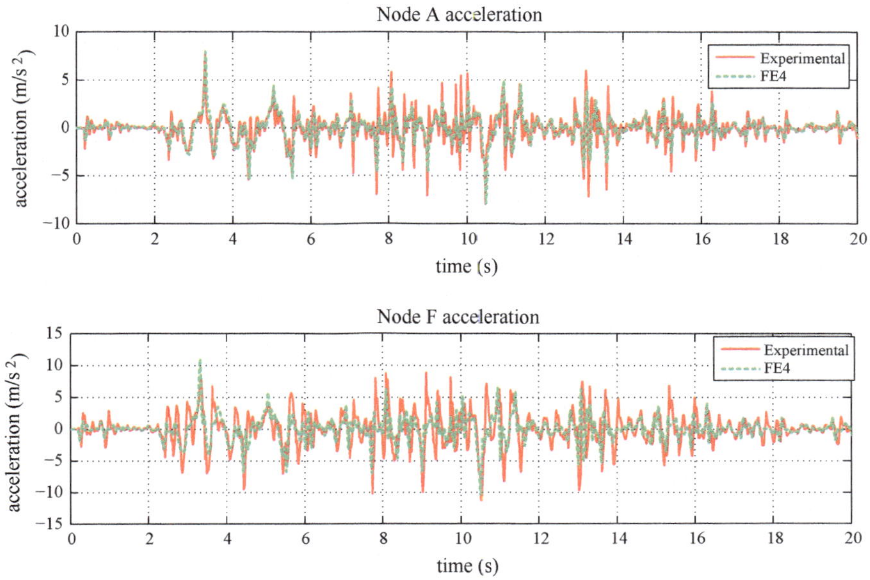

Fig. 9 Convergence plots for Setup FE4 in terms of approximation of the measured dynamic response of the damage/repaired (DRB-upper plots) and retrofitted (REB-lower plots) building offered at two characteristic locations

"The Hysteretic Multiscale Finite Element Method (HMsFEM)" [54]. The workings of the method have been demonstrated on the acceleration of the computation of polymer reinforced masonry structures in [57]. Following the construction stages outlined in the previous case study, [57] overview the analysis of an unreinforced and textile-retrofitted masonry wall, this time by means of a multiscale FE based analysis. For details on the methodological front, the interested reader is referred to [57] for a thorough outline.

The cantilever masonry wall presented in Fig. 10 is considered, comprising stone blocks, mortar and a single outer layer of textile-reinforcement in analogy to a single wall of the case study presented in section "Case Study-Polymeric Textiles for Seis mic Retrofitting of Masonry". The material properties are chosen in accordance to the previous case study, with addition of the properties of the smooth hysteretic law, which is now adopted for capturing the nonlinear behavior. The textile, whose configuration follows the material employed in the previous case-study is modeled as a homogenized anisotropic layer (Fig. 11). The elastic, plastic and hysteretic material properties of the constituents are summarized in Table 5.

The reference fine-meshed finite element model as illustrated in Fig. 10a comprises 2223 hex–elements with full integration and 3020 nodes. The multiscale model instead features only 10 RVEs and 44 nodes, with two types of coarse elements of 228 micro-elements each (Fig. 10b and c). The eigen-periods of the structure under consideration are $T_{in} = 0.19$ s and $T_{out} = 0.91$ s for the in-plane and out-of-plane eigen modes, respectively.

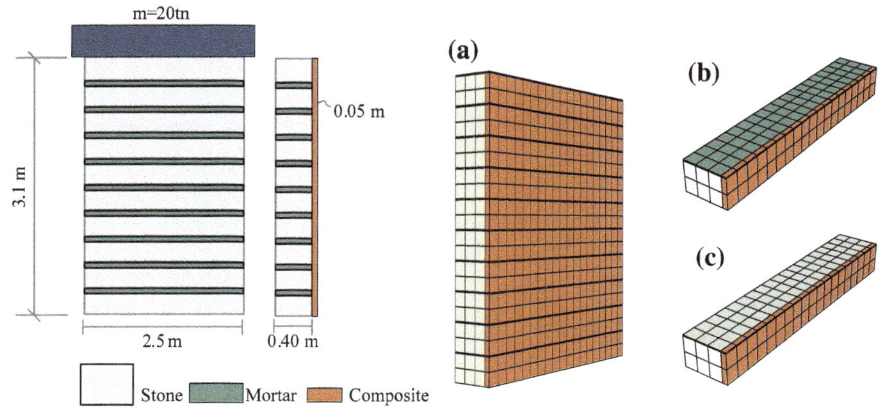

Fig. 10 Left: textile-retrofitted masonry wall; Right: **a** Finite element model, **b** Multiscale model RVE#1, **c** Multiscale model RVE#2

Table 5 Multiscale analysis models—constituent material properties

Parameters	Stone
Elastic/Plastic	$\rho_s = 1.8tn/m^3$, $E = 20.2$ GPa, $\sigma_c = 69.2$ MPa and $\sigma_t = 6.92$ MPa
Hysteretic shape	$\beta = 0.1$, $\gamma = 0.9$, $N = 25$
Deterioration	$c_\eta = 0.002$ and $c_s = 0.005$
Parameters	Mortar
Elastic/Plastic	$\rho_m = 1.2tn/m^3$, $E = 3494MPa$, $\sigma_c = 3$ MPa, $\sigma_t = 0.3$ MPa
	Drucker-Prager dilation angle: $\psi = 60^o$
Hysteretic shape	$\beta = \gamma = 0.5$, $N = 2$
Deterioration	$c_\eta = 0.002$ and $c_s = 0.05$
Pinching	$\zeta_1^0 = 1.0$, $\psi_0 = 0.05$, $\delta_\psi = 0.01$, $\mu = 0.0001$, $p = 1.2$, $q = 0.0001$
Parameters	Polymeric textile
Elastic	$\rho_t = 2.0tn/m^3$, $E_{11} = 40$ GPa, $E_{22} = E_{33} = 32$ GPa, $E_{12} = E_{23} = E_{13} = 4.5$ GPa
Plastic	{wrap-knitted fibers: $\sigma_t = 10$ MPa, $\sigma_s = 400$ MPa}, {mortar matrix: $\sigma_c = 30$ MPa}
Hysteretic Shape	$\beta = 0.5$, $\gamma = 0.5$, $N = 25$
Deterioration	$c_\eta = 0.002$ and $c_s = 0.005$

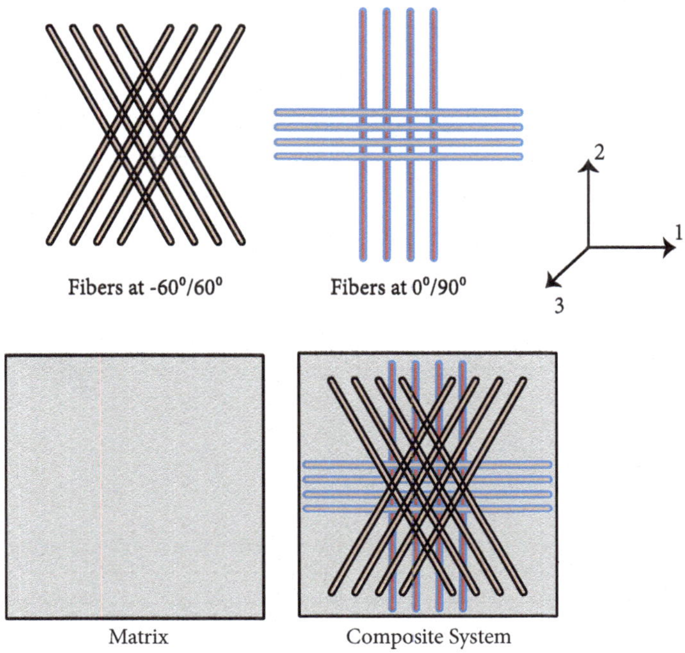

Fig. 11 Composition of the polymeric textile reinforcement

A series of time-history analyses is then performed at a constant time step of dt = 0.001s. The unreinforced (URM) and retrofitted (REB) stages, i.e., the wall with and without textile, are subjected to seven unscaled ground motion records obtained from the PEER strong motion database [73] (Friuli, 1976; Victoria-Mexico, 1980; Northridge, 1995; Imperial Valley(E06), 1979; Chi-Chi, 1999; Imperial Valley (E07), 1979; Coyote Lake, 1979). The multiscale approach is verified in terms of accuracy against a conventional FE model built in ABAQUS [74], rendering a very good approximation. The analysis time for the ABAQUS FE model is ca. 125 min, while the multiscale analysis necessitates only 14 min, thereby offering an 88% acceleration in computation.

The derived in-plane displacement time-histories, calculated at the free end, are contrasted in Figs. 12 (in-plane) and 13 (out-of-plane) for selected earthquake inputs in the URM and REB case. The peak displacements appear reduced for the REB case due to the strengthening functionality of the textile.

A benefit of the multiscale approach is the continual interaction between the fine and coarse scale, which allows for estimating micro-scale quantities (such as local strains), as well as examining phenomena of damage accumulation at the micro-mesh level. By carrying out such an analysis, it appears that despite reduction in the experienced peak displacements, damage accumulation still occurs. This is evident in the hysteretic energy accumulation plot for the bottom mortar layer in Fig. 14, albeit damage appears significantly reduced in the retrofitted case. Thus, although

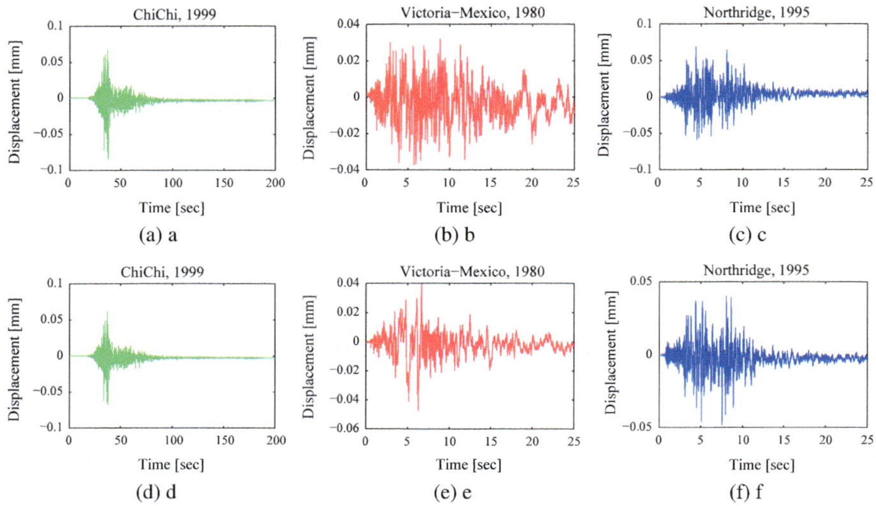

Fig. 12 In-plane displacement time histories for the unreinforced **a**, **b**, **c** and retrofitted **e**, **d**, **f** case

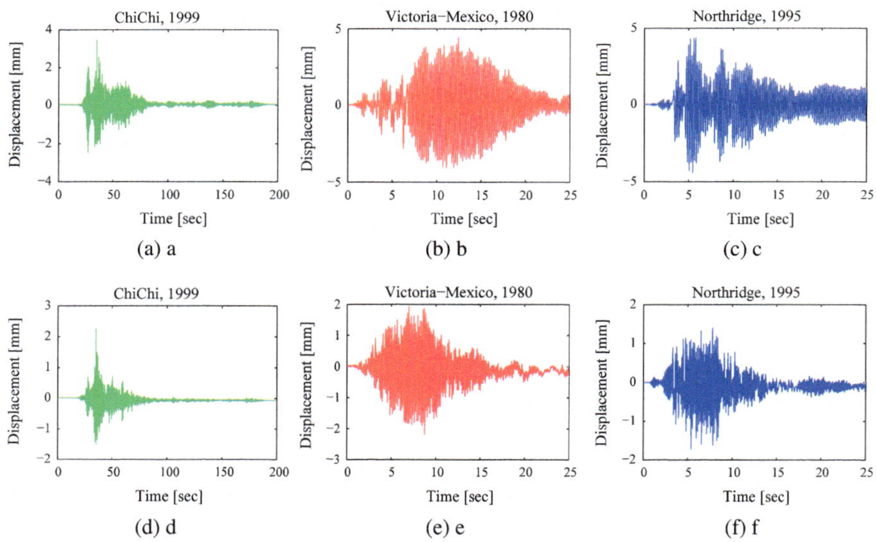

Fig. 13 Out-of-plane displacement time histories for the unreinforced **a**, **b**, **c** and retrofitted **d**, **e**, **f** case

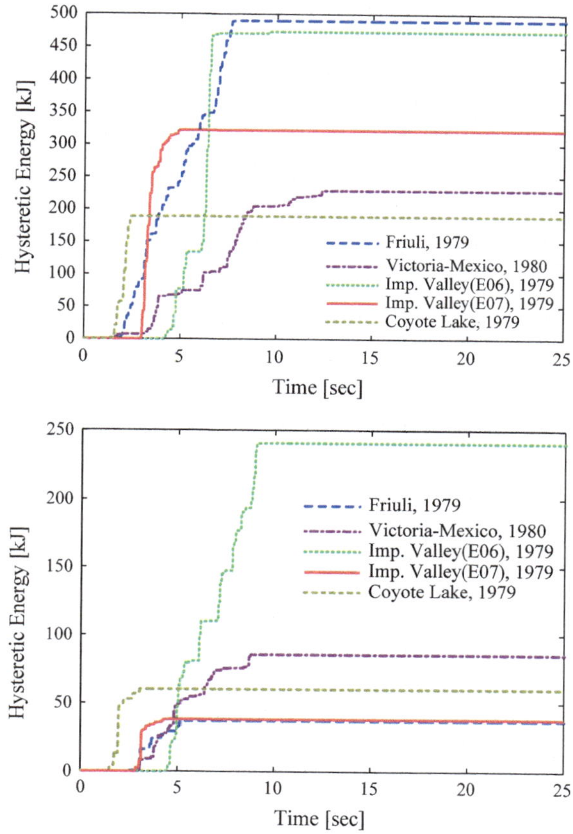

Fig. 14 Hysteretic energy accumulation at the bottom mortar layer, for the URM case (top); and the REB case (bottom)

the textile composite layer succeeds in increasing the overall strength and stiffness of the masonry wall, it need be combined with conventional measures to ensure compliance with requirements at damage limitation performance level.

The significant reduction in computation furnished by the HMsFEM scheme is particularly attractive with the context of an inverse analysis, as discussed in section "Case Study-Polymeric Textiles for Seismic Retrofitting of Masonry". Since the analysis time is reduced by an order of magnitude in this case, it would be feasible to consider implementation of identification techniques that are near-online, such as Kalman-type filters, as opposed to an offline optimization procedure, such as the GAs demonstrated earlier. This inverse problem formulation is over-viewed in [60], where the HMsFEM scheme is coupled with an unscented Kalman filter for system identification under availability of sparse measurements.

Conclusions

Polymeric textiles comprise a highly effective solution for shielding cultural heritage structures, such as masonry or natural-stone buildings, from natural hazards and extreme events. Motivated by an extensive experimental campaign on the application of a polymeric textile full-cover solution on a natural-stone building, this work outlines numerical strategies that are required for (i) the identification and characterization of the resulting composite structure in the elastic and plastic regime, and (ii) the accelerated and multiscale simulation of such systems, for facilitating optimization, reliability and risk assessment tasks.

- Firstly, the experimental campaign and the measurements gathered within the context of the EU funded project Polymast are overviewed, carried out for different stages of the building's life (unreinforced, damaged/repaired and retrofitted). Subsequently, the aggregated measurements serve for updating, or fine-tuning, computational models of the masonry-textile composite system.
- Initially, a finite-element based analysis is performed, adopting configurations of varying refinement, and therefore fidelity. The numerical representations are maintained at a mesoscopic scale, in order to facilitate the coupling with computationally intensive schemes, such as heuristic optimization. It is shown that inclusion of refinement in the representation, increases the flexibility of the model and allows for a better tuning to the true system and the experimental results. Nonetheless, the analysis proves computationally costly, and almost prohibitive when cast in the context of condition and risk assessment, i.e., when approached from a probabilistic assessment viewpoint.
- To alleviate this issue, particularly when nonlinear dynamic analysis is required, the Hysteretic Multiscale FEM (HMsFEM) methodology, developed in previous works of the authors [54], is demonstrated as a viable reduced order model. The benefits of the proposed procedure are two fold. (i) On one hand, a significant reduction is achieved in the overall computation, in this case resulting in an order of magnitude decrease in overall analysis time. (ii) On the other hand, and despite reduction in computation, precision and access to the information of the finer scale is maintained.

Naturally, the numerical scheme to be adopted should form a compromise between computational efficiency and complexity, which depends on the target of the simulation. The multiscale scheme discussed here is particularly meaningful when seeking to perform analyses which necessitate multiple iterations or multiple samples, as is often the case in risk and reliability assessment. Further benefits may be harnessed when this is coupled with monitoring technologies, which deliver not only global information, e.g., accelerations, but additionally local information, e.g., strains.

Acknowledgements Dr. Fuggini would like to gratefully acknowledge the support of European Community's Seventh Framework Programme [FP7/2007-2013] for access to Eucentre under grant agreement N 227887. Prof. Chatzi and Prof. Triantafyllou would like to gratefully acknowledge the support of the Swiss National Science Foundation under Research Grants #200021_146996, #200021_153379.

References

1. Thomas K (1996) Masonry walls: specification and design. Butterworth Heinemann, Ofxord
2. Salonikios T, Karakostas C, Lekidis V, Anthoine A (2003) Comparative inelastic pushover analysis of masonry frames. Eng Struct 25(12):1515–1523
3. Massart TJ, Peerlings RHJ, Geers MGD (2007) An enhanced multi-scale approach for masonry wall computations with localization of damage. Int J Numer Meth Eng 69(5):1022–1059
4. Oliveira DV, Lourenço PB, Roca P (2006) Cyclic behaviour of stone and brick masonry under uniaxial compressive loading. Mater Struct 39(2):247–257
5. Oliveira DV, Silva RA, Garbin E, Lourenço PB (2012) Strengthening of three-leaf stone masonry walls: an experimental research. Mater Struct 45(8):1259–1276
6. Mojsilović N (2011) Strength of masonry subjected to in-plane loading: a contribution. Int J Solids Struct 48(6):865–873
7. FEMA306 (1999) Evaluation of earthquake damaged concrete and masonry wall buildings-basic procedures, Technical Report, Federal Emergency Management Agency (FEMA), FEMA 306, Prepared by Applied Technology Council
8. Paulay T, Priestly MJN (2009) Seismic design of reinforced concrete and masonry buildings. Wiley
9. Triantafillou TC, Fardis MN (1997) Strengthening of historic masonry structures with composite materials. Materials and Structures 30(8):486–496
10. Krevaikas TD, Triantafillou TC (2005) Masonry confinement with fiber-reinforced polymers. J Compos Construct 9(2):128–135
11. Li T, Galati N, Tumialan J, Nanni A (2005) Analysis of unreinforced masonry concrete walls strengthened with glass fiber-reinforced polymer bars. ACI Struct J 102(4), 569–577, cited by 14
12. Triantafillou T (2011) Textile-Reinforced mortars (TRM). Springer: London, London, pp 113–127
13. Faella C, Martinelli E, Nigro E, Paciello S (2004) Tuff masonry walls strengthened with a new kind of c-frp sheet: experimental tests and analysis. In: Proceedings of the 13th world conference on earthquake engineering, Paper no. 923
14. Nurchi A, Valdes M (2005) Strengthening of stone masonry columns by means of cement-based composite wrapping. In: CCC 2005: 3rd International conference on composites in construction. Lyon, France, July 2005 (H. P, ed.), pp 1189–1196
15. Bischof P, Suter R, Chatzi E, Lestuzzi P (2014) On the use of cfrp sheets for the seismic retrofitting of masonry walls and the influence of mechanical anchorage. Polymers 6(7):1972
16. Messervey T, Zangani D, Fuggini C (2010) Sensor-embedded textiles for the reinforce-ment, dynamic characterisation, and structural health monitoring of masonry structures. In: Proceedings of the 5th EWSHM 2010, June 28 July 2, Sorrento, Italy, pp 1075–1282
17. Stempniewski L (2011) Polyfunctional technical textiles for the protectionand monitoring of masonry structures against earthquakes. Technical Report, Seventh Framework Programme, Capacities Specific Programme, Research Infrastructures, Project No.: 227887
18. Fuggini C, Chatzi E, Zangani D (2013) Combining genetic algorithms with a meso-scale approach for system identification of a smart polymeric textile. Comput-Aided Civil Infrastruct Eng 28(3):227–245
19. Soares C, de Freitas M, Arajo A, Pedersen P (1993) Identification of material properties of composite plate specimens. Compos Struct 25(14):277–285
20. Frederiksen P (1997) Experimental procedure and results for the identification of elastic constants of thick orthotropic plates. J Compos Mater 31:360–382
21. Rikards R, Chate A, Steinchen W, Kessler A, Bledzki A (1999) Method for identification of elastic properties of laminates based on experiment design. Compos Part B Eng 30(3):279–289
22. Morton J, Haig G (2011) Designers' guide to eurocode 6: design of masonry structures: EN 1996-1-1: General rules for reinforced and unreinforced masonry. ICE Publishing, London

23. Binda L, Pina-Henriques J, Anzani A, Fontana A, Loureno P (2006) A contribution for the understanding of load-transfer mechanisms in multi-leaf masonry walls: testing and modelling. Eng Struct 28(8):1132–1148
24. Anthoine A (1992) In-plane behaviour of masonry: a literature review. Report Eur 13840 En, commission of the european communities. Tech Report, JRC–Institute for Safety Technology, Ispra, Italy
25. Lourenco P (1996) Computational strategies for masonry structures. PhD Thesis, Delft University of Technology, Delft, The Netherlands
26. Tomazevic M, Lutman M (1996) Seismic behavior of masonry walls: Modeling of hysteretic rules. Journal of Structural Engineering 122(9):1048–1054
27. Chen S-Y, Moon F, Yi T (2008) A macroelement for the nonlinear analysis of in-plane unreinforced masonry piers. Eng Struct, 30(8), 2242–2252, Seismic reliability, analysis, and protection of historic buildings and heritage sites
28. Ferreira MAR, Lee HKH (2007) Multi-scale modeling: a bayesian perspective. Springer, Springer Series in Statistics, New York
29. Babuška I (1975) Homogenization approach in engineering. Technical Report ORO–3443-58; TN-BN–828 United States; NSA-33-022692
30. Nicot F, Darve F (2005) Hazards GNV of Structures A multi-scale approach to granular materials Mechanic Mater 37(9), 980–1006
31. Elsayed MS, Pasini D (2010) Multiscale structural design of columns made of regular octet-truss lattice material. Int J Solids Struct 47(1415):1764–1774
32. Mangipudi K, Onck P (2011) Multiscale modelling of damage and failure in two-dimensional metallic foams. J Mechanic Phys Solids 59(7):1437–1461
33. McDowell DL (2010) A perspective on trends in multiscale plasticity. Int J Plastic 26(9), 1280–1309. Special Issue In Honor of David L. McDowell
34. Hughes TJ, Scovazzi G, Bochev PB, Buffa A (2006) A multiscale discontinuous galerkin method with the computational structure of a continuous galerkin method. Comput Methods Appl Mech Eng 195(1922):2761–2787
35. Song J-H, Belytschko T (2009) Multiscale aggregating discontinuities method for micro-macro failure of composites. Compos Part B Eng 40(6), 417–426. Blast/Impact on engineered (nano)composite materials
36. Kanouté P, Boso DP, Chaboche JL, Schrefler BA (2009) Multiscale methods for composites: a review. Arch Comput Methods Eng 16(1):31–75
37. Belytschko T, de Borst R (2010) Multiscale methods in computational mechanics. Int J Numer Methods Eng 83(8–9):939–939
38. Xu X, Graham-Brady L (2005) A stochastic computational method for evaluation of global and local behavior of random elastic media. Comput Methods Appl Mech Eng 194(4244):4362–4385
39. Tootkaboni M, Graham-Brady L (2010) A multi-scale spectral stochastic method for homogenization of multi-phase periodic composites with random material properties. Int J Numer Methods Eng 83(1):59–90
40. Xu XF, Hu K, Beyerlein IJ, Deodatis G (2011) Statistical strength of hierarchical carbon nanotube composites. Int J Uncertain Quant 1(4):279–295
41. Massart T, Peerlings R, Geers M (2004) Mesoscopic modeling of failure and damage-induced anisotropy in brick masonry. Eur J Mech A/Solids 23(5):719–735
42. Massart T, Peerlings R, Geers M, Gottcheiner S (2005) Mesoscopic modeling of failure in brick masonry accounting for three-dimensional effects. Eng Fract Mech 72(8):1238–1253
43. Efendiev Y, Hou TY (2009) Multiscale finite element methods vol 4 of surveys and tutorials in the applied mathematical sciences. Springer
44. Zhang HW, Wu JK, Lv J (2012) A new multiscale computational method for elasto-plastic analysis of heterogeneous materials. Comput Mech 49(2):149–169
45. Karapitta L, Mouzakis H, Carydis P (2011) Explicit finite-element analysis for the in-plane cyclic behavior of unreinforced masonry structures. Earthquake Eng Struct Dynam 40(2):175–193

46. Wen Y (1976) Method of random vibration of hysteretic systems. J Eng Mech Div 102:249–263
47. Mayergoyz I (1998) Generalized preisach model of hysteresis. IEEE Trans Magnet 24(1):212–217
48. Erlicher S, Bursi OS (2008) Boucwen-type models with stiffness degradation: thermodynamic analysis and applications. J Eng Mech 134(10):843–855
49. Sivaselvan MV, Reinhorn AM (2000) Hysteretic models for deteriorating inelastic structures. J Eng Mech 126(6):633–640
50. Visintin A (1994) Differential models of hysteresis. In: Applied mathematical sciences, vol. 111, Springer
51. Ikhouane F, Rodellar J (2007) Systems with hysteresis: Analysis, identification and control using the Bouc-Wen model. Wiley, New York
52. Spanos PD, Kougioumtzoglou IA (2011) Harmonic wavelet-based statistical linearization of the Bouc-Wen hysteretic model. Taylor and Francis Group, pp 2649–2656
53. Triantafyllou SP, Koumousis VK (2012) A hysteretic quadrilateral plane stress element. vol. 82, 10–, pp 1675–1687
54. Triantafyllou S, Chatzi E (2014) A hysteretic multiscale formulation for nonlinear dynamic analysis of composite materials. Comput Mech 54(3):763–787
55. Zienkiewicz OC, Taylor RL, Zhu J (2005) The finite element method: its basis and fundamentals, 6th edn. Elsevier, Amsterdam
56. Triantafyllou S, Chatzi E (2014) Risk analysis of composite structures by subset estimation using the hysteretic multiscale finite element method, 157, pp 1564–1573
57. Triantafyllou SP, Chatzi EN (2015) Towards a multiscale scheme for nonlinear dynamic analysis of masonry structures with damage, Cham: Springer International Publishing pp 165–198
58. Nemat-Naser S (1982) On finite deformation elasto-plasticity. Int J Solids Struct 18(10):857–872
59. Washizu K (1983) Var Methods Elasticity Plastic. Pergamon Press, Oxford
60. Triantafyllou SP, Chatzi EN (2015) A hysteretic multiscale formulation for validating computational models of heterogeneous structures. J Strain Anal Eng Design
61. Iwan WD (1967) On a class of models for the yielding behavior of continuous and composite systems. J Appl Mech 34(3):612–617
62. Erlicher S (2003) Hysteretic degrading models for the low-cycle fatigue behaviour of structural elements: theory, numerical aspects and applications. PhD thesis, Department of Mechanical and Structural Engineering, University of Trento, Italy
63. Triantafyllou S, Koumousis V (2014) Hysteretic finite elements for the nonlinear static and dynamic analysis of structures. J Eng Mech 140(6), 04014025–1– 04014025–17
64. Lubliner J (2008) Plasticity theory. Dover Publications, New York
65. Foliente GC, Singh MP, Noori MN (1996) Equivalent linearization of generally pinching hysteretic and degrading systems. Earthquake Eng Struct Dynam 25:611–629
66. Zangani D (2010) Final report -polytect (polyfunctional technical textiles against natural hazards), project no. nmp2-ct-2006-026789. Technical Report D' Appolonia S.p.A
67. Peeters B, De Roeck G (Feb 2001) Stochastic system identification for operational modal analysis: a review. J Dynam Syst. measurem. Control 123:659–667
68. James GH, Carne TG, Lauffer JP, Nord AR (1992) Modal testing using natural excitation. In: Proceedings of 10th Int. Modal Analysis Conference, San Diego
69. Juang J-N, Pappa RS (Sept 1985) An eigensystem realization algorithm for modal parameter identification and model reduction. J Guid Control Dynam 8:620–627
70. Brincker R, Zhang L, Andersen P (2000) Modal identification from ambient responses using frequency domain decomposition. In: Proceedings of the 18th SEM International Modal Analysis Con-ference, San Antonio
71. Ansys academic research, release 16.2
72. Carrols DI open source code http://www.cuaerospace.com/carroll/ga.html
73. http://peer.berkeley.edu/peer_ground_motion_database/. Accessed 20 May 2014
74. Ri. abaqus version 6.11 [computer software]. Dassault systmes simulia, providence

Monitoring and Maintenance of Customized Structures for Underground Environments: The Case of Gran Sasso National Laboratory

Francesco Potenza

Abstract In recent years, the monitoring and maintenance of underground structures has received increasing attention. This chapter presents a structural analysis of two customized prototypes necessary for conducting experimental research within the underground halls of the Gran Sasso National Laboratory (LNGS). These structures consist primarily of a steel-framed structure built within a water tank. Different models representing the fluid–structure interaction are also discussed. The structural monitoring is carried out through dynamic experimental tests in different configurations (prototype empty or filled with water) and load conditions (release, hammer, and environmental). The experimental modal information is used to perform manual updating of the numerical finite element models. Finally, a possible structural monitoring system is proposed, which consists primarily of a distributed fiber optic sensing system for the vaults of the experimental halls of the LNGS.

Keywords Structural monitoring · Experimental test · Modal identification
Numerical modeling · Dynamic testing

Introduction

Underground structures play an essential role in the infrastructure of modern society, and thus require comprehensive risk analysis to inform choices regarding the adoption of suitable monitoring and protective systems. Seismically induced vibrations are among the most important dynamic effects that must be monitored; however, these are often neglected in the context of underground versus superficial structures. Indeed, significant damage has been found after high-intensity earthquakes in both external and underground structures [1]. The collapse of the Daikai metro in Kobe (Japan) during the Hyogoken-Nambu earthquake in 1995 [2],

F. Potenza (✉)
DICEAA—Department of Civil, Construction-Architectural and Environmental Engineering, University of L'Aquila, L'Aquila, Italy
e-mail: francesco.potenza@univaq.it

E. Ottaviano et al. (eds.), *Mechatronics for Cultural Heritage and Civil Engineering*, Intelligent Systems, Control and Automation: Science and Engineering 92,
https://doi.org/10.1007/978-3-319-68646-2_16

the damage observed in the Central Taiwan highway tunnel during the Chi-Chi earthquake in 1999 [3], and damage to the Bolu gallery in Turkey discovered after the Koceali seismic event in 1999 are notable examples. Damage to underground structures resulting from earlier earthquakes has also been documented [4], including the Kanto earthquake of 1923, the Izu Oshima Kinkai earthquake of 1978, and the Niigataken Chuetsu earthquake of 2004. It is also worth mentioning cases where tunnels designed to withstand seismic input have shown good performance during an earthquake, such as the Bay Area Rapid Transit (BART) system in San Francisco, CA (USA) during the 1989 Loma Prieta earthquake. The tunnel's structural behavior can be reasonably approximated as an elastic beam. The civil superficial structures are designed against the seismic load induced by ground motion acceleration, while the tunnels must be able to support the deformation induced by the displacement of the surrounding soil. Seismic input or the movement of a fault can cause instability or liquefaction of the surrounding soil, resulting in damage to underground structures [5]. Seismic analysis can be approached in one of two ways: *free-field deformation* and *soil–structure interaction*. The first strategy involves the analysis of seismically induced ground deformation in the absence of structures or excavations. In the second approach, these deformations can be altered if the seismic waves pass through a structure, which thus requires a suitable means of modeling the soil–structure interaction [2]. In tunnel design, the tensions produced by three structural behaviors must be analyzed and verified: (1) axial tension and compression induced by seismic waves parallel to the longitudinal axis of the tunnel, (2) longitudinal bending due to the components of the seismic wave in a transverse direction relative to the axis of the tunnel, and (3) distortion of the cross section (roundness) produced by normal seismic waves (sussultatory component). Comparisons between numerical and physical models in the context of these breaking mechanisms can be found in [6–10]. In [11], analysis of the deformations of a circular cross section is addressed, while a simplified analytical procedure for assessing the strain induced by a seismic motion on a cross section is proposed in [12]. In [13], the authors analyze the seismic response of circular tunnels embedded in a poroelastic medium, which considers the influence on the internal force due to the difference between the stiffness of the seismic joint and that of the linear pieces of the ring. In [14], a seismic analysis is pursued in which the tunnel is modeled as Timoshenko's beam on elastic soil. In the latter report, the seismic waves are widespread and parallel along the longitudinal axis of the tunnel, and the authors found a simplified closed-form solution capable of determining the displacements, deformations, and tensions. Numerical studies germane to the modeling of soil–structure interaction are included in [7, 15 , 16]. In other works, structural analysis is conducted using a two-dimensional finite element model, in which the modeling of the soil's constitutive behavior strongly affects the value of the internal tensions [17, 18]. Analysis of the residual load-bearing capacity of tunnels affected by earthquakes is discussed in [19, 20] . The first study proposes an empirical method for assessing the seismic capacity of the tunnels, while the second illustrates the application of non-linear static analysis (pushover) for seismic analysis and tunnel design. Analysis of the seismic local response within tunnels is poorly represented in the literature, with only a few works that address the problem experimentally, such as in [21, 22]. Finally, an additional area of research relates to structures immersed in water, i.e. Submerged Floating Tunnels (SFT), which investigates the effects of the soil's asynchronous motion, transferred vertically on the different water layers [23].

Gran Sasso National Laboratory (LNGS)

The Gran Sasso National Laboratory (LNGS) of the National Institute of Nuclear Physics (Fig. 1) represents the world's largest and most important underground research center, dedicated to the study of the fundamental constituents of matter and the laws that govern them. Inside the underground experimental halls are customized prototypes for the experimental study of phenomena that occur very rarely and are difficult to observe. The most fascinating questions that scientists there seek to answer relate to the birth of the universe, the behavior of the stars, and the nature of subatomic particles (neutrinos). Other experiments, like the last-generation DARKSIDE and XENON, are aimed at the direct detection of dark matter particles. The advanced technology developed for such studies could also be useful for other

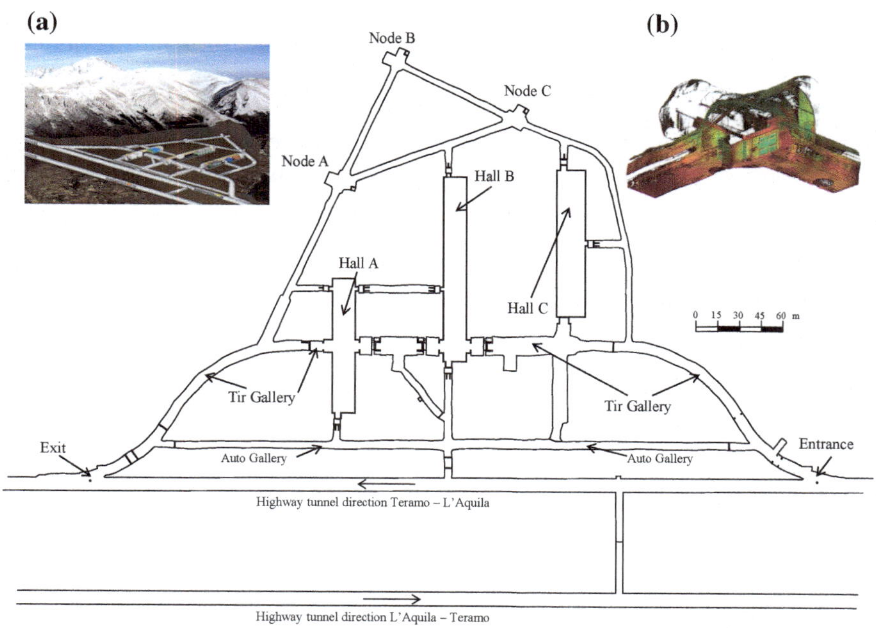

Fig. 1 General plan of the underground Gran Sasso National Laboratory. **a** 3D view inside the Gran Sasso mountain. **b** Laser scanner of node B

research fields, such as medical diagnosis. Currently, almost 1000 researchers from 32 countries associated with about 20 different experiments work at the LNGS.

The underground halls, located within the Gran Sasso mountain (the most important and highest massif of the central Apennines, fully located in the Abruzzo region), are positioned below about 1400 m of rock mass, and are accessible only through the highway tunnel, which serves as both entrance and exit. They have been equipped with all of the safety systems and apparatus needed for use by authorized persons. The natural rock cover provided by the mountain guarantees the ability to observe rare events by reducing the flux of cosmic rays. The inception of the LNGS is attributable to the famous Italian nuclear physicist Antonino Zichichi who obtained the public funding necessary for the realization of the underground labs. He called the underground experimental halls "*cosmic silence*". Today, management of the underground labs is carried out by a series of services, including the Research Division, Technical & General Service Division, Prevention & Protection Services, Public Affairs & Scientific Information, Administration and Directorate, all under the supervision of the LNGS director. The offices of these services are in external labs in Assergi, a small town located just outside the highway tunnel. A set of facilities including a canteen, sheds for storage, conference rooms, and small dorms is also located there. With these many features, the LNGS is clearly the most important laboratory of its kind.

The underground labs consist of three large halls (halls A, B, and C; see Fig. 1) as well as service tunnels, with a total area of about 18,000 m^2. To give an idea of the expansiveness of the labs, the dimensions of each hall are provided in Table 1. The vastness of the experimental halls facilitates large-scale prototype experiments which require continuous monitoring by specialized technicians. Two of the main service tunnels, the Tir and Auto galleries, can be seen in the general plan shown in Fig. 1. In the former, the peak of the transverse section is higher than that of the second, because it is dedicated to the passage of large vehicles, such as tirs or trucks, necessary for the transport of both heavy carpentry (i.e. steel structural elements) and chemical materials used in the latest-generation experiments (e.g. liquid argon), as well as buses used for guided tours on weekends. The Auto Gallery service tunnel, on the other hand, is used for a shuttle that transports researchers, technicians, and workers back and forth between the external and underground laboratories at intervals of approximately 20 min. The underground experimental tunnels, built in 1982, were designed by Pietro Lunardi and the main design choices are described in his published report [24]. From a structural and

Table 1 Dimensions of the main experimental halls

Experimental halls	Length [m]	Width [m]	Height [m]
A	100	15	16.50
B	126	15	16.50
C	100	18	18.00

geotechnical point of view, the most important issues considered during the design process regarded the large rock mass cover (about 1400 m) in relation to both the rock resistance and the large dimensions of the excavations, and the very thin coating facilities used to improve the stability of the rock provided also by special consolidation interventions. Indeed, tests carried out to estimate the limestone resistance showed its fragile nature, and the results of structural analysis revealed the possibility to overcome the load-bearing capacity of the same limestones during the excavations. For this reason, Lunardi envisioned an armed rock mass through the insertion of passive anchors that would attach the discontinuity points and increase the shear strength of the rock. The anchoring material is martensitic stainless steel (AISI 420), 30.6 mm in diameter and 6–8 m in length, arranged in a mesh of 2 m per side. In areas where the rock was more damaged, anchors 10–12 m in length, composed of a denser mesh, were adopted.

A detailed description of the rock mass and the area affected by the underground labs from a seismological and geological perspective can be found in the geological report by P.G. Catalano, "*I Laboratori Sotterranei, Relazione idrogeologica e sistemi di drenaggio e canalizzazione*". The report shows that the Gran Sasso mountain is intersected by three large and important faults, the *Valle Fredda*, *Fontari*, and *Overthrust*. The last of these crosses a zone of the underground labs called the Interferometric Station, which is composed of three minor tunnels, arranged triangularly with vertices defined as nodes A, B, and C (see Fig. 1). The aim of this configuration is to intercept two points of the Overthrust fault to monitor its small displacements. A Michelson interferometer was placed along the tunnels linking nodes A and B and nodes B and C to measure the distance between two fixed points. Two years ago, after 15 years of monitoring, the interferometer was removed to make room for new experimental proposals.

Numerical Modeling of the Experimental Prototypes

The underground labs represent an ideal environment in which to carry out experiments capable of detecting very rare phenomena. As mentioned above, the rock above the labs provides a natural filter to reduce the cosmic ray flux by a factor of one million. However, this is not adequate, because the large detectors need to be shielded from the natural radioactivity emitted by the rock inside the labs that could skew measurements or jam signals. Water tanks have provided an excellent means of shielding the detectors, which are fully immersed and protected by the demineralized water. Currently, in each underground hall there are experimental prototypes—GERDA, XENON, and DARKSIDE in halls A, B, and C, respectively—that make use of this configuration. Figure 2 presents two synthetic sketches of the layout of the XENON and DARKSIDE experiments. Here, the structural interactions between the main subsystems that constitute the complete prototype will be described using various finite element models.

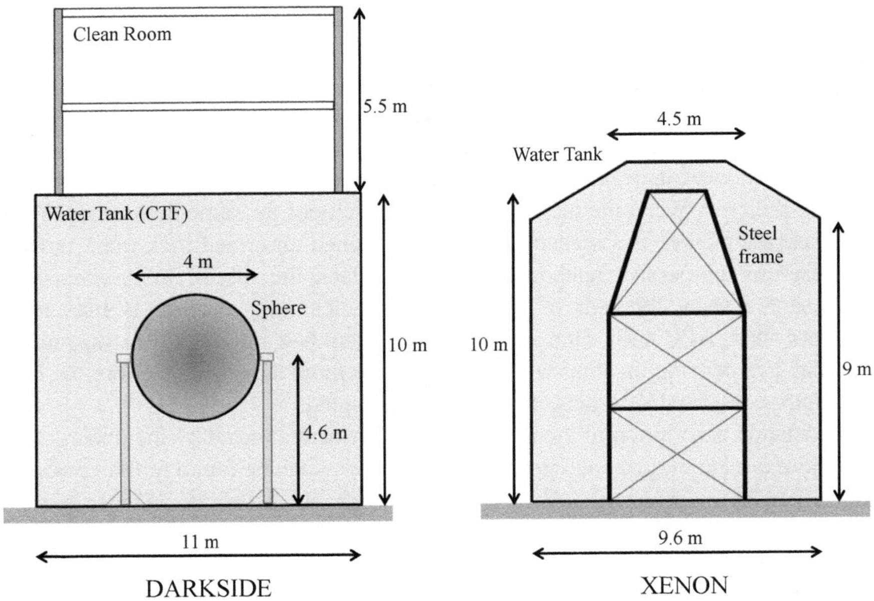

Fig. 2 Synthetic sketches of the experiments DARKSIDE and XENON

DARKSIDE Experiment

The substructures that comprise the DARKSIDE experiment include a water tank, built in 1994 and referred to as the Counting Test Facility (CTF), a sphere, within which is placed the detector and a clean room, formed by a steel frame collocated above the roof of the tank and completely connected to the same tank. The steel water tank has a circular base 11 m in diameter, with a total height of 10 m and a flat roof. The sphere, made of steel, is 4 m in diameter and is supported by a system of four columns with circular hollow transverse section (with an external diameter of 0.32 m and thickness of 12 mm). The arrangement and the total height of the columns put the sphere's center at a height of 4.60 m. The tank comprises a series of five cylindrical shells, the first two 8 mm thick and the remaining three 6 mm thick. The tank's bottom sheet is made of 8-mm-thick steel, and it is entirely positioned above a screed layer. The roof is composed of a grid of beams capable of supporting a distributed load of 4 KN/m^2. The structure that supports the sphere is braced by a system of horizontal and oblique elements arranged in identical fashion on each of the four sides. The bottom plates of the columns that support the sphere are incorporated within the tank's bottom sheet. The structural analysis aims to reveal the tank–sphere interaction that occurs through the liquid between the two substructures and with the tank bottom. In the following, the first two models are implemented to analyze the behavior of the tank and sphere separately, and are useful for simulating cases where the structures are empty or filled with liquid. Two additional models are used to evaluate the variation in the structural performance after coupling the tank and sphere and the construction of the clean room above the tank. The models can also represent the possible configurations of the prototype along its life cycle: (1) tank and sphere both empty, (2) tank empty and sphere full, (3) tank and sphere both full. All have been implemented as linear models.

The finite element models for the tank and sphere are illustrated in Fig. 3. In each, the geometry is the same as in the real case, while the description of the constitutive law of the material uses the nominal mechanical characteristics of the steel. The tank model, Fig. 3a, is completely realized using 1600 two-dimensional elements (shells). Boundary conditions have been applied to the bottom and top nodes. In the first, hinges have been inserted, while the top nodes linked together by a diaphragm. In the case of an empty tank, the mass assigned to each node derives from the self-weight of the finite elements. For the full tank, the mass of a water column with a height of 9.60 m is distributed on each node based on the area of influence of each element. The roof mass is concentrated in the central node on the top. The model of the sphere, shown in Fig. 3b, comprises 2000 two-dimensional elements used to describe the behavior of the sphere and its support. The four columns and the horizontal and oblique braces are modeled with mono-dimensional (beam) elements. The only boundary conditions are supports fixed to the bottom nodes.

To simulate the full sphere, a diaphragm has been inserted that links all nodes of the equatorial zone and the node in the sphere's center, which is assigned the total mass of water.

For all models, a specific weight of 8.9 KN/m^3 has been selected for the demineralized water. The dynamic behavior of the two models is evaluated using modal analysis. Table 2 shows the periods and shapes of the fundamental modes for each

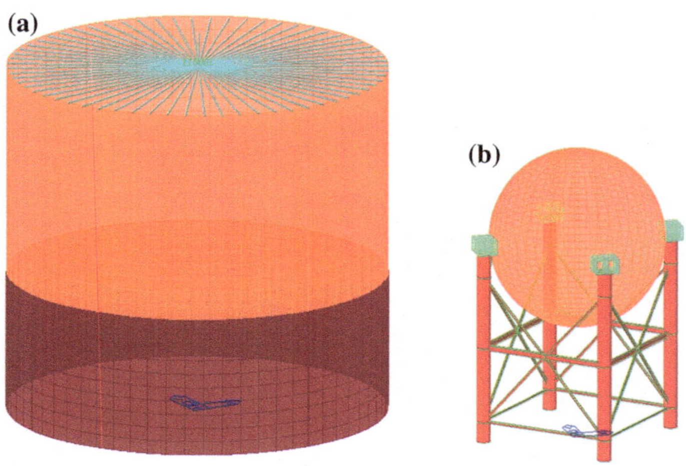

Fig. 3 Finite element models: water tank (**a**) and sphere (**b**)

Table 2 Fundamental modes for the tank and sphere

Tank				Sphere			
Empty		Full		Empty		Full	
Mode	T [*s*]	Mode	T [*s*]	Mode	T [*s*]	Mode	T [*s*]
T	0.0244	T	0.1400	T	0.0735	T	0.2254
T	0.0244	T	0.1400	T	0.0735	T	0.2254
–	–	–	–	R	0.0593	R	0.0588

model in different configurations (empty and full). The models representing empty structures show very low periods of 0.0244 s and 0.0735 s for the tank and sphere, respectively. Of course, the masses added to models for the water-filled structures make the modes more flexible. Indeed, the periods increase more than threefold, with the largest increase in the tank for which they are equal to 0.1400 s. The modal shapes are primarily translational (T): the total translational excited mass considering only these two modes is very close to 85%. It is worth highlighting the presence of a rotational (R) mode for the sphere for which the percentage of rotational excited mass is nearly 80%, while the percentage is less than 1% for the others modes.

The structural interaction between the tank and the sphere is achieved by coupling the two models previously described. The interaction can occur both from the bottom sheet and the internal liquid interposed between the two structures. With these considerations in mind, the following details have been incorporated in the models: (1) modeling of the bottom plate with evidence of particular meshing on the interaction zone between the base of the sphere's columns and tank bottom (Fig. 4a), and (2) a diaphragm that links the center and the equatorial nodes of the sphere with the corresponding nodes (i.e. same height) of the tank's lateral surface

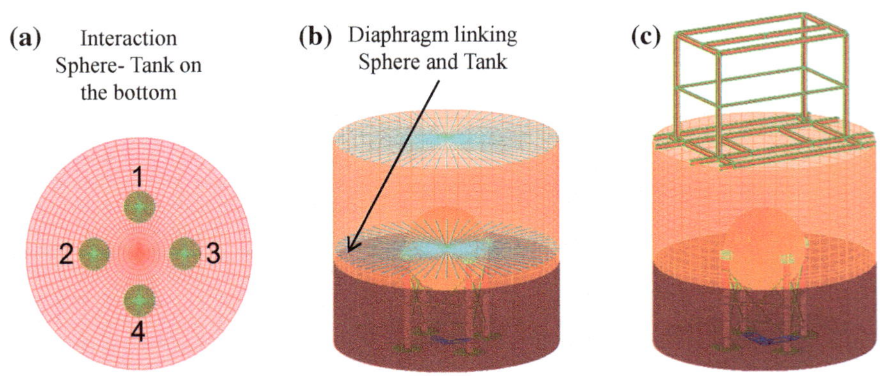

Fig. 4 Plate inserted at the base of the tank (**a**). Models representing the tank–sphere interaction (**b**) and the influence of the clean room above the tank (**c**)

Table 3 Fundamental modes for the tank–sphere models

Model A		Model B		Model C	
Shape	Period	Shape	Period	Shape	Period
Sphere	0.2096	Sphere	0.6076	Sphere–tank	0.1371
Sphere	0.2096	Sphere	0.6076	Sphere–tank	0.1371
Tank	0.0244	Tank	0.0244	–	–
Tank	0.0244	Tank	0.0244	–	–

(Fig. 4b). In addition, vertical linear springs with an elastic coefficient equal to 1 KN/cm^3 have been applied to the nodes of the bottom plate, while the nodes of the edges of the same plate are bounded by hinges. The nodes on the top are linked by a diaphragm. The constitutive law of the material and the mass distribution for both empty and full configurations are the same as in the previous decoupled models. The last model has been implemented considering the steel frame of the clean room collocated above the tank (Fig. 4c). This substructure is modeled using only mono-dimensional beams with the same material properties as the plate-shells used for the tank and sphere. Table 3 displays the fundamental modes for different configurations of the tank–sphere model (Fig. 3b). In models A and B, the interaction is achieved exclusively through the bottom plate. In the first model, the tank and sphere are both empty, and in the second, only the sphere is full of liquid. In model C, an additional interaction between the two substructures occurs through the liquid inside the tank due to the diaphragm linking the tank and sphere (Fig. 4b). In the first two models, the fundamental modal shapes are defined separately by the two substructures, with the tank's modes stiffer than those of the sphere. In these two cases the effect of the interaction makes the sphere's mode more flexible due to a different boundary condition at the bottom of the columns, as illustrated in Fig. 4a. The modes involving only the tank are the same in terms of shape and period for both models A and B and also for the model in Fig. 3a.

In model C, only two main and translational global modes are present, with identical periods of 0.1371. Other modes are not reported due to a very low participation mass coefficient. Finally, in the last model, the insertion of the clean room does not substantially alter the behavior described in model C. The effect of the steel frame above the tank separates the two periods slightly due to the imperfect symmetry of the frames (0.1392 s and 0.1384 s), and both become slightly more flexible.

This comprehensive numerical analysis highlights important changes in the dynamic behavior of the DARKSIDE prototype along its life cycle. There are no conservative models capable of taking into account all situations, but they can be more rigid or more flexible depending on the configuration. Indeed, the interaction makes the tank more flexible and the sphere more rigid relative to the corresponding independent cases.

XENON Experiment

The configuration of the XENON experiment is very similar to that of the DARKSIDE. The tank is cylindrical in shape, with a roof in the shape of a truncated cone. The steel frame collocated inside the tank that supports the cryostat has a quadratic layout plan and a total height of 10.30 m (Fig. 2). There are three elevations, with the first two approximately 3 m in height and the third about 4 m. In the latter, the columns are slightly inclined from the vertical. The model is intended to describe the fluid–structure interaction in order to assess the effect of the sloshing modes on the internal substructure.

Finite element models (Fig. 5a) were implemented using two-dimensional elements (plate) for the tank's lateral surface and for the elements of the internal steel frame, while the liquid was modeled using solid elements (brick). Nominal

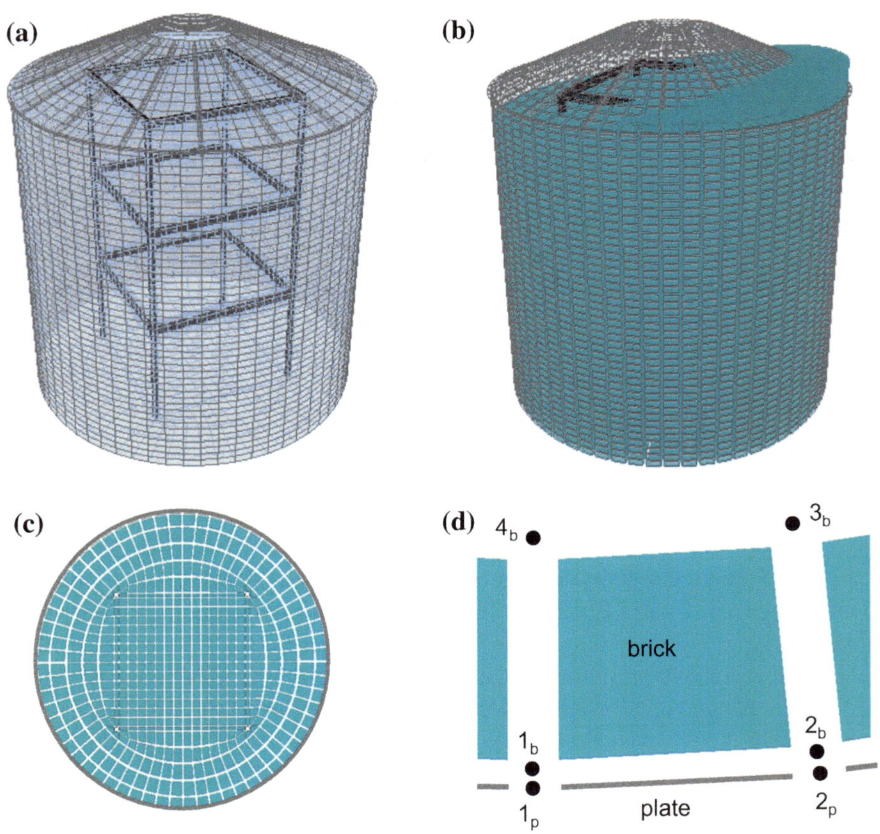

Fig. 5 Finite element model representing the XENON prototype (**a**). Modal shape involving only the brick elements, sloshing (**b**). Sectional planar view of the meshing (**c**). Internal boundary condition between brick and plate (**d**)

mechanical characteristics were used for the steel, while the elastic modulus, *E*, and the Poisson coefficient, ν, for the brick elements were calculated by setting the bulk modulus, K_ν, and the shear modulus, *G*, equal to 2.11 GPa and close to zero, respectively. Indeed, considering the material as elastic, linear, isotropic, and homogenous, the following expressions can be applied:

$$K_\nu = \frac{E}{[3(1-2\nu)]} \quad G = \frac{E}{[2(1-2\nu)]} \tag{1}$$

where 6.33e-5 GPa and 0.499995 are obtained for E and ν, respectively, on the basis of these hypotheses, a Poisson coefficient very close to 0.5 is representative of an incompressible material. The main choice of model is related to the internal boundary condition between brick and plate elements. Figure 5d shows, from a bird's-eye view, a generic brick element adjacent to a plate element belonging to the lateral surface of the tank. In this case, the nodes of the plate (1_p and 2_p) do not correspond to those of the brick (1_b and 2_b), but they have been separated along a radial direction a distance equal to 1 mm. The constraint inserted between these nodes is such that the displacements are only equal in radial directions. The same constraints have been applied between the plates of the internal steel frame and the brick elements. In this last case, displacements are equal only in the normal direction of the plate surface. The nodes on the base have been fixed in all six degrees of freedom. It is worth noting that to facilitate the internal meshing (Fig. 5c), the geometry along the height of the steel frame has been approximated, especially in the last elevation. This model is able to reproduce modal shapes very similar to the sloshing modes, as is clearly visible in Fig. 5b. One of the main modes, with a modal shape involving only the brick elements, is reported in the latter figure. In this case, the period is very high (about 3 s) and therefore very far from the range of periods observed in the previous case (about 0.13 s). Based on these results, the effects of the impulsive and convective motion can be decoupled and analyzed separately. In a spectral response analysis, the contribution of these modes could reasonably be neglected due to their high period values.

Dynamic Testing and Structural Monitoring

Among the different methods used to certify the suitability of a new structure, dynamic tests constitute an attractive and non-destructive procedure that is widely used, especially in the area of scientific research [25–27]. After construction of the sphere, 2 days of an experimental campaign were carried out to assess its dynamic behavior. The tests on the first and second day were carried out with the sphere empty and full, respectively. The general setup consists of two mono-directional accelerometers (Columbia SA-107LN, full-scale ±2 g) and a dSPACE 1103 PPC controller board that is used for data acquisition. The experimental setups, designed to extract the largest amount of information possible, are as follows. In setup 1, one

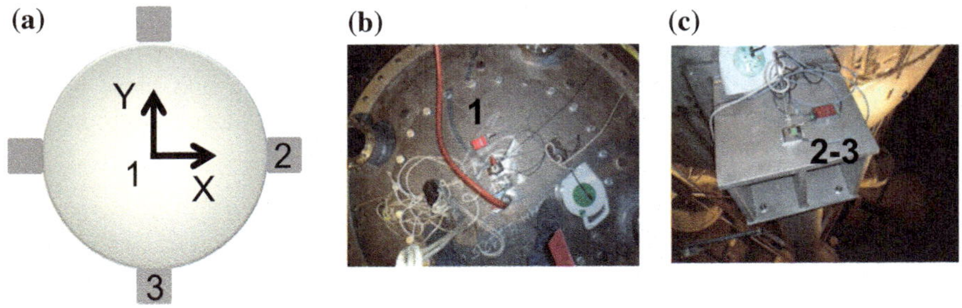

Fig. 6 Generic setup for the experimental tests (**a**). Photo of the accelerometers collocated in position 1 (north pole, **b**) and in positions 2 or 3 (column, **c**)

accelerometer was placed on top of the sphere (north pole, position 1 in Fig. 6a) and the other was placed above a column (position 3 in Fig. 6a), and both were positioned in X-direction. In setup 2, the direction of the accelerometer in position 1 was changed from X to Y, while both the position and direction were changed (position 2 in Fig. 6a and Y-direction). These setups are the same for both the empty and full configurations, and they are designed primarily to estimate the main modal frequencies for the translational mode in the X-direction (setup 1) and Y-direction (setup 2). In addition, the measurements recorded by the accelerometers in positions 2 and 3 have been processed to assess the frequencies associated with the torsional modes. Figure 6b and c shows two photos of the accelerometers in positions 1 and 2 or position 3, respectively. During each day of the experimental campaign, release, hammer, and environmental dynamic tests were performed. In the first case, a displacement in X and Y directions for setups 1 and 2, respectively, was quasi-statically imposed and then suddenly released. In the hammer test, a shot in the head of the column was given in X and Y directions for setups 1 and 2, respectively. The third experimental test involved 15 min of data acquisition in the presence of environmental noise. In Fig. 7, for the sake of brevity, the FFTs carried out on the registrations obtained in the empty configuration are shown only for the release and hammer tests for both setups 1 and 2. Nevertheless, these results provide enough information to understand the experimental dynamic structural behavior. Indeed, Fig. 7a and b clearly shows the main picks for the release and hammer tests, respectively, using the arrangement in setup 1. In these two graphs, the only pick visualized for the accelerometer placed in the north pole (continuous line) corresponds to 3 Hz, while for the accelerometer placed above the column (dashed line), two picks corresponding to 3 Hz and 11.33 Hz are evident. Consequently, it would seem reasonable to attribute the 3 Hz and 11.33 Hz frequencies to the first translational mode in X-direction and the first torsional mode, respectively. Similar results are found for the graphs in Fig. 7c and d for the release and hammer tests, respectively, carried out for setup 2. Indeed, the torsional mode appears to be 11.3 Hz, more or less, while the main mode in Y-direction is collocated in 2.67 Hz. Given the symmetry of the structures in elevation and the reliability of the nominal values of the steel's mechanical characteristics, differences between the frequencies

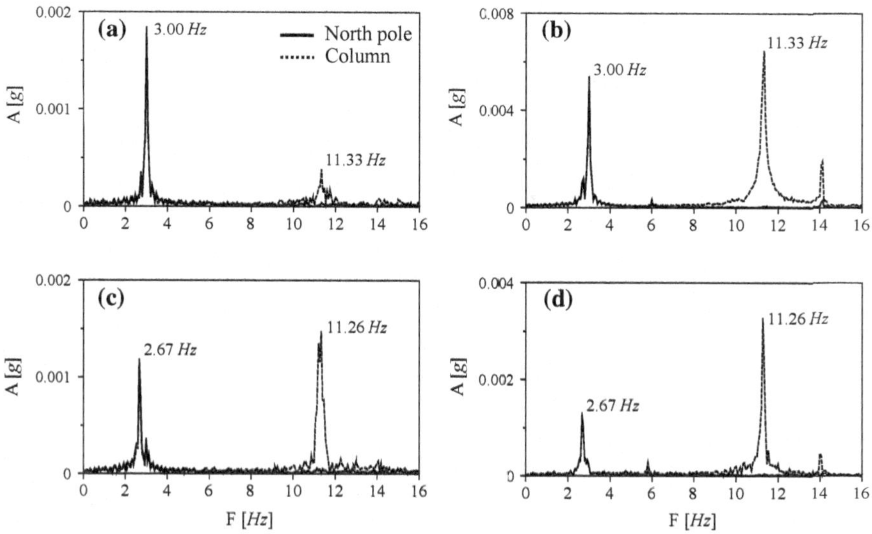

Fig. 7 FFTs of the experimental tests on the sphere: release in setups 1 (**a**) and 2 (**c**) and hammer in setups 1 (**b**) and 2 (**c**)

of the two main translational modes—which should be the same—may result from the constraints at the base of the columns. Thus manual updating of the model has been performed by varying the elastic coefficient of the vertical linear springs inserted under the nodes of the bottom plate. Model A, described above (both tank and sphere empty), has been improved, and good agreement is obtained between the main numerical and experimental modes whose model shape involves only the sphere (see Fig. 8, in which, for simplicity, only the sphere's deformation is reported because the tank remains un-deformed). It is worth noting that even if the difference for the torsional mode is quite high (approximately 20%), the result may

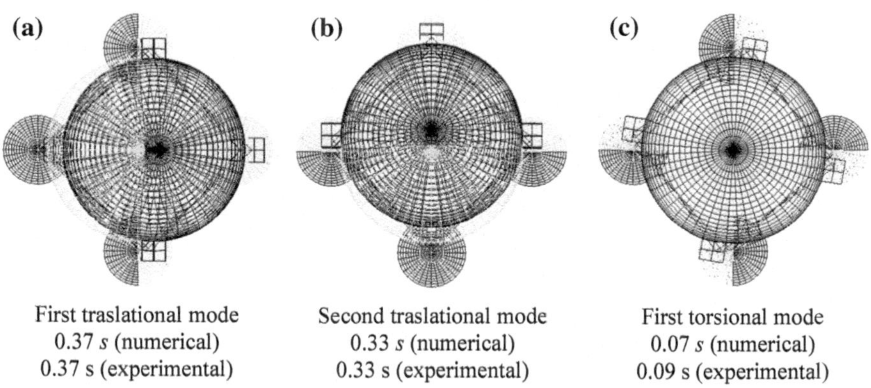

Fig. 8 Main numerical modes of the updated finite element model. View from above

be acceptable due to the very low period, which is difficult to change numerically and measure experimentally.

The structural monitoring is intended to be used to measure observable physical quantities on structures over a long period, from which information capable of describing the structural behavior can be extracted that could also be useful for the identification of damage [28–34]. Today, the key progress in the field of sensor technology, telecommunications, and information technology facilitates the development of the latest generation of instruments that improve the entire process. Among the different types of sensors (accelerometers, extensometers, inclinometers) fiber optic sensing represents an innovative instrumentation that is very sensitive to both environmental (e.g. temperature) and mechanical external forces. Fiber optic sensors have several advantages over traditional sensors, especially with regard to static structural monitoring. For instance, they are very compact, and sometimes can be embedded within the material. They are immune to electromagnetic interference, as there is no need for electric power, and because they are composed of glass, they do not induce electric currents. They exhibit high resolution and accuracy, as well as stability, and installation over long distances is possible. The most important types are fiber Bragg grating (FBG) systems, which are punctual sensors capable of measuring the deformation in points where they are collocated [35–38], and Brillouin optical time domain reflectometry (BOTDR), which measures deformations along the whole length of the fiber [39]. The latter technology could be applied to carry out static monitoring of the main vaults of the underground galleries of the LNGS. This system will make it possible to observe both transverse and longitudinal deformations due to the very low strain expectations. Moreover, the fiber optic could be integrated with a network of accelerometer sensors placed in appropriately selected points and capable of recording micro-tremor induced vibrations. The fusion of data measurement and the resulting information obtained should enable the assessment of any failure or relaxations of the ties in the vaults due to natural degradation phenomena. The accelerations recorded by sensors placed on the ground in each experimental hall could also be useful for assessing probabilistic earthquake hazard.

Conclusions

In this chapter, issues concerning the design, monitoring, and maintenance of underground structures have been analyzed. Particular attention has been paid to the effects induced by seismic excitation. The case study of the Gran Sasso National Laboratory of the National Institute of Nuclear Physics, the largest and most important underground research center in the world, has been presented. In this case the two main points discussed have been the possible choices for describing the behavior of customized structures used as experimental prototypes and the structural monitoring of the underground experimental hall vaults. In the first case, different finite element models, successively updated based on information derived from dynamic experimental tests, have been implemented, while in the second case, a monitoring system comprising fiber optic sensors and accelerometers has been proposed. Finally, an important open issue remains the difficult probabilistic assessment of earthquake intensity in the underground structure, due mainly to the lack of experimental recorded data.

Acknowledgements The author wishes to thank Eng. Paolo Martella (Head of Design Service at LNGS) for furnishing unique technical information on the laboratory infrastructures.

References

1. Ceci AM, Gattulli V, Potenza F (2013) Serviceability and damage scenario in irregular RC structures: post-earthquake observations and modelling predictions. J Perform Constr Facil 27 (1):98–115
2. Hashash YMA, Hook JJ, Schmidt B, I-Chiang Yao J (2001) Seismic design and analysis of underground structures. Tunn Undergr Space Technol **16**:247–293
3. Wang WL, Wang TT, Su JJ, Lin CH, Seng CR, Huang TH (2001) Assessment of damage in mountain tunnels due to the Taiwan Chi-Chi Earthquake. Tunn Undergr Space Technol 16:133–150
4. Yashiro K, Kojima Y, Shimizu M (2007) Historical earthquake damage to tunnels in Japan and case studies of railway tunnels in the 2004 Niigataken-Chuetsu Earthquake. Q Rep Railw Tech Res Inst 48:136–141
5. Azadi M, Hossein SMMM (2010) Analyses of the effect of seismic behaviour of shallow tunnels in liquefiable grounds. Tunn Undergr Space Technol 25:543–552
6. Boldini D, Graziani A (2012) Remarks on axisymmetric modelling of deep tunnels in argillaceous formations. II: fissured argillites. Tunn Undergr Space Technol 28:80–89
7. Hage Chehade F, Shahrour I (2008) Numerical analysis of the interaction between twin-tunnels: influence of the relative position and construction procedure. Tunn Undergr Space Technol 23:210–214
8. Hu J, Xu N (2011) Numerical analysis of failure mechanism of tunnel under different confining pressure. Procedia Eng 26:107–112
9. Yingjie L, Dingli Z, Qian F, Qingchun Y, Lu X (2014) A physical and numerical investigation of the failure mechanism of weak rocks surrounding tunnels. Comput Geotech 61:292–307
10. Talmon AM, Bezuijen A (2013) Calculation of longitudinal bending moment and shear force for Shanghai Yangtze river tunnel: application of lessons from Dutch research. Tunn Undergr Space Technol 35:161–171
11. Pakbaz MC, Yareevand A (2005) 2-D analysis of circular tunnel against earthquake loading. Tunn Undergr Space Technol 20:411–417
12. El Naggar H, Hinchberger DS, El Naggar MH (2008) Simplified analysis of seismic in-plane stresses in composite and joined tunnel linings. Soil Dyn Earthq Eng 28:1063–1077
13. Lu JF, Jeng DS, Lee TL (2007) Dynamic response of a piecewise circular tunnel embedded in a poroelastic medium. Soil Dyn Earthq Eng 27:875–891
14. Sanchez-Merino AL, Fernandez-Saez J, Navarro C (2009) Simplified longitudinal seismic response of tunnel linings subjected to surface waves. Soil Dyn Earthq Eng 29:579–582
15. Shahrour I, Khoshnoudian F, Sadek M, Mroueh H (2010) Elastoplastic analysis of the seismic response of tunnels in soft soils. Tunn Undergr Space Technol 25:478–482
16. Wang CJ (2011) Seismic racking of a dual-wall subway station box embedded in soft soil strata. Tunn Undergr Space Technol 26:83–91

17. Amorosi A, Boldini D (2009) Numerical modelling of the transverse dynamic behaviour of circular tunnels in clayey soils. Soil Dyn Earthq Eng 29:1059–1072
18. Anastasopoulos I, Gazetas G (2010) Analysis of cut-and cover tunnels against large tectonic deformation. Bull Earthq Eng 8:283–307
19. Hwang JH, Lu CC (2007) Seismic capacity assessment of old Sanyi railway tunnels. Tunn Undergr Space Technol 22:433–449
20. Liu J, Liu X (2008) Pushover analysis of Dakai subway station during the Osaka-Kobe earthquake in 1995. In: The 14th World Conference on Earthquake Engineering, Beijing, China, 12–17 October 2008
21. Chen J, Shi X, Li J (2010) Shaking table test of utility tunnel under non-uniform earthquake wave excitation. Soil Dyn Earthq Eng 30:1400–1416
22. Ciligir U, Gopal Madabhushi SP (2011) A model study on the effects of input motion on the seismic behaviour of tunnel. Soil Dyn Earthq Eng 31:452–462
23. Xiao J, Huang G (2010) Transverse earthquake response and design analysis of submerged floating tunnels with various shore connections. Procedia Eng 4:233–242
24. Lunardi P (1990) La ricerca e la tecnologia nella realizzazione di grandi cavità sotterranee: il laboratorio di fisica nucleare del Gran Sasso. In: International Conference Se vogliamo il verde sopra utilizziamo lo spazio profondo, Milano, Italy, pp 15–16 (in Italian)
25. Antonacci E, De Stefano A, Gattulli V, Lepidi M, Matta E (2012) Comparative study of vibration-based parametric identification techniques for a three-dimensional frame structure. Struct Control Health Monit 19(5):579–608
26. Diaferio M, Foti D, Giannoccaro NI (2015) Identification of the modal properties of a building of the Greek heritage. Key Eng Mater 628:150–159
27. Diaferio M, Foti D, Giannoccaro NI (2014) Non-destructive characterization and identification of the modal parameters of an old masonry tower. In: Proceedings of EESMS 2014—2014 IEEE Workshop on Environmental, Energy and Structural Monitoring Systems, Naples, Italy, 17–18 September 2014
28. Bhala S, Yang YW, Zhao J, Soh CK (2005) Structural health monitoring of underground facilities—technological issues and challenges. Tunn Undergr Space Technol 20:487–500
29. Potenza F, Federici F, Lepidi M, Gattulli V, Graziosi F, Colarieti A (2015) Long term structural monitoring of the damaged Basilica S. Maria di Collemaggio through a low-cost wireless sensor network. J Civ Struct Health Monit 5(5):655–676. https://doi.org/10.1007/s13349-015-0146-3
30. Federici F, Graziosi F, Faccio M, Gattulli V, Lepidi M, Potenza F (2012) An integrated approach to the design of wireless sensor networks for structural health monitoring. Int J Distrib Sens Netw, Article ID 594842, 1–16
31. Gattulli V, Lepidi M, Potenza F, Di Sabatino U (2016) Dynamics of masonry walls connected by a vibrating cable in a historic structure. Meccanica 51(11):2813–2826. https://doi.org/10.1007/s11012-016-0509-9
32. Yun HB, Park SH, Mehdawi N, Mokhtari S, Chopra M, Reddi L, Park KT (2014) Monitoring for close proximity tunneling effects on an existing tunnel using principal component analysis technique with limited sensor data. Tunn Undergr Space Technol 43:398–412
33. Lepidi M, Gattulli V, Vestroni F (2009) Damage identification in elastic suspended cables through frequency measurement. J Vib Control 15(6):867–896
34. Gattulli V, Graziosi F, Federici F, Potenza F, Colarieti A, Lepidi M (2013) Structural health monitoring of the Basilica S. Maria di Collemaggio. In: Proceedings of The Fifth International Conference on Structural Engineering, Mechanics and Computation (SEMC 2013), Cape Town, South Africa, 2–4 September 2013
35. Betz DC, Staudigel L, Trutzel MN, Kehlenbach M (2003) Structural monitoring using fiber-opitic bragg grating sensors. Struct Health Monit 2(2):145–152
36. Valvona F, Toti J, Gattulli V, Potenza F (2017) Effective seismic strengthening and monitoring of a masonry vault by using glass fiber reinforced cementitious matrix with embedded fiber bragg grating sensors. Compos B Eng 113:355–370. https://doi.org/10.1016/j.compositesb.2017.01.024

37. Todd MD, Johnson GA, Vohra ST (2001) Deployment of a fiber bragg grating-based measurement system in a structural health monitoring application. Smart Mater Struct 10 (3):534–539
38. Moyo P, Brownjohn JMW, Suresh R, Tjin SC (2005) Development of fiber Bragg grating sensors for monitoring civil infrastructures. Eng Struct 27(12):1828–1834
39. Gue CY, Wilcock M, Alhaddad MM, Elshafie MZEB, Soga K, Mair RJ (2015) The monitoring of an existing cast iron tunnel with distributed fiber optic sensing (DFOS). J Civ Struct Health Monit 5(5):573–586

Printed by Printforce, the Netherlands